Subsurface Ventilation

and

Environmental Engineering

T0396555

Subsurface Ventilation

and

Environmental Engineering

Malcolm J. McPherson

The Massey Professor of Mining Engineering,
Virginia Polytechnic Institute and State University
and
President,
Mine Ventilation Services,
Incorporated
USA

SPRINGER-SCIENCE+BUSINESS MEDIA, B.V.

First edition 1993

© 1993 Malcolm J. McPherson
Originally published by Chapman & Hall in 1993
Softcover reprint of the hardcover 1st edition 1993

Typeset in 10/12 pt Bembo by Thomson Press (India) Limited, New Delhi, India

ISBN 978-94-010-4677-0   ISBN 978-94-011-1550-6 (eBook)
DOI 10.1007/978-94-011-1550-6

A catalogue record for this book is available from the British Library

Library of Congress Cataloging-in-Publication Data

McPherson, Malcolm J.
    Subsurface ventilation and environmental engineering / Malcolm J.
McPherson.—1st ed.
        p.   cm.
    Includes bibliographical references and index.
    1. Mine ventilation.   2. Air—Pollution.   3. Mine fires.   4. Mine
explosions.   5. Underground areas—heating and ventilation.
6. Underground areas—Fires and fire prevention.   7. Explosions.
I. Title.
TN301.M37   1992
622'.42—dc20   92–6185
                                                          CIP

*This work has been undertaken in fulfilment of
a long-standing promise to my former teacher,
mentor and dear friend*

Professor Frederick Baden Hinsley

The book is dedicated to his memory

Contents

Acknowledgements

There are many people without whose contributions this book could not have been written. First, I thank Shirley, my wife, for her patience and understanding—not only through the long hours of midnight oil burning that took place during the writing but, more particularly, for the extended periods, stretching over many years, when she was left alone to look after the home and family while I was deep under the surface of some faraway country.

I am grateful to former colleagues in the Department of Mining Engineering, University of Nottingham, England, for sowing seeds of ideas that later produced practical designs and procedures, many of which are reflected in this book; especially Ian Longson with whom I rediscovered the fascinations of thermodynamic logic, Leslie H. Morris, Dr. Jim R. Brown and, most of all, Professor F. Baden Hinsley to whom this book is dedicated. I am also privileged in having worked with excellent students from whom I learned a great deal, at Nottingham, at the University of California, Berkeley, and now at Virginia Polytechnic and State University.

Despite having been involved in numerous research investigations, my knowledge of subsurface ventilation and environmental engineering has been advanced primarily by working on projects with mine ventilation engineers in many countries. Most of the case studies and examples in the book originated in such work. In particular, I am truly grateful for having had the opportunity of interacting with dedicated professional engineers in the United Kingdom, the countries of East and West Europe, South Africa, Australia, India, the United States of America and Canada.

I am indebted to the two ladies who shared typing the manuscript. First, my daughter Alison D. McPherson who also took great delight in correcting some of my mathematics, and, Lucy Musante, my secretarial assistant at Mine Ventilation Services, Inc.—the most skilled and dedicated secretary with whom I have ever worked. Most of the initial reviews of chapters were undertaken by staff of Mine Ventilation Services, namely Daniel J. Brunner, Justus Deen, Martha O'Leary and, most particularly, Keith G. Wallace who willingly volunteered for far more than his fair share of the work. Several chapters were reviewed by Dr. Felipe Calizaya, formerly at Berkeley and now at the Mackay School of Mines in Reno, Nevada.

Some of the analyses described in the book arose directly out of funded research. The physiological model in Chapter 17 was developed for the US Department of Energy via Sandia National Laboratories as part of an investigation into climatic conditions in a deep geological repository for nuclear waste. Some of the heat transfer and climatic simulation studies in Chapters 15 and 16, and investigations into the installation of booster fans outlined Chapter 9 were assisted by funding from the Generic Mineral Technology Center in Mine Systems Design and Ground Control, Office of Mineral Institutes, US Bureau of Mines under Grant No. G1125151. I am indebted to those organizations for financing the work.

Finally, but also foremost, I thank the Good Lord for guiding my career to a point when I could prepare this book.

Preface

This book has been written as a reference and text for engineers, researchers, teachers and students who have an interest in the planning and control of the environment in underground openings. While directed primarily to underground mining operations, the design procedures are also applicable to other complex developments of subsurface space such as nuclear waste repositories, commercial accommodation or vehicular networks. The book will, therefore, be useful for mining, civil, mechanical, and heating, ventilating and air-conditioning engineers involved in such enterprises. The chapters on airborne pollutants highlight means of measurement and control as well as physiological reaction. These topics will be of particular interest to industrial hygienists and students of industrial medicine.

One of the first technical applications of digital computers in the world's mining industries was for ventilation network analysis. This occurred during the early 1960s. However, it was not until low cost but powerful personal computers proliferated in engineering offices during the 1980s that the full impact of the computer revolution was realized in the day-to-day work of most mine ventilation engineers. This book reflects the changes in approach and design procedures that have been brought about by that revolution.

While the book is organized into six parts, it encompasses three broad areas. Following an introductory background to the subject, Chapters 2 and 3 provide the fundamentals of fluid mechanics and thermodynamics that are necessary for a complete understanding of large three-dimensional ventilation systems. Chapters 4 to 10, inclusive, offer a comprehensive treatment of subsurface airflow systems while Chapters 11 to 21 deal with the airborne hazards that are encountered in underground openings.

Each chapter is self-contained as far as is practicable. The interrelated features of the topics are maintained by means of copious cross-references. These are included in order that practicing engineers may progress through a design project and be reminded of the wider repercussions of decisions that might be made. However, numerous cross-references can be a little distracting. The student is advised to ignore them during an initial reading and unless additional information is sought.

Many of the chapters are subdivided into theoretical and descriptive sections. Again, these can be read separately although a full understanding of the purpose and range of application of design procedures can be gained only through a knowledge of both. When used as a refresher or text by practicing engineers, it is suggested that the relevant descriptive section be consulted first and reference made back to the corresponding analysis or derivation when necessary.

The use of the book as an aid to teaching and learning can be moulded to suit any given curriculum. For the full education of a subsurface ventilation and environmental engineer, Chapters 1 to 10 may be employed during a course on ventilation, i.e. airflow processes, leaving the chapters on gases, heat, dust, and fires and explosions for advanced courses. Where time is more restricted then the teacher may compile his or her own syllabus at any given level by choosing relevant sections from selected chapters.

In most countries, mining activities are regulated by specific state or national legislation. This book has been written for an international audience and reflects the author's experience of teaching and practice in a number of countries. While guideline threshold limit values are given, the reader is frequently reminded to consult the relevant local regulations for specific mandatory requirements and limitations on practical procedures. Système Internationale (SI) units are employed and a comprehensive list of conversion factors is provided.

Table of conversion factors between imperial and SI units

Quantity	Imperial to SI			SI to imperial		
Length	1 ft	= 0.3048	m	1 m	= 3.2808	ft
	1 yd	= 0.9144	m		= 1.0936	yd
	1 in	= 0.0254	m		= 39.3701	in
Area	1 ft²	= 0.0929	m²	1 m²	= 10.7639	ft²
	1 in²	= 0.000 645	m²		= 1550.003	in²
Acceleration	1 ft/s²	= 0.3048	m/s²	1 m/s²	= 3.2808	ft/s²
Force	1 lbf	= 4.448 2	N	1 N	= 0.2248	lbf
	1 tonf	= 9964.02	N			
Velocity	1 ft/s	= 0.3048	m/s	1 m/s	= 3.2808	ft/s
	1 ft/min	= 0.005 08	m/s		= 196.85	ft/min
Volume	1 ft³	= 0.028 32	m³	1 m³	= 35.315	ft³
	1 yd³	= 0.764 56	m³		= 1.308	yd³
	1 imperial gal	= 4.545	1	1 l	= 0.2200	imperial gal
	1 US gal	= 3.785	1	(0.001 m³)	= 0.2642	US gal

Table I (*Cont.*)

Quantity	Imperial to SI		SI to imperial	
Volume flow	1 ft³/s	= 0.028 32 m³/s	1 m³/s = 35.315	ft³/s
	1 ft³/min	= 0.000 472 m³/s	1 m³/s = 2118.9	ft³/min
	1 imperial gal/h	= 0.004 55 m³/h	1 m³/h = 220.0	imperial gal/h
	1 imperial gal/min	= 0.004 55 m³/min	1 m³/min = 220.0	imperial gal/min
		= 4.545 l/min	1 l/min = 0.220	imperial gal/min
		= 0.075 75 l/s	1 l/s = 13.20	imperial gal/min
	1 US gal/min	= 0.063 13 l/s	1 l/s = 15.84	US gal/min
Mass	1 lb	= 0.453 592 kg	1 kg = 2.204 62	lb
	1 imperial ton (2240 lb)	= 1.016 05 t	1 t = 1000 kg = 0.984 20	imperial ton
	1 short ton (2000 lb)	= 0.907 18 t	= 1.1023	short ton
Pressure, stress	1 lbf/ft²	= 47.880 N/m² = Pa	1 N/m² = Pa = 0.020 88	lbf/ft²
	1 lbf/in²	= 6894.76 N/m²	= 0.000 145	lbf/in²
	1 in wg	= 249.089 N/m²	= 0.004 015	in wg
	1 ft wg	= 2989.07 N/m²	= 0.000 3346	ft wg
	1 mm wg	= 9.807 N/m²	= 0.101 97	mm wg
	1 inHg	= 3386.39 N/m²	= 0.000 2953	inHg
	1 mmHg	= 133.32 N/m²	= 0.007 501	mmHg
		= 1.333 2 mb	= 0.01	mb
	Note: the millibar (1 mb = 100 N/m²) is included here as it is a familiar metric unit of pressure. It is not, however, an SI unit.			
Airway resistance	1 Atk	= 0.059 71 N s²/m⁸	1 N s²/m⁸ = 16.747	Atk
	1 PU	= 1.118 3 N s²/m⁸	1 N s²/m⁸ = 0.894 2	PU
Airway specific resistance	1 in wg per 10 000 ft³/min	= 22.366 N s²/m⁸	1 N s²/m⁸ = 0.044 7	in wg per 10 000 ft³/min

Quantity	Unit		Value	SI unit	Reference		Value	Unit
Friction factor	$1\,\mathrm{lbf\,min^2/ft^4}$	=	1.8554×10^6	$\mathrm{kg/m^3}$	$1\,\mathrm{kg/m^3}$	=	539.0×10^{-9}	$\mathrm{lbf\,min^2/ft^4}$
Density	$1\,\mathrm{lb/ft^3}$	=	16.0185	$\mathrm{kg/m^3}$	$1\,\mathrm{kg/m^3}$	=	$0.062\,43$	$\mathrm{lb/ft^3}$
	$1\,\mathrm{imperial\ ton/yd^3}$	=	1328.94	$\mathrm{kg/m^3}$		=	$0.000\,753$	$\mathrm{imperial\ ton/yd^3}$
	$1\,\mathrm{short\ ton/yd^3}$	=	1186.55	$\mathrm{kg/m^3}$		=	$0.000\,843$	$\mathrm{short\ ton/yd^3}$
Energy, work, heat	$1\,\mathrm{ft\,lbf}$	=	$1.355\,82$	J	$1\,\mathrm{J}$	=	$0.737\,56$	ft/lbf
	$1\,\mathrm{Btu}$	=	1055.06	J		=	$0.000\,948$	Btu
	$1\,\mathrm{cal}$	=	$4.186\,8$	J		=	$0.238\,89$	cal
	$1\,\mathrm{therm}$	=	105.506	MJ		=	$0.009\,478$	μtherm
	$1\,\mathrm{kWh}$	=	3.6	MJ		=	$0.000\,278$	Wh
Power	$1\,\mathrm{hp}$	=	745.700	W	$1\,\mathrm{W}$	=	$0.001\,341$	hp
Heatflow	$1\,\mathrm{ft\,lbf/min}$	=	0.0226	W		=	44.254	ft lbf/min
	$1\,\mathrm{Btu/min}$	=	17.584	W		=	$0.056\,87$	Btu/min
	$1\,\mathrm{RT}$	=	3517	W		=	$0.000\,2843$	RT
	Refrigeration (imperial) ton							
Specific energy,	$1\,\mathrm{ft\,lbf/lb}$	=	2.989	J/kg	$1\,\mathrm{J/kg}$	=	0.3345	ft lbf/lb
Calorific value	$1\,\mathrm{Btu/lb}$	=	2326	J/kg		=	$0.000\,430$	Btu/lb
	$1\,\mathrm{therm/imperial\ ton}$	=	0.1038	MJ/kg		=	9.634	μtherm/imperial ton
	$1\,\mathrm{therm/short\ ton}$	=	0.1163	MJ/kg		=	8.602	μtherm/short ton
Gas constants	$1\,\mathrm{ft\,lbf/lb\,{}^\circ R}$	=	$5.380\,3$	J/kg K	$1\,\mathrm{J/kg\,K}$	=	$0.185\,9$	ft lbf/lb °R
Specific heat, specific entropy	$1\,\mathrm{Btu/lb\,{}^\circ R}$	=	4186.8	J/kg K		=	$0.000\,2388$	Btu/lb °R
Specific volume	$1\,\mathrm{ft^3/lb}$	=	$0.062\,43$	$\mathrm{m^3/kg}$	$1\,\mathrm{m^3/kg}$	=	16.018	$\mathrm{ft^3/lb}$
	$1\,\mathrm{ft^3/imperial\ ton}$	=	$0.027\,87$	$\mathrm{m^3/t}$	$1\,\mathrm{m^3/t}$	=	35.881	$\mathrm{ft^3/imperial\ ton}$
	$1\,\mathrm{ft^3/short\ ton}$	=	$0.031\,21$	$\mathrm{m^3/t}$		=	32.037	$\mathrm{ft^3/short\ ton}$
				Note: 1 metric tonne (t) = 1000 kg				
Dynamic viscosity	$1\,\mathrm{lb/ft\,s}$	=	$1.488\,16$	$\mathrm{N\,s/m^2}$	$1\,\mathrm{N\,s/m^2}$	=	$0.671\,97$	lb/ft s
	$1\,\mathrm{poise}$	=	0.1	$\mathrm{N\,s/m^2}$		=	10	poise

Table I (*Cont.*)

Quantity	Imperial to SI			SI to imperial	
Kinematic viscosity	1 ft²/s	= 0.092903	m²/s	= 10.7639	ft²/s
	1 stokes	= 0.0001	m²/s	= 10000	stokes
Permeability	1 darcy	= 0.98693 × 10⁻¹²	m²	= 1.01324 × 10¹²	darcy
	1 md	= 0.98693 × 10⁻¹⁵	m²	= 1.01324 × 10¹⁵	md
Thermal conductivity	1 Btu ft/ft² h °R	= 1.73073	W/(m K) 1 W/(m K)	= 0.57779	Btu ft/ft² h °R
Thermal gradient	1 °F/ft	= 1.8227	°C/m 1 °C/m	= 0.5486	°F/ft
Moisture content	1 lb/lb	= 1	kg/kg 1 kg/kg	= 1	lb/lb
	1 gr/lb	= 0.0001429	kg/kg	= 7000	gr/lb
Radiation	1 rad	= 0.01	Gy 1 Gy	= 100	rad
	1 Curie	= 37 × 10⁹	Bq 1 Bq	= 27 × 10⁻¹²	Curie
	1 rem	= 0.01	Sv 1 Sv	= 100	rem
	1 Roentgen	= 2.58 × 10⁻⁴	C/kg 1 C/kg	= 3876	Roentgen

$$1 \text{ gray (Gy)} = 1 \text{ J/kg}$$
$$1 \text{ sievert (Sv)} = 1 \text{ J/kg}$$
$$1 \text{ becquerel (Bq)} = 1 \text{ disintegration/s}$$
$$1 \text{ coulomb (C)} = 1 \text{ A s}$$

Temperature:

$$K = {}^\circ C + 273.15$$
$$^\circ R = {}^\circ F + 459.67$$

For differential temperatures, 1 Centigrade degree = 1.8 Fahrenheit degrees. For actual temperature,

$$1.8 \times t(^\circ C) + 32 = {}^\circ F$$

and

$$\frac{t(^\circ F) - 32}{1.8} = {}^\circ C$$

1

Background to subsurface ventilation and environmental engineering

1.1 INTRODUCTION

Ventilation is sometimes described as the lifeblood of a mine, the intake airways being arteries that carry oxygen to the working areas and the returns veins that conduct pollutants away to be expelled to the outside atmosphere. Without an effective ventilation system, no underground facility that requires personnel to enter it can operate safely.

The slaughter of men, women and children that took place in the coal mines of Britain during the eighteenth and nineteenth centuries resulted in the theory and art of ventilation becoming the primary mining science. The success of research in this area has produced tremendous improvements in underground environmental conditions. Loss of life attributable to inadequate ventilation is now, thankfully, a relatively infrequent occurrence. Falls of ground rather than ventilation-related factors have become the most common cause of fatalities and injuries in underground mines. Improvements in ventilation have also allowed the productivity of mines to be greatly improved. Neither the very first nor the very latest powered machines could have been introduced underground without an adequate supply of air. Subsurface ventilation engineers are caught up in a continuing cycle. Their work allows rock to be broken in ever larger quantities and at greater depths. This, in turn, produces more dust, gases and heat, resulting in a demand for yet better environmental control.

This opening chapter takes a necessarily cursory look at the long history of mine ventilation and discusses the interactions between ventilation and the other systems that, jointly, comprise a complete mine or underground facility.

1.2 A BRIEF HISTORY OF MINE VENTILATION

Observations of the movements of air in underground passages have a long and fascinating history. Between 4000 and 1200 BC, European miners dug tunnels into chalk deposits searching for flint. Archaeological investigations at Grimes Graves in the south of England have shown that these early flint miners built brushwood fires at the working faces—presumably to weaken the rock. However, those Neolithic miners could hardly have failed to observe the currents of air induced by the fire. Indeed, the ability of fire to promote airflow was rediscovered by the Greeks, the Romans, in medieval Europe and during the Industrial Revolution in Britain.

The Laurium silver mines of Greece, operating in 600 BC, have layouts which reveal that the Greek miners were conscious of the need for a connected ventilating **circuit.** At least two airways served each major section of the mine and there is evidence that divided shafts were used to provide separate air intake and return connections to the surface. Underground mines of the Roman Empire often had twin shafts, and Pliny (AD 23–79) describes how slaves used palm fronds to waft air along tunnels.

Although metal mines were worked in Europe during the first 1500 years *anno Domini*, there remain few documented descriptions of their operations. The first great textbook on mining was written in Latin by Georgius Agricola, a physician in a thriving iron ore mining and smelting community of Bohemia in Central Europe. Agricola's *De Re Metallica*, produced in 1556, is profusely illustrated. A number of the prints show ventilating methods that include diverting surface winds into the mouths of shafts, wooden centrifugal fans powered by men and horses, bellows for auxiliary ventilation and air doors. An example of one of Agricola's prints is reproduced in Fig. 1.1.

Agricola was also well aware of the dangers of 'blackdamp', air that has suffered from a reduction in oxygen content—'miners are sometimes killed by the pestilential air that they breathe'—and of the explosive power of 'firedamp', a mixture of methane and air—'likened to the fiery blast of a dragon's breath'. *De Re Metallica* was translated into English in 1912 by Herbert C. Hoover and his wife, Lou. Hoover was a young American mining engineer who graduated from Stanford University and subsequently served as President of the United States during the term 1929–1933.

From the seventeenth century onwards, papers began to be presented to the Royal Society of the United Kingdom on the explosive and poisonous nature of mine atmospheres. The Industrial Revolution brought a rapid increase in the demand for coal. Conditions in many coal mines were quite horrific for the men, women, and children who were employed in them during the eighteenth and nineteenth centuries. Ventilation was induced either by purely natural effects, stagnating when air temperatures on the surface and underground were near equal, or by fire. The first ventilating furnaces of that era were built on surface but it was soon realized that burning coals suspended in a wire basket within the upcast shaft gave improved ventilation. Furthermore, the lower the basket, the better the effect. This quickly led to the construction of shaft bottom furnaces.

The only form of illumination until the early nineteenth century was the candle.

Figure 1.1 A print from Agricola's *De Re Metallica*. (*Reproduced by permission of Dover Publications.*)

A—Machine first described. B—This workman, treading with his feet, is compressing the bellows. C—Bellows without nozzles. D—Hole by which heavy vapours or blasts are blown out. E—Conduits. F—Tunnel. G—Second machine described. H—Wooden wheel. I—Its steps. K—Bars. L—Hole in same wheel. M—Pole. N—Third machine described. O—Upright axle. P—Its toothed drum. Q—Horizontal axle. R—Its drum which is made of rundles.

With historical hindsight we can see the conjunction of circumstances that caused the ensuing carnage: a seemingly insatiable demand for coal to fuel the steam engines of the Industrial Revolution, the working of seams rich in methane gas, inadequate ventilation, furnaces located in methane-laden return air and the open flames of candles. There are many graphic descriptions of methane and coal dust explosions, the suffering of mining communities, the heroism of rescue attempts and the strenuous efforts of mining engineers and scientists to find means of improving ventilation and providing illumination without the accompanying danger of igniting methane gas. Seemingly oblivious to the extent of the danger, miners would sometimes ignite pockets of methane intentionally, for amusement and to watch the blue flames flickering above their heads. Even the renowned engineer George Stephenson admitted to this practice during the inquiries of a government select committee on mine explosions in 1835. A common method of removing methane was to send a 'fireman' in before each shift, covered in sackcloths dowsed in water and carrying a candle on the end of a long rod. It was his task to burn out the methane before the miners went into the working faces.

John Buddle (1773–1843), an eminent mining engineer in the north of England, produced two significant improvements. First, he introduced 'dumb drifts' which bled sufficient fresh air from the base of a downcast shaft to feed the furnace. The return air, laden with methane, bypassed the furnace. The products of combustion entering the upcast shaft from the furnace were too cool to ignite the methane but still gave a good chimney effect in the shaft, thus inducing airflow around the mine. Buddle's second innovation was 'panel (or split) ventilation'. Until that time, air flowed sequentially through work areas, one after the other, continually increasing in methane concentration. Buddle originally divided the mine layout into discrete panels, with intervening barrier pillars, to counteract excessive floor heave. However, he found that by providing an intake and return separately to each panel the ventilating quantities improved markedly and methane concentrations decreased. He had discovered, almost by accident, the advantages of parallel layouts over series circuits. The mathematical proof of this did not come until Atkinson's theoretical analyses several decades later.

The quest for a safe form of illumination went on through the eighteenth century. Some of the earlier suggestions made by scientists of the time, such as using very thin candles, appear quite ludicrous to us today. One of the more serious attempts was the steel flint mill invented in 1733 by Carlisle Spedding, a well known mining engineer, again, in the north of England (Fig. 1.2). This device relied on a piece of flint being held against a rapidly revolving steel wheel. The latter was driven through a gear mechanism by a manually rotated handle. The complete device was strapped to the chest of a boy whose job was to produce a continuous shower of sparks in order to provide some illumination for the work place of a miner. The instrument was deemed safer than a candle but the light it produced was poor, intermittent, and still capable of igniting methane.

A crisis point was reached in 1812 when a horrific explosion at Felling, Gateshead, killed 92 miners. With the help of local clergymen, a society was formed to look

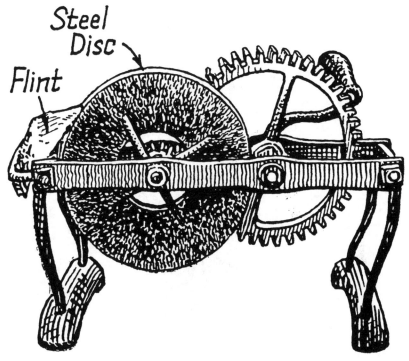

Figure 1.2 Spedding's Flint Mill. (*Reproduced by permission of Virtue and Co., Ltd.*)

into ways of preventing such disasters. Contact was made with Sir Humphrey Davy, President of the Royal Society, for assistance in developing a safe lamp. Davy visited John Buddle to learn more of conditions in the mines. As this was well before the days of electricity, he was limited to some form of flame lamp. Within a short period of experimentation he found that the flame of burning methane would not readily pass through a closely woven wire mesh. The Davy lamp had arrived (Fig. 1.3). Buddle's reaction is best expressed in a letter he wrote to Davy.

> I first tried it in a explosive mixture on the surface, and then took it into the mine...it is impossible for me to express my feelings at the time when I first suspended the lamp in the mine and saw it red hot...I said to those around me, 'We have at last subdued this monster.'

The lamp glowed 'red hot' because of the methane burning vigourously within it, yet the flames could not pass through the wire mesh to ignite the surrounding firedamp.

Davy lamps were introduced into British mines, then spread to other countries. Nevertheless, in the absence of effective legislation, candles remained in widespread use through the nineteenth century because of the better light that they produced.

Perhaps the greatest classical paper on mine ventilation was one entitled 'On the

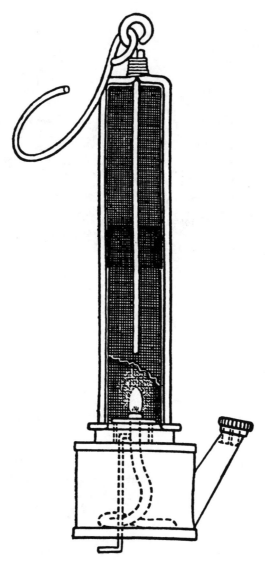

Figure 1.3   The original appearance of the Davy safety lamp. (*Reproduced by permission of Virtue and Co., Ltd.*)

theory of the ventilation of mines', presented by John Job Atkinson to the North of England Institute of Mining Engineers in December, 1854. Atkinson was a mining agent—an intermediary between management and the mine owners. He later became one of the first Inspectors of Mines. Atkinson appears to have been well educated in mathematics and languages, and was clearly influenced by the earlier work of French hydraulic engineers (Chapter 5). He seems to have had some difficulty in having his paper accepted. Officers of the Institute decided, perhaps understandably, that the 154 page paper was too long to be presented at a meeting. It was, however, published and a meeting of the Institute arranged to discuss it. Despite publicity referring to the importance of the subject, attendance at the meeting was poor and there was little discussion. In this paper, Atkinson proposed and expanded upon the principles on which most modern mine ventilation planning is still based. However, the analytical reasoning and mathematical analyses that he developed in great detail were simply too much for engineers of the day. The paper was consigned to the archives and it was some 60 years after Atkinson's death that his work was 're-discovered' and put into practice.

During Atkinson's productive years the first power-driven ventilators began to appear. These varied from enormous steam-driven piston and cylinder devices to elementary centrifugal fans.

The years around the turn of the century saw working conditions in mines coming under legislative control. Persons responsible for underground mining operations were required to obtain minimum statutory qualifications. Mine manager's examination papers concentrated heavily on ventilation matters until well into the twentieth century.

The 1920s saw further accelerated research in several countries. Improved instrumentation allowed organized ventilation surveys to be carried out to measure airflows and pressure drops for the purposes of ventilation planning, although there was no practical means of predicting airflows in other than simple circuits at that time. Atkinson's theory was confirmed in practice. The first successful axial fans were introduced in about 1930.

In 1943, Professor F. B. Hinsley produced another classical paper advancing understanding of the behaviour of airflow by using thermodynamic analyses. Hinsley also supervised the work at Nottingham University that led to the first practical use of analogue computers in 1952 to facilitate ventilation planning. This technique was employed widely and successfully for over a decade. The development of ventilation network analysis programs for digital computers in the early 1960s rendered the analogue devices obsolete. Initially, the network programs were written for, and required the power of, mainframe computers. These were employed throughout the 1970s. However, the 1980s saw a shift to desk-top computers and corresponding programs were developed. This is now the dominant method used for ventilation planning (Chapter 7).

The discipline of mine ventilation is an addictive subject for researchers of industrial history, full of lost discoveries and rediscoveries, excitement and despair, achievement and tragedy. It has been the subject of many papers and books. An excellent place to commence further reading is the text by Saxton serialized in Volume 146 of the *Mining Engineer*.

1.3   THE RELATIONSHIPS BETWEEN VENTILATION AND
OTHER SUBSURFACE SYSTEMS

1.3.1   The objectives of subsurface ventilation

The basic objective of an underground ventilation system is clear and simple. It is to provide airflows in sufficient quantity and quality to dilute contaminants to safe concentrations in all parts of the facility where personnel are required to work or travel. This basic requirement is incorporated into mining law in those countries that have such legislation. The manner in which 'quantity and quality' are defined varies from country to country depending on their mining history, the pollutants of greatest concern, the perceived dangers associated with those hazards and the political and social structure of the country. The overall requirement is that all persons must be able to work and travel within an environment that is safe and which provides reasonable comfort. An interpretation of the latter phase depends greatly on the geographical location of the mine and the background and expectations of the workforce. Personnel in a permafrost mine work in conditions that would be unacceptable to miners from an equatorial region, and vice versa—and neither set of conditions would be tolerated by factory or office workers. This perception of 'reasonable comfort' sometimes causes misunderstandings between subsurface ventilation engineers and those associated with the heating and ventilating industry for buildings.

While maintaining the essential objectives related to safety and health, subsurface environmental engineering has, increasingly, developed a wider purpose. In some circumstances, atmospheric pressure and temperature may be allowed to exceed the ranges that are acceptable for human tolerance. For example, in an underground repository for high level nuclear waste, a containment drift will be sealed against human access after emplacement of the waste canisters has been completed. However, the environment within the drift must still be maintained such that rock wall temperatures are controlled. This is necessary to enable the drift to be re-opened relatively quickly for retrieval of the nuclear waste at any subsequent time during the active life of the repository. Other forms of underground storage often require environmental control of pressure, temperature and humidity for the preservation of the stored material. Yet another trend is towards automated (manless) working faces and the possible use of underground space for *in situ* mineral processing. In such zones of future mines, environmental control will be required for the efficient operation of machines and processes, but not necessarily with an atmosphere acceptable to the unprotected human physiology.

1.3.2   Factors that affect the underground environment

During the development and operation of a mine or other underground facility, potential hazards arise from dust, gas emissions, heat and humidity, fires, explosions and radiation. Table 1.1 shows the factors that may contribute towards those hazards.

Table 1.1 Factors that feature in the creation and control of hazards in the subsurface environment

Factors that contribute to hazards			Methods of control	
Natural factors →	Design factors →	Hazard ←	Ancillary control ←	Airflow control
Depth below surface	Method of working	Dust	Dust suppression	Main fans
Surface climate	Layout of mine or facility	Gas emissions	Gas drainage	Booster fans Auxiliary ventilation
Geology		Heat and humidity	Refrigeration systems	Natural ventilation
Physical and chemical properties of rocks	Rate of rock fragmentation	Fires and explosions	Monitoring systems	Airlocks, stopping, air crossings,
	Mineral clearance	Radiation		Regulators
Gas content of strata	Type, size and siting of equipment			Number, size, lining, and layout of airways
Groundwater and other subsurface liquids	Vehicular traffic			
Age of airways	Stored materials			

These divide into features that are imposed by nature and those that are generated by design decisions on how to open up and operate the facility.

The major method of controlling atmospheric conditions in the subsurface is by airflow. This is produced, primarily, by main fans that are usually, but not necessarily, located on surface. National or state mining law may insist that main fans are sited on surface for gassy mines. While the main fan, or combination of main fans, handles all of the air that circulates through the underground network of airways, underground booster fans serve specific districts only. Auxiliary fans are used to pass air through ducts to ventilate blind headings. The distribution of airflow may further be controlled by ventilation doors, stoppings, air crossings and regulators.

It is often the case that it becomes impracticable or impossible to contend with all environmental hazards by ventilation alone. For example, increases in air temperature caused by compression of the air in the downcast shafts of deep mines may result in that air being too hot for personnel even before it enters the workings. No practical amount of increased airflow will solve that problem. Table 1.1 includes the ancillary control measures that may be advisable or necessary to supplement the ventilation system in order to maintain acceptable conditions underground.

1.3.3 The integration of ventilation planning into overall system design

The design of a major underground ventilation and environmental control system is a complex process with many interacting features. The principles of systems analyses should be applied to ensure that the consequences of such interaction are not overlooked (Chapter 9). However, ventilation and the underground environment must not be treated in isolation during planning exercises. They are, themselves, an integral part of the overall design of the mine or subsurface facility.

It has often been the case that the types, numbers and sizes of machines, the required rate of mineral production and questions of ground stability have dictated the layout of a mine without, initially, taking the demands of ventilation into account. This will result in a ventilation system that may lack effectiveness and, at best, will be more expensive in both operating and capital costs than would otherwise have been the case. A common error has been to size shafts that are appropriate for the hoisting duties but inadequate for the long-term ventilation requirement of the mine. Another frequent related problem is a ventilation infrastructure that was adequate for an initial layout but lacks the flexibility to handle fluctuating market demands for the mineral. Again, this can be very expensive to correct. The results of inadequate ventilation planning and system design are premature cessation of production, high costs of reconstruction, poor environmental conditions and, still too often, tragic consequences to the health and safety of the workforce. It is, therefore, most important that ventilation engineers should be incorporated as an integral part of a design team from the initial stages of planning a new mine or other underground facility.

FURTHER READING

Agricola, G. (1556) *De Re Metallica*. (Translated from Latin by H. C. Hoover and L. H. Hoover (1950), Dover Publications, New York.)

Atkinson, J. J. (1854) On the theory of the ventilation of mines. N. Engl. Inst. Min. Eng. (3), 118.

Hartley, Sir H. (1971) *Humphrey Davy*, S. R. Publishers.

Hinsley, F. B. (1967) The control of atmospheric conditions in mines. 11th Cadman Memorial Lecture. *Min. Eng.* (77), 289.

Hinsley, F. B. (1970) The development of the mechanical ventilation of coal mines. 1870–1940. *Univ. Nottingham Min. Mag.* **22**.

Mason, E. (1954) *Practical Coal Mining*. Virtue, London.

McPherson, M. J. (1964) Mine ventilation network problems, solution by digital computer. *Colliery Guardian* **209** (5392), 253–259.

Saxton, I. (1986–1987) Coal mine ventilation from Agricola to the 1980's. *Min. Eng.* **146**.

Scott, D. R., Hinsley, F. B. and Hudson, R. F. (1953) A calculator for the solution of ventilation network problems. *Trans. Inst. Min. Eng.* **112**, 623.

Wang, Y. J. and Hartman, H. L. (1967) Computer solution of three-dimensional mine ventilation networks with multiple fans and natural ventilation. *Int. J. Rock Mech. Min. Sci.* **2** (2), 129–154.

PART ONE

Basic Principles of Fluid Mechanics and Physical Thermodynamics

2

Introduction of fluid mechanics

2.1 INTRODUCTION

2.1.1 The concept of a fluid

A fluid is a substance in which the constituent molecules are free to move relative to each other. Conversely, in a solid, the relative positions of molecules remain essentially fixed under non-destructive conditions of temperature and pressure. While these definitions classify matter into fluids and solids, the fluids subdivide further into liquid and gases.

Molecules of any substance exhibit at least two types of forces; an attractive force that diminishes with the square of the distance between molecules, and a force of repulsion that becomes strong when molecules come very close together. In solids, the force of attraction is so dominant that the molecules remain essentially fixed in position while the resisting force of repulsion prevents them from collapsing into each other. However, if heat is supplied to the solid, the energy is absorbed internally causing the molecules to vibrate with increasing amplitude. If that vibration becomes sufficiently violent, then the bonds of attraction will be broken. Molecules will then be free to move in relation to each other —the solid melts to become a liquid.

When two moving molecules in a fluid converge on each other, actual collision is averted (at normal temperatures and velocities) because of the strong force of repulsion at short distances. The molecules behave as near perfectly elastic spheres, rebounding from each other or from the walls of the vessel. Nevertheless, in a liquid, the molecules remain sufficiently close together that the force of attraction maintains some coherence within the substance. Water poured into a vessel will assume the shape of that vessel but may not fill it. There will be a distinct interface (surface) between the water and the air or vapour above it. The mutual attraction between the water molecules is greater than that between a water molecule and molecules of the adjacent gas. Hence, the water remains in the vessel except for a few exceptional molecules that momentarily gain sufficient kinetic energy to escape through the interface (slow evaporation).

However, if heat continues to be supplied to the liquid then that energy is absorbed as an increase in the velocity of the molecules. The rising temperature of the liquid is, in fact, a measure of the internal kinetic energy of the molecules. At some critical temperature, depending on the applied pressure, the velocity of the molecules becomes so great that the forces of attraction are no longer sufficient to hold those molecules together as a discrete liquid. They separate to much greater distances apart, form bubbles of vapour and burst through the surface to mix with the air or other gases above. This is, of course, the common phenomenon of boiling or rapid evaporation. The liquid is converted into gas.

The molecules of a gas are identical to those of the liquid from which it evaporated. However, those molecules are now so far apart, and moving with such high velocity, that the forces of attraction are relatively small. The fluid can no longer maintain the coherence of a liquid. A gas will expand to fill any closed vessel within which it is contained.

The molecular spacing gives rise to distinct differences between the properties of liquids and gases. Two of these are, first, that the volume of gas with its large intermolecular spacing will be much greater than the same mass of liquid from which it evaporated. Hence, the density of gases (mass/volume) is much lower than that of liquids. Secondly, if pressure is applied to a liquid, then the strong forces of repulsion at small intermolecular distances offer such a high resistance that the volume of the liquid changes very little. For practical purposes most liquids may be regarded as incompressible. On the other hand, the far greater distances between molecules in a gas allow the molecules to be pushed closer together when subjected to compression. Gases, then, are compressible fluids.

Subsurface ventilation engineers need to be aware of the properties of both liquids and gases. In this chapter, we shall confine ourselves to incompressible fluids. Why is this useful when we are well aware that a ventilation system is concerned primarily with air, a mixture of gases and, therefore, compressible? The answer is that in a majority of mines and other subsurface facilities, the ranges of temperature and pressure are such that the variation in air density is fairly limited. Airflow measurements in mines are normally made to within 5% accuracy. A 5% change in air density occurs by moving through a vertical elevation of some 500 m in the gravitational field at the surface of the earth. Hence, the assumption of incompressible flow with its simpler analytical relationships gives acceptable accuracy in most cases. For the deeper and (usually) hotter facilities, the effects of pressure and temperature on air density must be taken into account through thermodynamic analyses. These are introduced in Chapter 3.

2.1.2   Volume flow, mass flow and the continuity equation

Most measurements of airflow in ventilation systems are based on the volume of air (m^3) that passes through a given cross-section of a duct or airway in unit time (1 s). The units of volume flow, Q, are, therefore, m^3/s. However, for accurate analyses when density variations are to be taken into account, it is preferable to work in mass flow—that is, the mass of air (kg) passing through the cross-section in 1 s. The units of mass flow, M, are then kg/s.

The relationship between volume flow and mass flow follows directly from the definition of density, ρ,

$$\rho = \frac{mass}{volume} \quad \frac{kg}{m^3} \tag{2.1}$$

and

$$\rho = \frac{mass\ flow}{volume\ flow} = \frac{M}{Q} \quad \frac{kg}{s} \frac{s}{m^3}$$

giving

$$M = Q\rho \quad kg/s \tag{2.2}$$

In any continuous duct or airway, the mass flows passing through all cross-sections along its length are equal, provided that the system is at steady state and there are no inflows or outflows of air or other gases between the two ends. If these conditions are met then

$$M = Q\rho = constant \quad kg/s \tag{2.3}$$

This is the simplest form of the continuity equation. It can, however, be written in other ways. A common method of measuring volume flow is to determine the mean velocity of air, u, over a given cross-section, and then to multiply by the area of that cross-section, A (Chapter 6):

$$Q = uA \quad \frac{m}{s} m^2 or \frac{m^3}{s} \tag{2.4}$$

Then the continuity equation becomes

$$M = \rho u A = constant \quad kg/s \tag{2.5}$$

As indicated in the preceding subsection, we can achieve acceptable accuracy in most situations within ventilation systems by assuming a constant density. The continuity equation then simplifies to

$$Q = uA = constant \quad m^3/s \tag{2.6}$$

This shows that, for steady-state conditions in a continuous airway, the velocity of the air varies inversely with cross-sectional area.

2.2 FLUID PRESSURE

2.2.1 The cause of fluid pressure

Section 2.1.1 described the dynamic behaviour of molecules in a liquid or gas. When a molecule rebounds from any confining boundary, a force equal to the rate of change of momentum of that molecule is exerted upon the boundary. If the area of the solid–fluid boundary is large compared with the average distance between molecular collisions then the statistical effect will be to give a uniform force distributed over that boundary. This is the case in most situations of importance in subsurface ventilation engineering.

Two further consequences arise from the bombardment of a very large number of molecules on a surface, each molecule behaving essentially as a perfectly elastic sphere. First, the force exerted by a static fluid will always be normal to the surface. We shall discover later that the situation is rather different when the dynamic forces of a moving fluid stream are considered (section 2.3). Secondly, at any point within a static fluid, the pressure is the same in all directions. Hence, static pressure is a scalar rather than a vector quantity.

Pressure is sometimes carelessly confused with force or thrust. The quantitative definition of pressure, P, is clear and simple

$$P = \frac{\text{force}}{\text{area}} \quad \frac{\text{N}}{\text{m}^2} \tag{2.7}$$

In the SI unit system, force is measured in newtons (N) and area in square metres. The resulting unit of pressure, the N/m^2, is usually called a pascal (Pa) after the French philosopher, Blaise Pascal (1623–1662).

2.2.2   Pressure head

If a liquid of density ρ is poured into a vertical tube of cross-sectional area, A, until the level reaches a height h, the volume of liquid is

$$\text{volume} = hA \quad \text{m}^3$$

Then from the definition of density (mass/volume), the mass of the liquid is

$$\text{mass} = \text{volume} \times \text{density}$$
$$= hA\rho \quad \text{kg}$$

The weight of the liquid will exert a force, F, on the base of the tube equal to mass $\times$ gravitational acceleration (g):

$$F = hA\rho g \quad \text{N}$$

However, as pressure = force/area, the pressure on the base of the tube is

$$P = \frac{F}{A} = \rho g h \quad \frac{\text{N}}{\text{m}^2} \text{ or Pa} \tag{2.8}$$

Hence, if the density of the liquid is known then the pressure may be quoted as h, the **head** of liquid. This concept is used in liquid-type manometers (section 2.2.4) which, although in declining use, are likely to be retained for many purposes owing to their simplicity.

Equation (2.8) can also be used for air and other gases. In this case, it should be remembered that the density will vary with height. A mean value may be used with little loss in accuracy for most mine shafts. However, here again, it is recommended that the more precise methodologies of thermodynamics be employed for elevation differences of more than 500 m.

2.2.3 Atmospheric pressure and gauge pressure

The blanket of air that shrouds the earth extends to approximately 40 km above the surface. At that height, its pressure and density tend towards zero. As we descend towards the earth, the number of molecules per unit volume increases, compressed by the weight of the air above. Hence, the pressure of the atmosphere also increases. However, the pressure at any point in the lower atmosphere is influenced not only by the column of air above it but also by the action of convection, wind currents and variations in temperature and water vapour content. Atmospheric pressure near the surface, therefore, varies with both place and time.

At the surface of the earth, atmospheric pressure is of the order of 100 000 Pa. For practical reference this is often translated into 100 kPa although the basic SI units should always be used in calculations. Older units used in meteorology for atmospheric pressure are the bar (10^5 Pa) and the millibar (100 Pa).

For comparative purposes, reference is often made to standard atmospheric pressure. This is the pressure that will support a 0.760 m column of mercury having a density of 13.5951×10^3 kg/m^3 in a standard earth gravitational field of 9.8066 m/s^2.

Then, from equation (2.8)

$$\text{one standard atmosphere} = \rho \times g \times h$$
$$= 13.5951 \times 10^3 \times 9.8066 \times 0.760$$
$$= 101.324 \times 10^3 \quad \text{Pa}$$

or

$$101.324 \quad \text{kPa}$$

The measurement of variations in atmospheric pressure is important during ventilation surveys (Chapter 6), for psychrometic measurements (Chapter 14), and also for predicting the emission of stored gases into a subsurface ventilation system (Chapter 12). However, for many purposes, it is required to measure differences in pressure. One common example is the difference between the pressure within a system such as a duct and the local atmosphere pressure. This is referred to as **gauge pressure**.

$$\text{absolute pressure} = \text{atmospheric pressure} + \text{gauge pressure} \qquad (2.9)$$

If the pressure within the system is below that of the local ambient atmospheric pressure then the negative gauge pressure is often termed the **suction pressure** or **vacuum** and the sign ignored. Care should be taken when using equation (2.9) as the gauge pressure may be positive or negative. However, the absolute pressure is always positive. Although many quoted measurements are pressure differences, it is the absolute pressures that are used in thermodynamic calculations. We must not forget to convert when necessary.

2.2.4 Measurement of air pressure

Barometers

Equation (2.8) showed that the pressure at the bottom of a column of liquid is equal to the product of the head (height) of the liquid, its density and the local value of gravitational acceleration. This principle was employed by Evangelista Torricelli (1608–1647), the Italian who invented the mercury barometer in 1643. Torricelli poured mercury into a glass tube, about one metre in length, closed at one end, and upturned the tube so that the open end dipped into a bowl of mercury. The level in the tube would then fall until the column of mercury, h, produced a pressure at the base that just balanced the atmospheric pressure acting on the open surface of mercury in the bowl.

The atmospheric pressure could then be calculated as (see equation (2.8))

$$p = \rho g h \quad \text{Pa}$$

where, in this case, ρ is the density of mercury.

Modern versions of the Torricelli instrument are still used as standards against which other types of barometer may be calibrated. Barometric (atmospheric) pressures are commonly quoted in millimetres (or inches) of mercury. However, for precise work, equation (2.8) should be employed using the density of mercury corresponding to its relevant temperature. Accurate mercury barometers have a thermometer attached to the stem of the instrument for this purpose and a sliding micrometer to assist in reading the precise height of the column. Furthermore, the local value of gravitational acceleration should be ascertained as this depends on latitude and altitude.

The space above the mercury in the barometer will not be a perfect vacuum as it contains mercury vapour. However, this exerts a pressure of less than 0.000 16 kPa at 20 °C and is quite negligible compared with the surface atmospheric pressure of near 100 kPa. This, coupled with the fact that the high density of mercury produces a barometer of reasonable length, explains why mercury rather than any other liquid is used. A water barometer would need to be about 10.5 m in height.

Owing to their fragility and slowness in reacting to temperature changes, mercury barometers are unsuitable for underground surveys. An aneroid barometer consists of a closed vessel which has been evacuated to a near-perfect vacuum. One or more elements of the vessel are flexible. These may take the form of a flexing diaphragm or the vessel itself may be shaped as a helical or spiral spring. The near-zero pressure within the vessel remains constant. However, as the surrounding atmospheric pressure varies, the appropriate element of the vessel will flex. The movement may be transmitted mechanically, magnetically or electrically to an indicator and/or re-corder.

Low cost aneroid barometers may be purchased for domestic or sporting use. Most altimeters are, in fact, aneroid barometers calibrated in metres (or feet) head of air. For the high accuracy required in ventilation surveys (Chapter 6) precision aneroids are available.

Another principle that can be employed in pressure transducers, including barom-eters, is the piezoelectric property of quartz. The natural frequency of a quartz beam varies with the applied pressure. As electrical frequency can be measured with great precision, this allows the pressure to be determined with good accuracy.

Differential pressure instruments

Differences in air pressure that need to be measured frequently in subsurface ventila-tion engineering rarely exceed 7 or 8 kPa and are often of the order of only a few pascals. The traditional instrument for such low pressure differences is the manometer. This relies on the displacement of liquid to produce a column, or head, that balances the differential pressure being measured. The most rudimentary manometer is the simple glass U-tube containing water, mercury or other liquid. A pressure difference applied across the ends of the tube causes the liquid levels in the two limbs to be displaced in opposite directions. A scale is used to measure the vertical distance between the levels and equation (2.8) used to calculate the required pressure differ-ential. Owing to the past widespread use of water manometers, the millimetre (or inch) of water column came to be used commonly as a measure of small pressure differentials, much as a head of mercury has been used for atmospheric pressures. However, it suffers from the same disadvantages in that it is not a primary unit but depends on the liquid density and local gravitational acceleration.

When a liquid other than water is used, the linear scale may be increased or decreased, dependent on the density of the liquid, so that it still reads directly in head of water. A pressure head in one fluid can be converted to a head in any other fluid provided that the ratio of the two densities is known:

$$p = \rho_1 g h_1 = \rho_2 g h_2 \quad \text{Pa}$$

or

$$h_2 = \frac{\rho_1}{\rho_2} h_1 \quad \text{m} \tag{2.10}$$

For high precision, the temperature of the liquid in a manometer should be obtained and the corresponding density determined. Equation (2.10) is then used to correct the reading, h_1, where ρ_1 is the actual liquid density and ρ_2 is the density at which the scale is calibrated.

Many variations of the manometer have been produced. Inclining one limb of the U-tube shortens its practicable range but gives greater accuracy of reading. Careful levelling of inclined manometers is required and they are no longer used in subsurface pressure surveys. Some models have one limb of the U-tube enlarged into a water reservoir. The liquid level in the reservoir changes only slightly compared with the balancing narrow tube. In the direct lift manometer, the reservoir is connected by flexible tubing to a short sight glass of variable inclination which may be raised or lowered against a graduated scale. This manipulation enables the meniscus to be adjusted to a fixed mark on the sight glass. Hence the level in the reservoir remains unchanged. The addition of a micrometer scale gives this instrument both a good range and high accuracy.

One of the problems in some water manometers is a misformed meniscus, particularly if the inclination of the tube is less than 5° from the horizontal. This difficulty may be overcome by employing a light oil, or other liquid that has good wetting properties on glass. Alternatively, the two limbs may be made large enough in diameter to give horizontal liquid surfaces whose position can be sensed electronically or by touch probes adjusted through micrometers.

U-tube manometers may feature as part of the permanent instrumentation of main and booster fans. Provided that the connections are kept firm and clean, there is little that can go wrong with these devices. Compact and portable inclined gauges are available for rapid readings of pressure differences across doors and stoppings in underground ventilation systems. However, in modern pressure surveying (Chapter 6) manometers have been replaced by the diaphragm gauge. This instrument consists essentially of a flexible diaphragm, across which is applied the differential pressure. The strain induced in the diaphragm is sensed electrically, mechanically or by magnetic means and transmitted to a visual indicator or recorder.

In addition to its portability and rapid reaction, the diaphragm gauge has many advantages for the subsurface ventilation engineer. First, it reflects directly a true pressure (force/area) rather than indirectly through a liquid medium. Secondly, it reacts relatively quickly to changes in temperature and does not require precise levelling. Thirdly, diaphragm gauges may be manufactured over a wide variety of ranges. A ventilation survey team may typically carry gauges ranging from 0–100 Pa to 0–5 kPa (or to encompass the value of the highest fan pressure in the system). One disadvantage of the diaphragm gauge is that its calibration may change with time and usage. Recalibration against a laboratory precision manometer is recommended prior to an important survey.

Other appliances are used occasionally for differential pressures in subsurface pressure surveys. Piezoelectric instruments are likely to increase in popularity. The aerostat principle eliminates the need for tubing between the two measurement points and leads to a type of differential barometer. In this instrument, a closed and rigid air vessel is maintained at a constant temperature and is connected to the outside atmosphere via a manometer or diaphragm gauge. As the inside of the vessel remains at near-constant pressure, any variations in atmospheric pressure cause a reaction on the manometer or gauge. Instruments based on this principle require independent calibration as slight movements of the diaphragm or liquid in the manometer result in the inside pressure not remaining truly constant.

2.3   FLUIDS IN MOTION

2.3.1   Bernoulli's equation for ideal fluids

As a fluid stream passes through a pipe, duct or other continuous opening, there will, in general, be changes in its velocity, elevation and pressure. In order to follow such changes it is useful to identify the differing forms of energy contained within a given mass of the fluid. For the time being, we will consider that the fluid is **ideal**; that is, it has no viscosity and proceeds along the pipe with no shear forces and no frictional losses. Secondly, we will ignore any thermal effects and consider mechanical energy only.

Suppose we have a mass, m, of fluid moving at velocity, u, at an elevation, Z, and a barometric pressure P. There are three forms of mechanical energy that we need to consider. In each case, we shall quantify the relevant term by assessing how much work we would have to do in order to raise that energy quantity from zero to its actual value in the pipe, duct or airway.

Kinetic energy

If we commence with the mass, m, at rest and accelerate it to velocity u in t seconds by applying a constant force F, then the acceleration will be uniform and the mean velocity is

$$\frac{0+u}{2} = \frac{u}{2} \quad \text{m/s}$$

Then

$$\text{distance travelled} = \text{mean velocity} \times \text{time}$$

$$= \frac{u}{2}t \quad \text{m}$$

Furthermore, the acceleration is defined as

$$\frac{\text{increase in velocity}}{\text{time}} = \frac{u}{t} \quad \text{m/s}^2$$

The force is given by

$$F = \text{mass} \times \text{acceleration}$$

$$= m\frac{u}{t} \quad \text{N}$$

and the work done to accelerate from rest to velocity u is

$$\text{WD} = \text{force} \times \text{distance} \quad \text{N m}$$

$$= m\,\frac{u}{t} \times \frac{u}{2}\,t$$

$$= m\,\frac{u^2}{2}\quad \text{N m or J}\tag{2.11}$$

The kinetic energy of the mass m is, therefore $mu^2/2$ joules.

Potential energy

Any base elevation may be used as the datum for potential energy. In most circumstances of underground ventilation engineering, it is differences in elevation that are important.

If our mass m is located on the base datum then it will have a potential energy of zero relative to that datum. We then exert an upward force, F, sufficient to counteract the effect of gravity:

$$F = \text{mass} \times \text{acceleration}$$

$$= mg\quad \text{N}$$

where g is the gravitational acceleration. In moving upward to the final elevation of Z metres above the datum, the work done is

$$\text{WD} = \text{force} \times \text{distance}$$

$$= mgZ\quad \text{J}\tag{2.12}$$

This gives the potential energy of the mass at elevation Z.

Flow work

Suppose we have a horizontal pipe, open at both ends and of cross-sectional area A as shown in Fig. 2.1. We wish to insert a plug of fluid, volume v and mass m, into the pipe. However, even in the absence of friction, there is a resistance due to the pressure of the fluid, P, that already exists in the pipe. Hence, we must exert a force, F, on the plug of fluid to overcome that resisting pressure. Our intent is to find the work done on the plug of fluid in order to move it a distance s into the pipe.

The force, F, must balance the pressure, P, which is distributed over the area, A:

$$F = PA\quad \text{N}$$

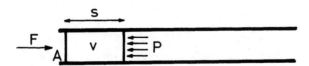

Figure 2.1   Flow work done on a fluid entering a pipe.

$$\text{work done} = \text{force} \times \text{distance}$$

$$= PAs \quad \text{J}$$

However, the product As is the swept volume v, giving

$$\text{WD} = Pv$$

Now, by definition, the density is

$$\rho = \frac{m}{v} \quad \frac{\text{kg}}{\text{m}^3}$$

or

$$v = \frac{m}{\rho}$$

Hence, the work done in moving the plug of fluid into the pipe is

$$\text{WD} = \frac{Pm}{\rho} \quad \text{J} \tag{2.13}$$

or P/ρ joules per kilogram.

As fluid continues to be inserted into the pipe to produce a continuous flow, then each individual plug must have this amount of work done on it. That energy is retained within the fluid stream and is known as the **flow work**. The appearance of pressure, P, within the expression for flow work has resulted in the term sometimes being labelled 'pressure energy'. This is very misleading as flow work is entirely different to the 'elastic energy' stored when a closed vessel of fluid is compressed. Some authorities also object to the term 'flow work' and have suggested 'convected energy' or, simply, 'Pv work'. Note that in Fig. 2.1 the pipe is open at both ends. Hence the pressure, P, inside the pipe does not change with time (the fluid is not compressed) when plugs of fluid continue to be inserted in a frictionless manner. When the fluid exits the system, it will carry kinetic and potential energy, and the corresponding flow work with it.

Now we are in a position to quantify the total mechanical energy of our mass of fluid, m. From expressions (2.11), (2.12) and (2.13)

$$\text{total mechanical} = \text{kinetic} + \text{potential} + \text{flow}$$
$$\text{energy} \qquad \text{energy} \qquad \text{energy} \qquad \text{work}$$

$$= \frac{mu^2}{2} + mZg + m\frac{P}{\rho} \quad \text{J} \tag{2.14}$$

If no mechanical energy is added to or subtracted from the fluid during its traverse through the pipe, duct or airway, and in the absence of frictional effects, the total mechanical energy must remain constant throughout the airway. Then equation (2.14) becomes

$$m\left(\frac{u^2}{2} + Zg + \frac{P}{\rho}\right) = \text{constant} \quad \text{J} \tag{2.15}$$

Another way of expressing this equation is to consider two stations, 1 and 2 along the pipe, duct or airway. Then

$$m\left(\frac{u_1^2}{2} + Z_1g + \frac{P_1}{\rho_1}\right) = m\left(\frac{u_2^2}{2} + Z_2g + \frac{P_2}{\rho_2}\right)$$

Now as we are still considering the fluid to be incompressible,

$$\rho_1 = \rho_2 = \rho \quad \text{(say)}$$

giving

$$\frac{u_1^2 - u_2^2}{2} + (Z_1 - Z_2)g + \frac{P_1 - P_2}{\rho} = 0 \quad \frac{\text{J}}{\text{kg}} \tag{2.16}$$

Note that dividing by m on both sides has changed the units of each term from J to J/kg. Furthermore, if we multiplied throughout by ρ then each term would take the units of pressure. Bernoulli's equation has, traditionally, been expressed in this form for incompressible flow.

Equation (2.16) is of fundamental importance in the study of fluid flow. It was first derived by Daniel Bernoulli (1700–1782), a Swiss mathematician, and is known throughout the world by his name.

As fluid flows along any opening, Bernoulli's equation allows us to track the interrelationships between the variables. Velocity u, elevation Z, and pressure P may all vary, but their combination as expressed in Bernoulli's equation remains true. It must be remembered, however, that it has been derived here on the assumptions of ideal (frictionless) conditions, constant density and steady-state flow. We shall see later how the equation must be amended for the real flow of compressible fluids.

2.3.2 Static, total and velocity pressures

Consider the level duct shown on Fig. 2.2. Three gauge pressures are measured. To facilitate visualization, the pressures are indicated as liquid heads on U-tube manom-

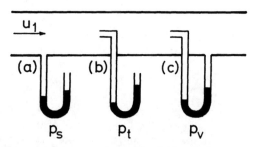

Figure 2.2 (a) Static, (b) total and (c) velocity pressures.

eters. However, the analysis will be conducted in terms of true pressure (N/m^2) rather than head of fluid.

In position (a), one limb of the U-tube is connected perpendicular through the wall of the duct. Any drilling burrs on the inside have been smoothed out so that the pressure indicated is not influenced by the local kinetic energy of the air. The other limb of the manometer is open to the ambient atmosphere. The gauge pressure indicated is known as the static pressure, p_s.

In position (b) the left tube has been extended into the duct and its open end turned so that it faces directly into the fluid stream. As the fluid impacts against the open end of the tube, it is brought to rest and the loss of its kinetic energy results in a local increase in pressure. The pressure within the tube then reflects the sum of the static pressure and the kinetic effect. Hence the manometer indicates a higher reading than in position (a). The corresponding pressure, p_t, is termed the total pressure.

The increase in pressure caused by the kinetic energy can be quantified by using Bernoulli's equation (2.16). In this case $Z_1 = Z_2$, and $u_2 = 0$. Then

$$\frac{P_2 - P_1}{\rho} = \frac{u_1^2}{2}$$

The local increase in pressure caused by bringing the fluid to rest is then

$$p_v = P_2 - P_1 = \rho \frac{u_1^2}{2} \quad \text{Pa} \qquad (2.17)$$

This is known as the velocity pressure and can be measured directly by connecting the manometer as shown in position (c). The left connecting tube of the manometer is at gauge pressure p_t and the right tube at gauge pressure p_s. It follows that

$$p_v = p_t - p_s$$

or

$$p_t = p_s + p_v \quad \text{Pa} \qquad (2.18)$$

In applying this equation, care should be taken with regard to sign as the static pressure, p_s, will be negative if the barometric pressure inside the duct is less than that of the outside atmosphere.

If measurements are actually made using a liquid in glass manometer as shown on Fig. 2.2 then the reading registered on the instrument is influenced by the head of fluid in the manometer tubes above the liquid level. If the manometer liquid has a density ρ_1, and the superincumbent fluid in both tubes has a density ρ_d, then the head indicated by the manometer, h, should be converted to true pressure by the equation

$$p = (\rho_1 - \rho_d)gh \quad \text{Pa} \qquad (2.19)$$

Reflecting back on equation (2.8) shows that this is the usual equation relating fluid head and pressure with the density replaced by the difference in the two fluid densities. In ventilation engineering, the superincumbent fluid is air, having a very low density compared with liquids. Hence, the ρ_d term in equation (2.19) is usually neglected.

However, if the duct or pipe contains a liquid rather than a gas then the full form of equation (2.19) should be employed.

A further situation arises when the fluid in the duct has a density, ρ_d, that is significantly different from that of the air (or other fluid), ρ_a, which exists above the liquid in the right-hand tube of the manometer in Fig. 2.2(a). Then

$$p = (\rho_1 - \rho_d)gh - (\rho_d - \rho_a)gh_2 \quad \text{Pa} \tag{2.20}$$

where h_2 is the vertical distance between the liquid level in the right side of the manometer and the connection into the duct.

Equations (2.19) and (2.20) can be derived by considering a pressure balance on the two sides of the U-tube above the lower of the two liquid levels.

2.3.3 Viscosity

Bernoulli's equation was derived in section 2.3.1 on the assumption of an ideal fluid, i.e. that flow could take place without frictional resistance. In subsurface ventilation engineering almost all of the work input by fans (or other ventilating devices) is utilized against frictional effects within the airways. Hence, we must find a way of amending Bernoulli's equation for the frictional flow of real fluids.

The starting point in an examination of 'frictional flow' is the concept of viscosity. Consider two parallel sheets of fluid a very small distance, dy, apart but moving at different velocities u and $u + du$ (Fig. 2.3). An equal but opposite force, F, will act upon each layer, the higher velocity sheet tending to pull its slower neighbour along and, conversely, the slower sheet tending to act as a brake on the higher velocity layer. If the area of each of the two sheets in near contact is A, then the shear stress is defined as τ (Greek 'tau') where

$$\tau = \frac{F}{A} \quad \frac{\text{N}}{\text{m}^2} \tag{2.21}$$

Among his many accomplishments, Isaac Newton (1642–1727) proposed that, for parallel motion of streamlines in a moving fluid, the shear stress transmitted across the fluid in a direction perpendicular to the flow is proportional to the rate of change of velocity, du/dy (velocity gradient):

$$\tau = \frac{F}{A} = \mu \frac{du}{dy} \quad \frac{\text{N}}{\text{m}^2} \tag{2.22}$$

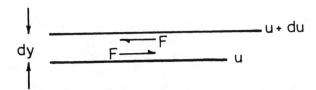

Figure 2.3 Viscosity causes equal but opposite forces to be exerted on adjacent laminae of fluid.

where the constant of proportionality, μ, is known as the coefficient of dynamic viscosity (usually referred to simply as dynamic viscosity). The dynamic viscosity of a fluid varies with its temperature. For air, it may be determined from

$$\mu_{air} = (17.0 + 0.045\, t) \times 10^{-6} \quad \frac{N\,s}{m^2}$$

and for water,

$$\mu_{water} = \left(\frac{64.72}{t + 31.766} - 0.2455 \right) \times 10^{-3} \quad \frac{N\,s}{m^2}$$

where t = temperature (°C) in the range 0–60 °C

The units of viscosity are derived by transposing equation (2.22):

$$\mu = \tau \frac{dy}{du} \quad \frac{N}{m^2}\, m\, \frac{s}{m} \text{ or } \frac{N\,s}{m^2}$$

A term which commonly occurs in fluid mechanics is the ratio of dynamic viscosity to fluid density. This is called the kinematic viscosity, v (Greek 'nu')

$$v = \frac{\mu}{\rho} \quad \frac{N\,s\,m^3}{m^2\,kg} \text{ or } N\,m\,\frac{s}{kg}$$

As $1\,N = 1\,kg \times 1\,(m/s^2)$, these units become

$$kg\, \frac{m\,m\,s}{s^2\,kg} = \frac{m^2}{s}$$

It is the transmission of shear stress that produces frictional resistance to motion in a fluid stream. Indeed, a definition of an 'ideal fluid' is one that has zero viscosity.

Following from our earlier discussion on the molecular behaviour of fluids (section 2.1.1), there would appear to be at least two effects that produce the phenomenon of viscosity. One is the attractive forces that exist between molecules—particularly those of liquids. This will result in the movement of some molecules tending to drag others along, and for the slower molecules to inhibit motion of faster neighbours. The second effect may be visualized by glancing again at Fig. 2.3. If molecules from the faster moving layer stray sideways into the slower layer then the inertia that they carry will impart kinetic energy to that layer. Conversely, migration of molecules from the slower to the faster layer will tend to retard its motion.

In liquids, the molecular attraction effect is dominant. Heating a liquid increases the internal kinetic energy of the molecules and also increases the average intermolecular spacing. Hence, as the attractive forces diminish with distance, the viscosity of a liquid decreases with respect to temperature. In a gas, the molecular attractive force is negligible. The viscosity of gases is much less than that of liquids and is caused by the molecular inertia effect. In this case, the increased velocity of molecules caused by heating will tend to enhance their ability to transmit inertia across streamlines and, hence, we may expect the viscosity of gases to increase with respect to temperature. This is, in fact, the situation observed in practice.

In both of these explanations of viscosity, the effect works between consecutive layers equally well in both directions. Hence, dynamic equilibrium is achieved with both the higher and lower velocity layers maintaining their net energy levels. Unfortunately, no real process is perfect in fluid mechanics. Some of the useful mechanical energy will be transformed into the much less useful heat energy. In a level airway, the loss of mechanical energy is reflected in an observable drop in pressure. This is often termed the 'frictional pressure drop'.

Recalling that Bernoulli's equation was derived for mechanical energy terms only in section 2.3.1, it follows that for the flow of real fluids, the equation must take account of the frictional loss of mechanical energy. We may rewrite equation (2.16) as

$$\frac{u_1^2}{2} + Z_1 g + \frac{P_1}{\rho} = \frac{u_2^2}{2} + Z_2 g \frac{P_2}{\rho} + F_{12} \qquad \frac{J}{kg} \qquad (2.23)$$

where F_{12} = energy converted from the mechanical form to heat (J/kg). The problem now turns to one of quantifying the frictional term F_{12}. For that, we must first examine the nature of fluid flow.

2.3.4 Laminar and turbulent flow; Reynolds' number

In our everyday world, we can observe many examples of the fact that there are two basic kinds of fluid flow. A stream of oil poured out of a can flows smoothly and in a controlled manner while water, poured out at the same rate, would break up into cascading rivulets and droplets. This example seems to suggest that the type of flow depends on the fluid. However, a light flow of water falling from a circular outlet has a steady and controlled appearance, but if the flow rate is increased the stream will assume a much more chaotic form. The type of flow seems to depend on the flow rate as well as the type of fluid.

Throughout the nineteenth century, it was realized that these two types of flows existed. The German engineer G. H. L. Hagen (1797–1884) found that the type of flow depended upon the velocity and viscosity of the fluid. However, it was not until the 1880s that Professor Osborne Reynolds of Manchester University in England established a means of characterizing the type of flow regime through a combination of experiments and logical reasoning.

Reynolds' laboratory tests consisted of injecting a filament of coloured dye into the bell mouth of a horizontal glass tube that was submerged in still water within a large glass-walled tank. The other end of the tube passed through the end of the tank to a valve which was used to control the velocity of water within the tube. At low flow rates, the filament of dye formed an unbroken line in the tube without mixing with the water. At higher flow rates the filament of dye began to waver. As the velocity in the tube continued to be increased the wavering filament suddenly broke up to mix almost completely with the water.

In the initial type of flow, the water appeared to move smoothly along streamlines, layers or laminae, parallel to the axis of the tube. We call this **laminar flow**. Appropriately, we refer to the completely mixing type of behaviour as **turbulent**

flow. Reynolds' experiments had, in fact, identified a third regime—the wavering filament indicated a transitional region between fully laminar and fully turbulent flow. Another observation made by Reynolds was that the break-up of the filament always occurred, not at the entrance, but about 30 diameters along the tube.

The essential difference between laminar and turbulent flow is that in the former, movement across streamlines is limited to the molecular scale, as described in section 2.3.3. However, in turbulent flow, packets of fluid move sideways in small turbulent eddies. These should not be confused with the even larger and more predictable oscillations that may occur with respect to time and position such as the vortex action caused by fans, pumps or obstructions in the airflow. The turbulent eddies appear random in the complexity of their motion. However, as with all 'random' phenomena, the term is used generically to describe a process that is too complex to be described by current mathematical knowledge. The development of supercomputers and numerical simulation shows promise of yielding methods of analysis and predictive models of turbulent flow. At the present time, however, most calculations involving turbulent flow still depend upon empirical factors.

The flow of air in the vast majority of 'ventilated' places underground is turbulent in nature. However, the sluggish movement of air or other fluids in zones behind stoppings or through fragmented strata may be laminar. It is, therefore, important that the subsurface ventilation engineer be familiar with both types of flow.

Returning to Osborne Reynolds, he found that the development of full turbulence depended not only on velocity, but also on the diameter of the tube. He reasoned that if we were to compare the flow regimes between differing geometrical configurations and various fluids we must have some combination of geometric and fluid properties that quantified the degree of similitude between any two systems. Reynolds was also familiar with the concepts of 'inertial (kinetic) force', $\rho(u^2/2)$ (newtons per square metre of cross section) and 'viscous force', $\tau = \mu \, du/dy$ (newtons per square metre of shear surface). Reynolds argued that the dimensionless ratio of 'inertial forces' to 'viscous forces' would provide a basis of comparing fluid systems:

$$\frac{\text{inertial force}}{\text{viscous force}} = \rho \frac{u^2}{2} \frac{1}{\mu} \frac{dy}{du} \tag{2.24}$$

Now, for similitude to exist, all steady-state velocities, u, or changes in velocity, du, within a given system are proportional to each other. Furthermore, all lengths are proportional to any chosen characteristic length, L. Hence, in equation (2.24) we can replace du by u, and dy by L. The constant, 2, can also be dropped as we are simply looking for a combination of variables that characterize the system. That combination now becomes

$$\rho u^2 \frac{1}{\mu} \frac{L}{u}$$

or

$$\frac{\rho u L}{\mu} = \text{Re} \tag{2.25}$$

As equation (2.24) is dimensionless then so, also, must this latter expression be dimensionless. This can easily be confirmed by writing down the units of the component variables.

The result we have reached here is of fundamental importance to the study of fluid flow. The dimensionless group $\rho u L/\mu$ is known universally as Reynolds' number, Re. In subsurface ventilation engineering, the characteristic length is normally taken to be the hydraulic mean diameter of an airway, d, and the characteristic velocity is usually the mean velocity of the airflow. Then

$$\mathrm{Re} = \frac{\rho u d}{\mu}$$

At Reynolds' numbers of less than 2000, viscous forces prevail and the flow will be laminar. The Reynolds number over which fully developed turbulence exists is less well defined. The onset of turbulence will occur at Reynolds' numbers of 2500 to 3000 assisted by any vibration, roughness of the walls of the pipe or any momentary perturbation in the flow.

Example   A ventilation shaft of diameter 5 m passes an airflow of 200 m³/s at a mean density of 1.2 kg/m³ and a mean temperature of 18 °C. Determine the Reynolds number for the shaft.

Solution   For air at 18 °C

$$\mu = (17.0 + 0.045 \times 18) \times 10^{-6}$$
$$= 17.81 \times 10^{-6}\,\mathrm{N\,s/m^2}$$

Air velocity

$$u = \frac{Q}{A} = \frac{200}{\pi 5^2/4}$$

$$= 10.186\,\mathrm{m/s}$$

$$\mathrm{Re} = \frac{\rho u d}{\mu}$$

$$= \frac{1.2 \times 10.186 \times 5}{17.81 \times 10^{-6}} = 3.432 \times 10^6$$

This Reynolds' number indicates that the flow will be turbulent.

2.3.5   Frictional losses in laminar flow; Poiseuille's equation

Now that we have a little background on the characteristics of laminar and turbulent flow, we can return to Bernoulli's equation corrected for friction (equation (2.23))

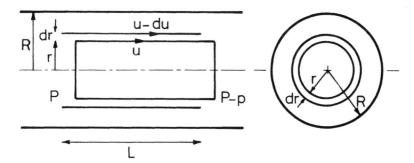

Figure 2.4 Viscous drag opposes the motive effect of applied pressure difference.

and attempt to find expressions for the work done against friction, F_{12}. First, let us deal with the case of laminar flow.

Consider a pipe of radius R as shown in Fig. 2.4. As the flow is laminar, we can imagine concentric cylinders of fluid telescoping along the pipe with zero velocity at the walls and maximum velocity in the centre. Two of these cylinders of length L and radii r and $r + dr$ are shown. The velocities of the cylinders are u and $u - du$ respectively.

The force propagating the inner cylinder forward is produced by the pressure difference across its two ends, p, multiplied by its cross-sectional area, πr^2. This force is resisted by the viscous drag of the outer cylinder, τ, acting on the 'contact' area $2\pi r L$. As these forces must be equal at steady state conditions,

$$2\pi r \tau L = \pi r^2 p$$

However, $\tau = -\mu\, du/dr$ (equation (2.22) with a negative du), giving

$$-\mu \frac{du}{dr} = \frac{r}{2} \frac{p}{L}$$

or

$$du = -\frac{p}{L} \frac{r}{2\mu} dr \quad \text{m/s} \tag{2.26}$$

For a constant diameter tube, the pressure gradient along the tube p/L is constant. So, also, is μ for the Newtonian fluids that we are considering. (A Newtonian fluid is defined as one in which viscosity is independent of velocity.)

Equation (2.26) can, therefore, be integrated to give

$$u = -\frac{p}{L} \frac{1}{2\mu} \frac{r^2}{2} + C \tag{2.27}$$

At the wall of the tube, $r = R$ and $u = 0$. This gives the constant of integration to be

$$C = \frac{p}{L} \frac{R^2}{4\mu}$$

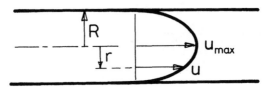

Figure 2.5 The velocity profile for laminar flow is parabolic.

Substituting back into equation (2.27) gives

$$u = \frac{1}{4\mu}\frac{p}{L}(R^2 - r^2) \quad \text{m/s} \tag{2.28}$$

Equation (2.28) is a general equation for the velocity of the fluid at any radius and shows that the velocity profile across the tube is parabolic (Fig. 2.5). Along the centre-line of the tube, $r = 0$ and the velocity reaches a maximum of

$$u_{max} = \frac{1}{4\mu}\frac{p}{L}R^2 \quad \frac{\text{m}}{\text{s}} \tag{2.29}$$

The velocity terms in the Bernoulli equation are mean velocities. It is, therefore, preferable that the work done against viscous friction should also be expressed in terms of a mean velocity, u_m. We must be careful how we define **mean velocity** in this context. Our convention is to determine it as

$$u_m = Q/A \quad \text{m/s} \tag{2.30}$$

where Q = volume airflow (m³/s) and A = cross-sectional area (m²). We could define another mean velocity by integrating the parabolic equation (2.28) with respect to r and dividing the result by R. However, this would not take account of the fact that the volume of fluid in each concentric shell of thickness dr increases with radius. In order to determine the true mean velocity, consider the elemental flow dQ through the annulus of cross-sectional area $2\pi r\, dr$ at radius r and having a velocity of u (Fig. 2.4):

$$dQ = u2\pi r\, dr$$

Substituting for u from equation (2.28) gives

$$dQ = \frac{2\pi}{4\mu}\frac{p}{L}(R^2 - r^2)r\, dr$$

$$Q = \frac{2\pi}{4\mu}\frac{p}{L}\int_0^R (R^2 r - r^3)\, dr$$

Integrating gives

$$Q = \frac{\pi R^4}{8\mu}\frac{p}{L} \quad \text{m}^3/\text{s} \tag{2.31}$$

This is known as the Poiseuille equation or, sometimes, the Hagen–Poiseuille Equation. J. L. M. Poiseuille (1799–1869) was a French physician who studied the flow of blood in capillary tubes.

For engineering use, where the dimensions of a given pipe and the viscosity of fluid are known, Poiseuille's equation may be written as a pressure drop–quantity relationship.

$$p = \frac{8\mu L}{\pi R^4} Q$$

or

$$p = R_L Q \quad \text{Pa} \tag{2.32}$$

where

$$R_L = \frac{8\mu L}{\pi R^4} \quad \text{N s/m}^5$$

and is known as the laminar resistance of the pipe: equation (2.32) shows clearly that in laminar flow the frictional pressure drop is proportional to the volume flow for any given pipe and fluid.

Combining equations (2.30) and (2.31) gives the required mean velocity

$$u_m = \frac{\pi R^4}{8\mu} \frac{p}{L} \frac{1}{\pi R^2}$$

$$= \frac{R^2}{8\mu} \frac{p}{L} \quad \text{m/s} \tag{2.33}$$

or

$$p = \frac{8\mu u_m}{R^2} L \quad \text{Pa} \tag{2.34}$$

This latter form gives another expression for the frictional pressure drop in laminar flow.

To see how we can use this equation in practice, let us return to the frictional form of Bernoulli's equation (see equation (2.23))

$$\frac{u_1^2 - u_2^2}{2} + (Z_1 - Z_2)g + \frac{P_1 - P_2}{\rho} = F_{12} \quad \frac{\text{J}}{\text{kg}}$$

Now for incompressible flow along a level pipe of constant cross-sectional area, $Z_1 = Z_2$ and $u_1 = u_2 = u_m$. Then

$$\frac{P_1 - P_2}{\rho} = F_{12} \quad \frac{\text{J}}{\text{kg}} \tag{2.35}$$

However, $P_1 - P_2$ is the same pressure difference as p in equation (2.34). Hence the work done against friction is

$$F_{12} = \frac{8\mu u_m}{\rho R^2} L \quad \frac{J}{kg} \tag{2.36}$$

Bernoulli's equation for incompressible laminar frictional flow now becomes

$$\frac{u_1^2 - u_2^2}{2} + (Z_1 - Z_2)g + \frac{P_1 - P_2}{\rho} = \frac{8\mu u_m}{\rho R^2} L \quad \frac{J}{kg} \tag{2.37}$$

If the pipe is of constant cross-sectional area, then $u_1 = u_2 = u_m$ and the kinetic energy term disappears. On the other hand, if the cross-sectional area and, hence, the velocity, varies along the pipe then u_m may be established as a weighted mean. For large changes in cross-sectional area, the full length of pipe may be subdivided into increments for analysis.

Example   A pipe of diameter 2 cm rises through a vertical distance of 5 m over the total pipe length of 2000 m. Water of mean temperature 15 °C flows up the tube to exit at atmospheric pressure of 100 kPa. If the required flowrate is 1.6 litres per minute, find the resistance of the pipe, the work done against friction and the head of water that must be applied at the pipe entrance.

Solution   It is often the case that measurements made in engineering are not in SI units. We must be careful to make the necessary conversions before commencing any calculations.

$$\text{flow rate } Q = 1.6 \text{ litres/minute}$$

$$= \frac{1.6}{1000 \times 60} = 2.667 \times 10^{-5} \, m^3/s$$

cross-sectional area of pipe $A = \pi d^2/4$

$$= \pi \times (0.02)^2/4 = 3.142 \times 10^{-4} \, m^2$$

Mean velocity $u = \dfrac{Q}{A}$

$$= \frac{2.667 \times 10^{-5}}{3.142 \times 10^{-4}} = 0.08488 \, m/s$$

(We have dropped the subscript m. For simplicity the term u from now on will refer to the mean velocity defined as Q/A.) From section 2.3.3, the viscosity of water at 15 °C is

$$\mu = \left(\frac{64.72}{15 + 31.766} - 0.2455 \right) \times 10^{-3}$$

$$= 1.138 \times 10^{-3} \, N\,s/m^2$$

Before we can begin to assess frictional effects we must check whether the flow is laminar or turbulent. We do this by calculating the Reynolds number

$$Re = \frac{\rho u d}{\mu}$$

where ρ = density of water (taken as $1000\,kg/m^3$):

$$Re = \frac{1000 \times 0.08488 \times 0.02}{1.138 \times 10^{-3}}$$

$$= 1491 \quad \text{(dimensionless)}$$

As Re is below 2000, the flow is laminar and we should use the equations based on viscous friction.

Laminar resistance of pipe (from equation (2.32))

$$R_L = \frac{8\mu L}{\pi R^4}$$

$$= \frac{8 \times 1.1384 \times 10^{-3} \times 2000}{\pi \times (0.01)^4}$$

$$= 580 \times 10^6 \, N\,s/m^5$$

Frictional pressure drop in the pipe (equation (2.32))

$$p = R_L Q$$

$$= 580 \times 10^6 \times 2.667 \times 10^{-5}$$

$$= 15\,461\,Pa$$

Work done against friction (equation (2.36))

$$F_{12} = \frac{8\mu u L}{R^2}$$

$$= \frac{8 \times 1.1384 \times 10^{-3} \times 0.08488 \times 2000}{1000 \times (0.01)^2}$$

$$= 15.461\,J/kg$$

This is the amount of mechanical energy transformed to heat in joules per kilogram of water. Note the similarity between the values for frictional pressure drop, p, and work done against friction, F_{12}. We have illustrated, by this example, a relationship between p and F_{12} that will be of particular significance in comprehending the behaviour of airflows in ventilation systems, namely

$$\frac{p}{\rho} = F_{12}$$

In fact, having calculated p as $15\,461\,Pa$, the value of F_{12} may be quickly evaluated as

$$\frac{15\,461}{1000} = 15.461 \text{ J/kg}$$

To find the pressure at the pipe inlet we may use Bernoulli's equation corrected for frictional effects (see equation (2.23)):

$$\frac{u_1^1 - u_2^2}{2} + (Z_1 - Z_2)g + \frac{P_1 - P_2}{\rho} = F_{12} \quad \frac{\text{J}}{\text{kg}}$$

In this example

$$u_1 = u_2$$
$$Z_1 - Z_2 = -5 \text{ m}$$

and

$$P_2 = 100 \text{ kPa} = 100\,000 \text{ Pa}$$

giving

$$-5 \times 9.81 + \frac{P_1 - 100\,000}{1000} = 15.461 \text{ J/kg}$$

This yields the absolute pressure at the pipe entry as

$$P_1 = 164.5 \times 10^3 \text{ Pa}$$

or

$$164.5 \text{ kPa}$$

If the atmospheric pressure at the location of the bottom of the pipe is also 100 kPa, then the gauge pressure, p_g, within the pipe at that same location

$$p_g = 164.5 - 100 = 64.5 \text{ kPa}$$

This can be converted into a head of water, h_1, from equation (2.8):

$$p_g = \rho g h_1$$

$$h_1 = \frac{64.5 \times 10^3}{1000 \times 9.81}$$

$$= 6.576 \text{ m of water}$$

Thus, a header tank with a water surface maintained 6.576 m above the pipe entrance will produce the required flow of 1.6 litres/minute along the pipe.

The experienced engineer would have determined this result quickly and directly after calculating the frictional pressure drop to be 15 461 Pa. The frictional head loss

$$h_1 = \frac{p}{\rho g} = \frac{15\,461}{1000 \times 9.81} = 1.576 \text{ m of water}$$

The head of water at the pipe entrance must overcome the frictional head loss as well as the vertical lift of 5 m. Then

$$h_1 = 5 + 1.576 = 6.576 \, \text{m of water}$$

2.3.6 Frictional losses in turbulent flow

The previous section showed that the parallel streamlines of laminar flow and Newton's perception of viscosity enabled us to produce quantitative relationships through purely analytical means. Unfortunately, the highly convoluted streamlines of turbulent flow caused by the interactions between both localized and propagating eddies have, so far, proved resistive to completely analytical techniques. Numerical methods using the memory capacities and speeds of supercomputers allow the flow to be simulated by a large number of small packets of fluids, each one influencing the behaviour of those around it. These mathematical models may be used to simulate turbulent flow in given geometrical systems, or to produce statistical trends. However, the vast majority of engineering applications involving turbulent flow still rely on a combination of analysis and empirical factors. The construction of physical models for observation in wind tunnels or other fluid flow test facilities remains a common means of predicting the behaviour and effects of turbulent flow.

The Chézy–Darcy equation

The discipline of hydraulics was studied by philosophers of the ancient civilizations. However, the beginnings of our present treatment of fluid flow owe much to the hydraulic engineers of eighteenth and nineteenth century France. During his reign, Napoleon Bonaparte encouraged the research and development necessary for the construction of water distribution and drainage systems in Paris.

Antoine de Chézy (1719–1798) carried out a series of experiments on the river Seine and on canals in about 1769. He found that the mean velocity of water in open ducts was proportional to the square root of the channel gradient, cross-sectional area of flow and inverse of the wetted perimeter:

$$u \propto \sqrt{\frac{A}{\text{per}} \frac{h}{L}}$$

where h = vertical distance dropped by the channel in a length L (h/L = hydraulic gradient), per = wetted perimeter (m) and $\propto$ means 'proportional to'. Inserting a constant of proportionality, c, gives

$$u = c \sqrt{\frac{A}{\text{per}} \frac{h}{L}} \quad \text{m/s} \tag{2.38}$$

where c is known as the Chézy coefficient.

Equation (2.38) has become known as Chézy's equation for channel flow. Subsequent analysis shed further light on the significance of the Chézy coefficient.

When a fluid flows along a channel, a mean shear stress τ is set up at the fluid–solid boundaries. The drag on the channel walls is then

$$\tau \text{ per } L$$

where per is the 'wetted' perimeter. This must equal the pressure force causing the fluid to move, pA, where p is the difference in pressure along length L:

$$\tau \text{ per } L = Ap \quad \text{N} \tag{2.39}$$

(A similar equation was used in section 2.3.5 for a circular pipe.) However,

$$p = \rho g h \quad \text{Pa}$$

(equation (2.8)) giving

$$\tau = \frac{A}{\text{per}} \rho g \frac{h}{L} \quad \frac{\text{N}}{\text{m}^2} \tag{2.40}$$

If the flow is fully turbulent, the shear stress or drag, τ, exerted on the channel walls is also proportional to the inertial (kinetic) energy of the flow expressed in joules per cubic metre:

$$\tau \propto \rho \frac{u^2}{2} \quad \frac{\text{J}}{\text{m}^3} = \frac{\text{Nm}}{\text{m}^3} \text{ or } \frac{\text{N}}{\text{m}^2}$$

or

$$\tau = f\rho \frac{u^2}{2} \quad \frac{\text{N}}{\text{m}^2} \tag{2.41}$$

where f is a dimensionless coefficient which, for fully developed turbulence, depends only on the roughness of the channel walls.

Equating equations (2.40) and (2.41) gives

$$f\frac{u^2}{2} = \frac{A}{\text{per}} g \frac{h}{L}$$

or

$$u = \sqrt{\frac{2g}{f}} \sqrt{\frac{A}{\text{per}} \frac{h}{L}} \quad \text{m/s} \tag{2.42}$$

Comparing this with equation (2.38) shows that Chézy's coefficient, c, is related to the roughness of the channel:

$$c = \sqrt{\frac{2g}{f}} \quad \text{m}^{1/2}/\text{s} \tag{2.43}$$

The development of flow relationships was continued by Henri Darcy (1803–1858), another French engineer, who was interested in the turbulent flow of water in pipes. He adapted Chézy's work to the case of circular pipes and ducts running full. Then $A = \pi d^2/4$, per $= \pi d$ and the fall in elevation of Chézy's channel became the head loss,

h (metres of fluid) along the pipe length L. Equation (2.42) now becomes

$$u^2 = \frac{2g}{f} \frac{\pi d^2}{4} \frac{1}{\pi d} \frac{h}{L}$$

or

$$h = \frac{4fLu^2}{2gd} \quad \text{metres of fluid} \tag{2.44}$$

This is the well-known Chézy–Darcy equation, sometimes also known simply as the Darcy equation or the Darcy–Weisbach equation.

The head loss, h, can be converted to a frictional pressure drop, p, by the now familiar relationship, $p = \rho g h$, to give

$$p = \frac{4fL}{d} \frac{\rho u^2}{2} \quad \text{Pa} \tag{2.45}$$

or a frictional work term

$$F_{12} = \frac{p}{\rho} = \frac{4fL}{d} \frac{u^2}{2} \quad \frac{\text{J}}{\text{kg}} \tag{2.46}$$

The Bernoulli equation for frictional and turbulent flow becomes

$$\frac{u_1^2 - u_2^2}{2} + (Z_1 - Z_2)g + \frac{P_1 - P_2}{\rho} = \frac{4fL}{d} \frac{u^2}{2} \quad \frac{\text{J}}{\text{kg}} \tag{2.47}$$

where u is the mean velocity.

The most common form of the Chézy–Darcy equation is that given as equation (2.44). Leaving the constant 2 uncancelled provides a reminder that the pressure loss due to friction is a function of kinetic energy, $u^2/2$. However, some authorities have combined the 4 and the f into a different coefficient of friction $\lambda(=4f)$ while others, presumably disliking Greek letters, replaced the symbol λ by (would you believe it?) f. We now have a confused situation in the literature of fluid mechanics where f may mean the original Chézy–Darcy coefficient of friction, or four times that value. When reading the literature, care should be taken to confirm the nomenclature used by the relevant author. Throughout this book, f is used to mean the original Chézy–Darcy coefficient as used in equation (2.44).

In order to generalize our results to ducts or airways of non-circular cross-section, we may define a hydraulic radius as

$$r_h = \frac{A}{\text{per}} \quad \text{m} \tag{2.48}$$

$$= \frac{\pi d^2}{4 \pi d} = \frac{d}{4}$$

Reference to the 'hydraulic mean diameter' denotes $4A/\text{per}$. This device works well for turbulent flow but must not be applied to laminar flow where the resistance to flow is caused by viscous action throughout the body of the fluid rather than concentrated around the perimeter of the walls.

Substituting for d in equation (2.45) gives

$$p = fL \frac{\text{per}}{A} \rho \frac{u^2}{2} \quad \text{Pa} \tag{2.49}$$

This can also be expressed as a relationship between frictional pressure drop, p, and volume flow, Q. Replacing u by Q/A in equation (2.49) gives

$$p = \frac{fL}{2} \frac{\text{per}}{A^3} \rho Q^2 \quad \text{Pa}$$

or

$$p = R_t \rho Q^2 \quad \text{Pa} \tag{2.50}$$

where

$$R_t = \frac{fL}{2} \frac{\text{per}}{A^3} \quad \text{m}^{-4} \tag{2.51}$$

This is known as the rational turbulent resistance of the pipe, duct or airway and is a function only of the geometry and roughness of the opening.

The coefficient of friction, f

It is usually the case that a significant advance in research opens up new avenues of investigation and produces a flurry of further activity. So it was following the work of Osborne Reynolds. During the first decade of this century, fluid flow through pipes was investigated in great detail by engineers such as Thomas E. Stanton (1865–1931) and J. R. Pannel in the United Kingdom and L. Prandtl in Germany. A major cause for concern was the coefficient of friction, f.

There were two problems. First, how could one predict the value of f for any given pipe without actually constructing the pipe and conducting a pressure–flow test on it? Secondly, it was found that f was not a true constant but varied with Reynolds' number for very smooth pipes and, particularly, at low values of Reynolds' number. The latter is not too surprising as f was introduced initially as a constant of proportionality between shear stress at the walls and inertial force of the fluid (equation (2.41)) for fully developed turbulence. At the lower Reynolds' numbers we may enter the transitional or even laminar regimes.

Figure 2.6 illustrates the types of results that were obtained. A very smooth pipe exhibited a continually decreasing value of f. This is labelled as the turbulent smooth pipe. However, for rougher pipes, the values of f broke away from the smooth pipe curve at some point and, after a transitional region, settled down to a constant value, independent of Reynolds' number. This phenomenon was quantified empirically through a series of classical experiments conducted in Germany by Johann Nikuradse, a former student of Prandtl.

Nikuradse took a number of smooth pipes of diameter 2.5, 5 and 10 cm, and coated the inside walls uniformly with grains of graded sand. The roughness of

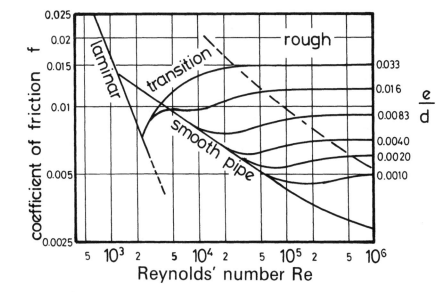

Figure 2.6 Variation of f with respect to Re as found by Nikuradse.

each tube was then defined as e/d where e was the diameter of the sand grains and d the diameter of the tube. The advantages of dimensionless numbers had been well learned from Reynolds. The corresponding f–Re relationships are illustrated on Fig. 2.6.

The investigators of the time were then faced with an intriguing question. How could a pipe of given roughness and passing a turbulent flow be 'smooth' (i.e. follow the smooth pipe curve) at certain Reynolds' numbers but become 'rough' (constant f) at higher Reynolds' numbers? The answer lies in our initial concept of turbulence—the formation and maintenance of small, interacting and propagating eddies within the fluid stream. These necessitate the existence of cross-velocities with vector components perpendicular to the longitudinal axis of the tube. At the walls there can be no cross-velocities except on a molecular scale. Hence, there must be a thin layer close to each wall through which the velocity increases from zero (actually at the wall) to some finite velocity sufficiently far away from the wall for an eddy to exist. Within that thin layer the streamlines remain parallel to each other and to the wall, i.e. laminar flow.

Although this laminar sublayer is very thin, it has a marked effect on the behaviour of the total flow in the pipe. All real surfaces (even polished ones) have some degree of roughness. If the peaks of the roughness, or asperities, do not protrude through the laminar sublayer then the surface may be described as 'hydraulically smooth' and the wall resistance is limited to that caused by viscous shear within the fluid. On the other hand, if the asperities protrude well beyond the laminar sublayer then the disturbance to flow that they produce will cause additional eddies to be formed, consuming mechanical energy and resulting in a higher resistance to flow. Further-more, as the velocity and, hence, the Reynolds number increases, the thickness of the laminar sublayer decreases. Any given pipe will then be hydraulically smooth if the asperities are submerged within the laminar sublayer and hydraulically rough if the asperities project beyond the laminar sublayer. Between the two conditions there will be a transition zone where some, but not all, of the asperities protrude through the laminar sublayer. The hypothesis of the existence of a laminar sublayer explains the behaviour of the curves in Fig. 2.6.

Nikuradse's work marked a significant step forward in that it promised a means of predicting the coefficient of friction and, hence, the resistance of any given pipe passing turbulent flow. However, there continued to be difficulties. In real pipes, ducts or underground airways, the wall asperities are not all of the same size, nor are they uniformly dispersed. In particular, mine airways show great variation in their roughness. Concrete lining in ventilation shafts may have a uniform e/d value as low as 0.001. On the other hand, where shaft tubbing or regularly spaced airway supports are used, the turbulent wakes on the downstream side of the supports create a dependence of airway resistance on their distance apart. Furthermore, the immediate wall roughness may be superimposed upon larger-scale sinuosity of the airways and, perhaps, the existence of cross-cuts or other junctions. The larger-scale vortices produced by these macro effects may be more energy demanding than the smaller eddies of normal turbulent flow and, hence, produce a much higher value of f. Many airways also have wall roughnesses that exhibit a directional bias, produced by the mechanized or drill and blast methods of driving the airway, or natural cleavage of the rock.

For all of these reasons, there may be a significant divergence between Nikuradse's curves and results obtained in practice, particularly in the transitional zone. Further experiments and analytical investigations were carried out in the late 1930s by C. F. Colebrook in England. The equations that were developed were somewhat awkward to use. However, the concept of 'equivalent sand grain roughness' was further developed by the American engineer Lewis F. Moody in 1944. The ensuing chart, shown on Fig. 2.7, is known as the Moody diagram and is now widely employed by practicing engineers to determine coefficients of friction.

Equations describing $f-Re$ relationships

The literature is replete with relationships that have been derived through combinations of analysis and empiricism to describe the behaviour of the coefficient of friction, f, with respect to Reynolds' number on the Moody chart. No attempt is made here at a comprehensive discussion of the merits and demerits of the various relationships. Rather, a simple summary is given of those equations that have been found to be most useful in ventilation engineering.

Laminar flow

The straight line that describes laminar flow on the log–log plot of Fig. 2.7 is included in the Moody chart for completeness. However, Poiseuille's equation (2.31)

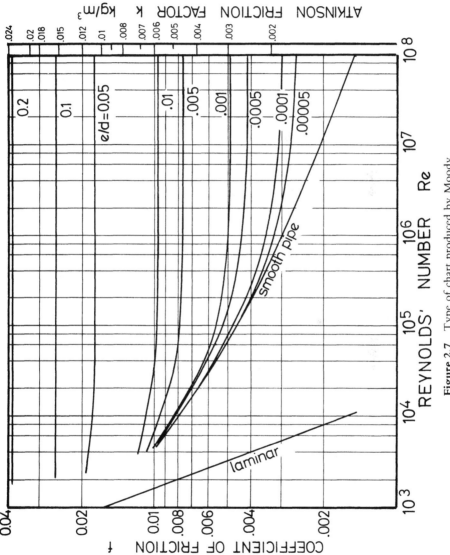

Figure 2.7 Type of chart produced by Moody.

can be used directly to establish frictional pressure losses for laminar flow without using the chart. The corresponding f–Re relationship is easily established.

Combining equations (2.34) and (2.45) gives

$$p = \frac{8\mu u L}{R^2} = \frac{4fL}{d}\frac{\rho u^2}{2} \quad \text{Pa}$$

Substituting $R = d/2$ gives

$$f = 16\frac{\mu}{\rho u d}$$

or

$$f = \frac{16}{\text{Re}} \quad \text{dimensionless} \tag{2.52}$$

Smooth pipe turbulent curve

Perhaps the most widely accepted equation for the smooth pipe turbulent curve is that produced by both Nikuradse and the Hungarian engineer T. Von Kármán:

$$\frac{1}{\sqrt{f}} = 4\log_{10}(\text{Re}\sqrt{f}) - 0.4 \tag{2.53}$$

This suffers from the disadvantage that f appears on both sides. Blasius suggested the approximation

$$f = \frac{0.0791}{\text{Re}^{0.25}} \tag{2.54}$$

This gives a good approximation for Reynolds' numbers in the range 3000 to 10^5.

Rough pipes

When fully developed rough pipe turbulence has been established, the viscous forces are negligible compared with inertial forces. The latter are proportional to the shear stress at the walls (equation (2.41)). Hence, in this condition f becomes independent of Reynolds' number and varies only with e/d. A useful equation for this situation was suggested by Von Kármán.

$$f = \frac{1}{4[2\log_{10}(d/e) + 1.14]^2} \tag{2.55}$$

The most general of the f–Re relationships in common use is the Colebrook–White equation. This has been expressed in a variety of ways, including

$$\frac{1}{\sqrt{4f}} = 1.74 - 2\log_{10}\left(2\frac{e}{d} + \frac{18.7}{\text{Re}\sqrt{4f}}\right) \tag{2.56}$$

and

$$\frac{1}{\sqrt{f}} = -4 \log_{10}\left(\frac{e/d}{3.7} + \frac{1.255}{\mathrm{Re}\sqrt{f}}\right) \tag{2.57}$$

Here again, f, appears on both sides making these equations awkward to use in practice. It was, in fact, this difficulty that led Moody into devising his chart.

The advantage of the Colebrook–White equation is that it is applicable to both rough and smooth pipe flow and for the transitional region as well as fully developed turbulence. For hydraulically smooth pipes, $e/d = 0$, and the Colebrook–White equation simplifies to the Nikuradse relationship of equation (2.53). On the other hand, for high Reynolds' numbers, the term involving Re in equation (2.57) may be ignored. The equation then simplifies to

$$f = \left[4 \log_{10}\left(\frac{e/d}{3.7}\right)\right]^{-2} \tag{2.58}$$

This gives the same results as Von Kármán's rough pipe equation (2.55) for fully developed turbulence.

Example A vertical shaft is 400 m deep, 5 m in diameter and has wall roughenings of height 5 mm. An airflow of 150 m³/s passes at a mean density of 1.2 kg/m³. Taking the viscosity of the air to be 17.9 × 10⁻⁶ N s/m² and ignoring changes in kinetic energy, determine

1. the coefficient of friction, f
2. the turbulent resistance, R_t (m⁻⁴)
3. the frictional pressure drop, p (Pa)
4. the work done against friction, F_{12} (J/kg)
5. the barometric pressure at the shaft bottom if the shaft top pressure is 100 kPa

Solution For a 400 m deep shaft, we can assume incompressible flow (section 2.1.1):

$$\text{cross-sectional area, } A = \frac{\pi \times 5^2}{4} = 19.635 \text{ m}^2$$

$$\text{perimeter, per} = 5\pi = 15.708 \text{ m}$$

$$\text{air velocity, } u = \frac{Q}{A} = \frac{150}{19.635} = 7.639 \text{ m/s}$$

In order to determine the regime of flow, we must first find the Reynolds' number (equation (2.25)):

$$\mathrm{Re} = \frac{\rho u d}{\mu}$$

$$= \frac{1.2 \times 7.639 \times 5}{17.9 \times 10^{-6}} = 2.561 \times 10^6$$

1. *Coefficient of friction, f.* At this value of Re, the flow is fully turbulent (section 2.3.4). We may then use the Moody chart to find the coefficient of friction, f. However, for this we need the equivalent roughness

$$\frac{e}{d} = \frac{5 \times 10^{-3}}{5} = 0.001$$

Hence at $e/d = 0.001$ and Re $= 2.561 \times 10^6$ on Fig. 2.7 we can estimate $f = 0.0049$. (Iterating equation (2.57) gives $f = 0.004\,94$. As the friction coefficient is near constant at this Reynolds' number, we could use equation (2.55) to give $f = 0.004\,90$ or equation (2.58) which gives $f = 0.004\,91$.)

2. *Turbulent resistance, R_t (equation (2.51)).*

$$R_t = \frac{fL\,\text{per}}{2A^3} = \frac{0.0049 \times 400 \times 15.708}{2(19.635)^3}$$

$$= 0.002\,036\,\text{m}^{-4}$$

3. *Frictional pressure drop, p (equation (2.50)).*

$$p = R_t \rho Q^2$$
$$= 0.002\,036 \times 1.2 \times (150)^2 = 54.91\,\text{Pa}$$

4. *Work done against friction, F_{12} (equation (2.46)).*

$$F_{12} = \frac{p}{\rho} = \frac{54.91}{1.2} = 45.76\,\text{J/kg}$$

5. *Barometric pressure at shaft bottom, P_2.* This is obtained from Bernoulli's equation (2.47) with no change in kinetic energy.

$$(Z_1 - Z_2)g + \frac{P_1 - P_2}{\rho} = F_{12}$$

giving

$$P_2 = (Z_1 - Z_2)g\rho - F_{12}\rho + P_1$$
$$= 400 \times 9.81 \times 1.2 - 54.91 + 100\,000$$
$$P_2 = 104\,654\,\text{Pa or } 104.654\,\text{kPa}$$

FURTHER READING

Blasius, H. (1913) Däs ahnlichtkeitgesetz bei Reibungsvorgängen in flussigkeiten. *Forsch. Geb. Ing.* **131**.
Colebrook, C. F. and White, C. M. (1937) Experiments with fluid friction in roughened pipes. *Proc. R. Soc. London, Ser. A* **161**, 367.
Colebrook, C. F. and White, C. M. (1939) Turbulent flow in pipes with particular reference

to the transition region between the smooth and rough pipe laws. *Proc. Inst. Civ. Eng.* **II**, 133.

Daugherty, R. L. and Franzini, J. B. (1977) *Fluid Mechanics, with Engineering Applications*, 7th edn., McGraw-Hill, New York.

Lewitt, E. H. (1959) *Hydraulics and Fluid Mechanics*, 10th edn., Pitman, London.

Massey, B. S. (1968) *Mechanics of Fluids*, Van Nostrand, New York.

Moody, L. F. (1944) Friction factors for pipe flow. *Trans. Am. Soc. Mech. Eng.* **66**, 671–684.

Nikuradse, J. (1933) Strömungsgesetze in rauhen Rohren. *VDI-Forschungsh.* **361**.

Prandtl, L. (1933) Neuere Ergebnisse der Turbulenz-forschung. *Z. VDI* (77), 105.

Reynolds, O. (1883) The motion of water and the law of resistance in parallel channels. *Proc. R. Soc. London* **35**.

Rohsenhow, W. M. and Choi, H. (1961) *Heat, Mass and Momentum Transfer*, International Series in Engineering, Prentice-Hall, Englewood Cliffs, NJ.

Von Karman, T. (1939) *Trans. ASME* **61**, 705.

3

Fundamentals of steady flow thermodynamics

3.1 INTRODUCTION

The previous chapter emphasized the behaviour of incompressible fluids in motion. Accordingly, the analyses were based on the mechanisms of fluid **dynamics**. In expanding these to encompass compressible fluids and to take account of **thermal** effects, we enter the world of thermodynamics.

This subject divides into two major areas. Chemical and statistical thermodynamics are concerned with reactions involving mass and energy exchanges at a molecular or atomic level, while physical thermodynamics takes a macroscopic view of the behaviour of matter subjected to changes of pressure, temperature and volume but not involving chemical reactions. Physical thermodynamics subdivides further into the study of 'closed' systems, within each of which remains a fixed mass of material such as a gas compressed within a cylinder, and 'open' systems through which material flows. A subsurface ventilation system is, of course, an open system with air continuously entering and leaving the facility. In this chapter, we shall concentrate on open systems with one further restriction—that the mass flow of air at any point in the system does not change with time. We may, then, define our particular interest as one of steady flow physical thermodynamics.

Thermodynamics began to be developed as an engineering discipline after the invention of a practicable steam engine by Thomas Newcomen in 1712. At that time, heat was conceived to be a massless fluid named 'caloric' that had the ability to flow from a hotter to a cooler body. Improvements in the design and efficient operation of steam engines, particularly by James Watt (1736–1819) in Scotland, highlighted deficiencies in this concept. During the middle of the nineteenth century the caloric theory was demolished by the work of James P. Joule (1818–1889) in England, H. L. F. Helmholtz (1821–1894) and Rudolph J. E. Clausius (1822–1888) in Germany, and Lord Kelvin (1824–1907) and J. C. Maxwell (1831–1879) of Scotland.

The application of thermodynamics to mine ventilation systems was heralded by the publication of a watershed paper in 1943 by Frederick B. Hinsley (1900–1988). His work was motivated by consistent deviations that were observed when mine ventilation surveys were analysed using incompressible flow theory, and by Hinsley's recognition of the similarity between plots of pressure against specific volume constructed from measurements made in mine downcast and upcast shafts, and indicator diagrams produced by compressed air or heat engines. The new thermodynamic theory was particularly applicable to the deep and hot mines of South Africa. Mine ventilation engineers of that country have contributed greatly to theoretical advances and practical utilization of the more exact thermodynamic methods.

3.2 PROPERTIES OF STATE, WORK AND HEAT

3.2.1 Thermodynamic properties; state of a system

In Chapter 2 we introduced the concepts of fluid density and pressure. In this chapter we shall consider the further properties of temperature, internal energy, enthalpy and entropy. These will be introduced in turn and where appropriate. For the moment, let us confine ourselves to temperature.

Reference to the temperature of substances is such an everyday occurrence that we seldom give conscious thought to the foundations upon which we make such measurements. The most common basis has been to take two fixed temperatures such as those of melting ice and boiling water at standard atmospheric pressure, to ascribe numerical values to those temperatures and to define a scale between the two fixed points. Anders Celsius (1701–1744), a Swedish astronomer, chose to give values of 0 and 100 to the temperatures of melting ice and boiling water respectively, and to select a linear scale between the two. The choice of a linear scale is, in fact, quite arbitrary but leads to a convenience of measurement and simpler relationships between temperature and other thermodynamic properties. The scale thus defined is known as the Celsius (or Centigrade) scale. The older Fahrenheit scale was named after Gabriel Fahrenheit (1686–1736), the German scientist who first used a mercury-in-glass thermometer. Fahrenheit's two 'fixed' but rather inexact points were 0 for a mixture of salt, ice and water, and 96 for the average temperature of the human body. A linear scale was then found to give Fahrenheit temperatures of 32 for melting ice and 212 for boiling water. These were later chosen as the two fixed points for the Fahrenheit scale but, unfortunately, the somewhat unusual numeric values were retained. In the SI system of units, temperatures are normally related to degrees Celsius.

However, through a thermodynamic analysis, another scale of temperature can be defined that does not depend on the melting or boiling points of any substance. This is called the **absolute** or **thermodynamic** temperature scale. N. L. Sadi Carnot (1796–1832), a French military engineer, showed that a theoretical heat engine operating between fixed inlet and outlet temperatures becomes more efficient as the difference between those two temperatures increases. Absolute zero on the thermodynamic temperature scale is defined theoretically as that outlet temperature at which an ideal heat engine operating between two fixed temperature reservoirs would become 100% efficient, i.e. operate without producing any reject heat. Absolute zero temperature is a theoretical datum that can be approached but never quite attained. We can then choose any other fixed point and interval to define a unit or **degree** on the absolute temperature scale. The SI system of units employs the Celsius degree as the unit of temperature and retains 0 °C and 100 °C for melting ice and boiling water. This gives absolute zero as -273.15 °C. Thermodynamic temperatures quoted on the basis of absolute zero are always positive numbers and are measured in kelvins (after Lord Kelvin). A difference of one kelvin is equivalent to a difference of one Celsius degree. Throughout this book, absolute temperatures are identified by the symbol T and temperatures shown as t or θ denote degrees Celsius.

$$T = t(°C) + 273.15 \quad K \tag{3.1}$$

The kelvin units, K, should be shown without a degree (°) sign. Thermodynamic calculations should always be conducted using the absolute temperature, T in kelvins. However, as a kelvin is identical to a degree Celsius, temperature differences may be quoted in either units.

   The **state** of any point within a system is defined by the thermodynamic properties of the fluid at that point. If air is considered to be a pure substance of fixed composition then any two independent properties are sufficient to define its thermodynamic state. In practice, the two properties are often pressure and temperature as these can be measured directly. If the air is not of fixed composition as, for example, in airways where evaporation or condensation of water occurs, then at least one more property is required to define its thermodynamic state (Chapter 14).

   The **intensive** or **specific** properties of state (quoted on the basis of unit mass) define completely the thermodynamic state of any point within a system or sub-system and are independent of the processes which led to the establishment of that state.

3.2.2   Work and heat

Both work and heat involve the transfer of energy. In SI units, the fundamental numerical equivalence of the two is recognized by their being given the same units, joules, where

$$1 \text{ joule} = 1 \text{ newton} \times 1 \text{ metre (N m)}$$

   Work usually (but not necessarily) involves mechanical movement against a resisting force. An equation used repeatedly in Chapter 2 was

$$\text{work done} = \text{force} \times \text{distance}$$

or

$$dW = F\,dL \quad \text{N m or J} \tag{3.2}$$

and is the basis for the definition of a joule. Work may be added as mechanical energy from an external source such as a fan or pump. Additionally, it was shown in section 2.3.1 that 'flow work', Pv (J) must be done to introduce a plug of fluid into an open system. However, it is only at entry (or exit) of the system that the flow work can be conceived as a measure of force × distance. Elsewhere within the system the flow work is a point function. It is for this reason that some engineers prefer not to describe it as a work term.

Heat is transferred when an energy exchange takes place because of a temperature difference. When two bodies of differing temperatures are placed in contact then heat will 'flow' from the hotter to the cooler body. (In fact, heat can be transferred by convection or radiation without physical contact.) It was this conception of heat flowing that gave rise to the caloric theory. Our modern hypothesis is that heat transfer involves the excitation of molecules in the receiving substance, increasing their internal kinetic energy at the expense of those in the emitting substance.

Equation (3.2) showed that work can be described as the product of a driving potential (force) and distance. It might be expected that there is an analogous relationship for heat, dq, involving the driving potential of temperature and some other property. Such a relationship does, in fact, exist and is quantified as

$$dq = T\,ds \quad J \tag{3.3}$$

The variable, s, is named **entropy** and is a property that will be discussed in more detail in section 3.4.2.

It is important to realize that neither work nor heat is a property of a system. Contrary to popular phraseology which still retains reminders of the old caloric theory, no system 'contains' either heat or work. The terms become meaningful only in the context of energy transfer across the boundaries of the system. Furthermore, the magnitude of the transfer depends on the process path or particular circumstances existing at that time and place on the boundary. Hence, to be precise, the quantities dW and dq should actually be denoted as inexact differentials δW and δq.

The rate at which energy transfers take place is commonly expressed in one of two ways. First, on the basis of unit mass of the fluid, i.e. J/kg. This was the method used to dimension the terms of the Bernoulli equation in section 2.3.1. Secondly, an energy transfer may be described with reference to time, joules per second. This latter method produces the definition of **power**:

$$\text{power} = \frac{dW}{\text{time}} \text{ or } \frac{dq}{\text{time}} \quad \frac{J}{s} \tag{3.4}$$

where the unit, J/s, is given the name watt after the Scots engineer, James Watt.

Before embarking on any analyses involving energy transfers, it is important to define a sign convention. Many textbooks on thermodynamics use the rather confusing convention that heat transferred to a system is **positive**, but work transferred to the system is **negative**. This strange irrationality has arisen from the historical development of physical thermodynamics being motivated by the study of heat engines, these consuming heat energy (supplied in the form of a hot vapour or burning fuel) and producing a mechanical work output. In subsurface ventilation engineering, work input from fans is mechanical energy transferred to the air and, in most cases, heat is transferred from the surrounding strata or machines also to the air. Hence, in this engineering discipline it is convenient as well as being mathematically consistent to regard all energy transfers to the air as being **positive**, whether those energy transfers are work or heat. This is the sign convention utilized throughout this book.

3.3 SOME BASIC RELATIONSHIPS

3.3.1 Gas laws and gas constants

An ideal gas is one in which the volume of the constituent molecules is zero and where there are no intermolecular forces. Although no real gas conforms exactly to that definition, the mixture of gases that constitute air behaves in a manner that differs negligibly from an ideal gas within the ranges of temperature and pressure found in subsurface ventilation engineering. Hence, the thermodynamic analyses outlined in this chapter will assume ideal gas behaviour.

Some 20 years before Isaac Newton's major works, Robert Boyle (1627–1691) developed a vacuum pump and found, experimentally, that gas pressure, P, varied inversely with the volume, v, of a closed system for constant temperature.

$$P \propto \frac{1}{v} \quad \text{(Boyle's law)} \tag{3.5}$$

where $\propto$ means 'proportional to'.

In the following century and on the other side of the English Channel in France, Jacques A. C. Charles (1746–1823) discovered, also experimentally, that

$$v \propto T \quad \text{(Charles' law)} \tag{3.6}$$

for constant pressure, where T = absolute temperature.

Combining Boyle's and Charles' laws gives

$$Pv \propto T$$

where both temperature and pressure vary. Alternatively, inserting a constant of proportionality, R',

$$Pv = R'T \quad \text{J} \tag{3.7}$$

To make the equation more generally applicable, we can replace R' by mR where m is the mass of gas, giving

$$Pv = mRT \quad \text{J} \tag{3.8}$$

or

$$P\frac{v}{m} = RT \quad \text{J/kg}$$

But v/m is the volume of 1 kg, i.e. the **specific** volume of the gas, V (m³/kg). Hence,

$$PV = RT \quad \text{J/kg} \tag{3.9}$$

This is known as the general gas law and R is the gas constant for that particular gas or mixture of gases, having dimensions of J/(kg K). The specific volume, V, is simply the reciprocal of density.

$$V = 1/\rho \quad \text{m}^3/\text{kg} \tag{3.10}$$

Hence, the general gas law can also be written as

$$\frac{P}{\rho} = RT \quad \text{J/kg} \tag{3.11}$$

or

$$\rho = \frac{P}{RT} \quad \frac{\text{kg}}{\text{m}^3}$$

giving an expression for the density of an ideal gas. As R is a constant for any perfect gas, it follows from equation (3.9) that the two end states of any process involving an ideal gas are related by the equation

$$\frac{P_1 V_1}{T_1} = \frac{P_2 V_2}{T_2} = R \quad \frac{\text{J}}{\text{kg K}} \tag{3.12}$$

Another feature of the gas constant, R, is that although it takes a different value for each gas, there is a useful and simple relationship between the gas constants of all ideal gases.

Avogadro's law states that equal volumes of ideal gases at the same temperature and pressure contain the same number of molecules. Applying these conditions to equation (3.8) for all ideal gases gives

$$mR = \text{constant}$$

Furthermore, if the same volume of each gas contains an equal number of molecules, it follows that the mass, m, is proportional to the weight of a single molecule, that is, the molecular weight of the gas, M. Then

$$MR = \text{constant} \tag{3.13}$$

The product MR is a constant for all ideal gases and is called the universal gas constant, R_u. In SI units, the value of R_u is 8314.36 J/K. The dimensions are sometimes defined as J/(kg mol K) where one mole is the amount of gas contained in a mass M (kg), i.e. its molecular weight expressed in kilograms.

The gas constant for any ideal gas can now be found provided that its molecular weight is known:

$$R = \frac{8314.36}{M} \quad \frac{\text{J}}{\text{kg K}} \tag{3.14}$$

For example, the equivalent molecular weight of dry air is 28.966, giving its gas constant as

$$R = \frac{8314.36}{28.966} = 287.04 \, \frac{J}{kg \, K} \qquad (3.15)$$

Example   At the top of a mine downcast shaft the barometric pressure is 100 kPa and the air temperature is 18.0 °C. At the shaft bottom, the corresponding measurements are 110 kPa and 27.4 °C respectively. The airflow measured at the shaft top is 200 m^3/s. If the shaft is dry, determine

1. the air densities at the shaft top and shaft bottom,
2. the mass flow of air, and
3. the volume flow of air at the shaft bottom.

Solution   Use subscripts 1 and 2 for the top and bottom of the shaft respectively.

1.
$$\rho_1 = \frac{P_1}{RT_1}$$

$$= \frac{100\,000}{287.04 \times (273.15 + 18)} = 1.1966 \, kg/m^3$$

In any calculation, the units of measurement must be converted to the basic SI unless ratios are involved. Hence 100 kPa = 100 000 Pa.

$$\rho_2 = \frac{P_2}{RT_2}$$

$$= \frac{110\,000}{287.04 \times (273.15 + 27.4)} = 1.2751 \, kg/m^3$$

2.
$$\text{mass flow } M = Q_1 \rho_1$$
$$= 200 \times 1.1966 = 239.3 \, kg/s$$

3.
$$Q_2 = \frac{M}{\rho_2}$$

$$= \frac{239.3}{1.2751} = 187.7 \, m^3/s$$

Example   Calculate the volume of 100 kg of methane at a pressure of 75 kPa and a temperature of 42 °C.

Solution   The molecular weight of methane (CH_4) is

$$12.01 + (4 \times 1.008) = 16.04$$

$$R(\text{methane}) = \frac{8314.36}{16.04} = 518.4 \text{ J/(kg K)}$$

from equation (3.14). Therefore for the volume of 100 kg

$$v = \frac{mRT}{P} \quad \text{from equation (3.8)}$$

$$= \frac{100 \times 518.4 \times (273.15 + 42)}{75\,000} = 217.8 \text{ m}^3$$

3.3.2 Internal energy and the first law of thermodynamics

Suppose we have 1 kg of gas in a closed container as shown in Fig. 3.1. For simplicity, we shall assume that the vessel is at rest with respect to the earth and is located on a base horizon. The gas in the vessel has neither macro kinetic energy nor potential energy. However, the molecules of the gas are in motion and possess a molecular or 'internal' kinetic energy. The term is usually shortened to **internal energy**. In the fluid mechanics analyses of Chapter 2 we dealt only with mechanical energy and there was no need to involve internal energy. However, if we are to study thermal effects then we can no longer ignore this form of energy. We shall denote the specific (per kilogram) internal energy as U (J/kg).

Now suppose that, by rotation of an impeller within the vessel, we add work δW to the closed system and we also introduce an amount of heat δq. The gas in the vessel still has zero macro kinetic energy and zero potential energy. The energy

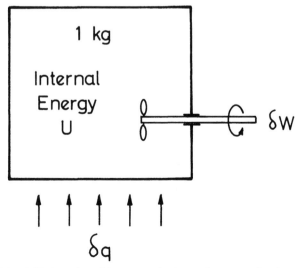

Figure 3.1 Added work and heat raise the internal energy of a closed system.

that has been added has simply caused an increase in the internal energy.

$$dU = \delta W + \delta q \quad \text{J/kg} \tag{3.16}$$

The change in internal energy is determined only by the net energy that has been transferred across the boundary and is independent of the form of that energy (work or heat) or the process path of the energy transfer. Internal energy is, therefore, a thermodynamic property of state.

Equation (3.16) is sometimes known as the non-flow energy equation and is a statement of the first law of thermodynamics. This equation also illustrates that the first law is simply a quantified restatement of the general law of conservation of energy.

3.3.3   Enthalpy and the steady flow energy equation

Let us now return to steady flow through an open system. Bernoulli's equation (2.15) for a frictionless system included mechanical energy terms only and took the form

$$\frac{u^2}{2} + Zg + PV = \text{constant} \quad \frac{\text{J}}{\text{kg}} \tag{3.17}$$

(where specific volume, $V = 1/\rho$). In order to expand this equation to include all energy terms, we must add internal energy, giving

$$\frac{u^2}{2} + Zg + PV + U = \text{constant} \quad \frac{\text{J}}{\text{kg}} \tag{3.18}$$

One of the features of thermodynamics is the regularity with which certain groupings of variables appear. So it is with the group $PV + U$. This sum of internal energy and the PV product is particularly important in flow systems. It is given the name **enthalpy** and the symbol H

$$H = PV + U \quad \text{J/kg} \tag{3.19}$$

Now pressure, P, specific volume, V, and specific internal energy, U, are all thermodynamic properties of state. It follows, therefore, that enthalpy, H, must also be a thermodynamic property of state and is independent of any previous process path.

The energy equation (3.18) can now be written as

$$\frac{u^2}{2} + Zg + H = \text{constant} \quad \frac{\text{J}}{\text{kg}} \tag{3.20}$$

Consider the continuous airway shown on Fig. 3.2. If there were no energy additions between stations 1 and 2 then the energy equation (3.20) would give

$$\frac{u_1^2}{2} + Z_1 g + H_1 = \frac{u_2^2}{2} + Z_2 g + H_2 \quad \frac{\text{J}}{\text{kg}}$$

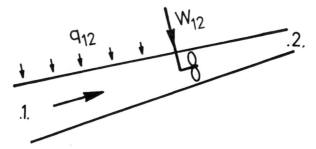

Figure 3.2 Heat and work added to a steady flow system.

However, Fig. 3.2 shows that a fan adds W_{12} of mechanical energy and that strata heat transfers q_{12} of thermal energy to the air. These terms must be added to the total energy at station 1 in order to give the total energy at station 2

$$\frac{u_1^2}{2} + Z_1 g + H_1 + W_{12} + q_{12} = \frac{u_2^2}{2} + Z_2 g + H_2 \quad \frac{J}{kg} \tag{3.21}$$

This is usually rearranged as follows:

$$\frac{u_1^2 - u_2^2}{2} + (Z_1 - Z_2)g + W_{12} = H_2 - H_1 - q_{12} \quad \frac{J}{kg} \tag{3.22}$$

Equation (3.22) is of fundamental importance in steady flow processes and is given a special name, the **steady flow energy equation**.

Note that the steady flow energy equation does not contain a term for friction, nor does it require any such term. It is a total energy balance. Any frictional effects will reduce the mechanical energy terms and increase the internal energy but will have no influence on the overall energy balance. It follows that equation (3.22) is applicable both to ideal (frictionless) processes and also to real processes that involve viscous resistance and turbulence. Such generality is admirable but does, however, give rise to a problem for the ventilation engineer. In airflow systems, mechanical energy is expended against frictional resistance only—we do not normally use the airflow to produce a work output through a turbine or any other device. It is, there-fore, important that we are able to quantify work done against friction. This seems to be precluded by equation (3.22) as no term for friction appears in that equation.

To resolve the difficulty, let us recall Bernoulli's mechanical energy equation (2.23) corrected for friction

$$\frac{u_1^2 - u_2^2}{2} + (Z_1 - Z_2)g + \frac{P_1 - P_2}{\rho} = F_{12} \quad \frac{J}{kg}$$

If we, again, add work, W_{12}, between stations 1 and 2 then the equation becomes

$$\frac{u_1^2 - u_2^2}{2} + (Z_1 - Z_2)g + W_{12} = \frac{P_2 - P_1}{\rho} + F_{12} \quad \frac{J}{kg} \tag{3.23}$$

(Although any heat, q_{12}, that may also be added does not appear explicitly in a mechanical energy equation it will affect the balance of mechanical energy values arriving at station 2.)

We have one further step to take. As we are now dealing with gases, we should no longer assume that density, ρ, remains constant. If we consider the complete process from station 1 to station 2 to be made up of a series of infinitesimally small steps then the pressure difference across each step becomes dP and the flow work term becomes dP/ρ or $V\,dP$ (as $V = 1/\rho$). Summing up, or integrating all such terms for the complete process gives

$$\frac{u_1^2 - u_2^2}{2} + (Z_1 - Z_2)g + W_{12} = \int_1^2 V\,dP + F_{12} \quad \frac{J}{kg} \tag{3.24}$$

This is sometimes called the steady flow mechanical energy equation. It is, in fact, simply a statement of Bernoulli's equation made applicable to compressible flow.

Finally, compare equations (3.22) and (3.24). Their left-hand sides are identical. Hence, we can write an expanded version of the steady flow energy equation:

$$\boxed{\frac{u_1^2 - u_2^2}{2} + (Z_1 - Z_2)g + W_{12} = \int_1^2 V\,dP + F_{12} = H_2 - H_1 - q_{12} \quad \frac{J}{kg}} \tag{3.25}$$

or, in differential form,

$$- u\,du - g\,dz + dW = V\,dP + dF = dH - dq$$

This important three-part equation is the starting point for the application of thermodynamics to subsurface ventilation systems and, indeed, to virtually all steady flow compressible systems. In most applications within mechanical engineering the potential energy term $(Z_1 - Z_2)g$ is negligible and can be dropped to simplify the equation. However, the large elevation differences that may be traversed by subsurface airflows often result in this term being dominant and it must be retained for mine ventilation systems.

Students of the subject and ventilation engineers dealing with underground networks of airways should become completely familiarized with equation (3.25). It is absolutely fundamental to a complete understanding of the behaviour of airflow in subsurface ventilation systems.

3.3.4   Specific heats and their relationship to gas constant

When heat is supplied to any unconstrained substance, there will, in general, be two effects. First, the temperature of the material will rise and, secondly, if the material is not completely constrained, there will be an increase in volume. (An exception is the contraction of water when heated between 0 and 4 °C.) Hence, the heat is utilized both in raising the temperature of the material and in doing work against the surrounding as it expands.

The specific heat of a substance has been described in elementary texts as that amount of heat which must be applied to 1 kg of the substance in order to raise its temperature through one Celsius degree. There are some difficulties with this simplistic definition. First, the temperature of any solid, liquid or gas can be increased by doing work on it and without applying any heat—for example by rubbing friction or by rotating an impeller in a liquid or gas. Secondly, if we allow the substance to expand against a resisting pressure then the work that is done on the surroundings necessarily requires that more energy must be applied to raise the 1 kg through the same 1 °C than if the system were held at constant volume. The specific heat must have a higher value if the system is at constant pressure than if it is held at constant volume. If both pressure and volume are allowed to vary in a partially constrained system, then yet another specific heat will be found. As there are an infinite number of variations in pressure and volume between the conditions of constant pressure and constant volume, it follows that there are an infinite number of specific heats for any given substance. This does not seem to be a particularly useful conclusion. However, we can apply certain restrictions which make the concept of specific heat a valuable tool in thermodynamic analyses.

First, extremely high stresses are produced if a liquid or solid is prevented from expanding during heating. For this reason, the specific heats quoted for liquids and solids are normally those pertaining to constant pressure, i.e. allowing free expansion. In the case of gases we can, indeed, have any combination of changes in the pressure and volume. However, we confine ourselves to the two simple cases of constant volume and constant pressure.

Let us take the example of a vessel of fixed volume containing 1 kg of gas. We add heat, δq, until the temperature has risen through δT. Then, from the first law (equation (3.16)) with $\delta W = 0$ for fixed volume,

$$\delta q = dU$$

From our definition of specific heat,

$$C_v = \left(\frac{\delta q}{\delta T}\right)_v = \left(\frac{dU}{\delta T}\right)_v$$

As this is the particular specific heat at constant volume, we use the subscript v. At any point during the process, the specific heat takes the value pertaining to the corresponding temperature and pressure. Hence, we should define C_v more generally as

$$C_v = \left(\frac{\partial U}{\partial T}\right)_v \quad \frac{J}{kg\,K} \tag{3.26}$$

However, for an ideal gas the specific heat is independent of either pressure or temperature and we may write

$$C_v = \frac{dU}{dT} \quad \frac{J}{kg\,K} \tag{3.27}$$

In section 3.3.2, the concept of internal energy was introduced as a function of the internal kinetic energy of the gas molecules. As it is the molecular kinetic energy that governs the temperature of the gas, it is reasonable to deduce that internal energy is a function of temperature only. This can, in fact, be proved mathematically through an analysis of Maxwell's equations assuming an ideal gas. Furthermore, as any defined specific heat remains constant, again, for an ideal gas, equation (3.27) can be integrated directly between any two end points to give

$$U_2 - U_1 = C_v(T_2 - T_1) \quad \text{J/kg} \tag{3.28}$$

Now let us examine the case of adding heat, δq, while keeping the pressure constant, i.e. allowing the gas to expand. In this case, from equation (3.19)

$$q = PV + U = H \quad \text{J/kg} \tag{3.29}$$

Therefore, the specific heat becomes

$$C_p = \left(\frac{\delta q}{\delta T}\right)_p = \left(\frac{\delta H}{\delta T}\right)_p \quad \frac{\text{J}}{\text{kg K}} \tag{3.30}$$

In the limit, this defines C_p as

$$C_p = \left(\frac{\partial H}{\partial T}\right)_p \quad \frac{\text{J}}{\text{kg K}} \tag{3.31}$$

Here again, for a perfect gas, specific heats are independent of pressure and temperature and we can write

$$C_p = \frac{dH}{dT} \quad \frac{\text{J}}{\text{kg K}} \tag{3.32}$$

or

$$H_2 - H_1 = C_p(T_2 - T_1) \quad \text{J/kg} \tag{3.33}$$

We shall use this latter equation extensively in Chapter 8 when we apply thermodynamic theory to subsurface ventilation systems. The same equation also shows that for an ideal gas, enthalpy, H, is a function of temperature only, C_p being a constant.

There is another feature of C_p that often causes conceptual difficulty. We introduced equations (3.29) and (3.30) by way of a constant pressure process. However, we did not actually enforce the condition of constant pressure in those equations. Furthermore, we have twice stated that for an ideal gas the specific heats are independent of either pressure or temperature. It follows that C_p can be used in equation (3.33) for ideal gases even when the pressure varies. For flow processes, the term **specific heat at constant pressure** can be rather misleading.

A useful relationship between the specific heats and gas constant is revealed if we substitute the general gas law $PV = RT$ (equation (3.9)) into equation (3.29):

$$H = RT + U \quad \text{J/kg}$$

Differentiating with respect to T gives

$$\frac{dH}{dT} = R + \frac{dU}{dT} \quad \frac{\text{J}}{\text{kg K}}$$

Table 3.1 Thermodynamic properties of gases at atmospheric pressures and a temperature of 26.7°C

Gas	Molecular weight M	Gas constant $(J/(kg\ K))$ $R = 8314.36/M$	Specific heats $(J/(kg\ K))$		Isentropic index $\gamma = C_p/C_v$	R/C_p $= (\gamma - 1)/\gamma$
			C_p	$C_v = C_p - R$		
Air (dry)	28.966	287.04	1005	718.0	1.400	0.2856
Water vapour	18.016	461.5	1884	1422	1.324	0.2450
Nitrogen	28.015	296.8	1038	741.2	1.400	0.2859
Oxygen	32.000	259.8	916.9	657.1	1.395	0.2833
Carbon dioxide	44.003	188.9	849.9	661.0	1.286	0.2223
Carbon monoxide	28.01	296.8	1043	746.2	1.398	0.2846
Methane	16.04	518.4	2219	1700	1.305	0.2336
Helium	4.003	2077	5236	3159	1.658	0.3967
Hydrogen	2.016	4124	14361	10237	1.403	0.2872
Argon	39.94	208.2	524.6	316.4	1.658	0.3968

i.e.

$$C_p = R + C_v \quad J/(kg\,K)$$

or

$$R = C_p - C_v \quad J/(kg\,K) \tag{3.34}$$

The names 'heat capacity' or 'thermal capacity' are sometimes used in place of specific heat. These terms are relics remaining from the days of the caloric theory when heat was throught to be a fluid without mass that could be 'contained' within a substance. We have also shown that temperatures can be changed by work as well as heat transfer so that the term specific heat is itself open to challenge.

Two groups of variables involving specific heats and gas constant that frequently occur are the ratio of specific heats (isentropic index),

$$\gamma = \frac{C_p}{C_v} \quad \text{(dimensionless)} \tag{3.35}$$

and

$$\frac{R}{C_p} \quad \text{(dimensionless)}$$

Using equation (3.34), it can easily be shown that

$$\frac{R}{C_p} = \frac{\gamma - 1}{\gamma} \quad \text{(dimensionless)} \tag{3.36}$$

Table 3.1 gives data for those gases that may be encountered in subsurface ventilation engineering. As no real gases follow ideal gas behaviour exactly, the values of the 'constants' in the table vary slightly with pressure and temperature. For this reason, there may be minor differences in published values. Those given in Table 3.1 are referred to low (atmospheric) pressures and a temperature of 26.7 °C.

3.3.5   The second law of thermodynamics

Heat and work are mutually convertible. Each joule of thermal energy that is converted to mechanical energy in a heat engine produces one joule of work. Similarly each joule of work expended against friction produces one joule of heat. This is another statement of the first law of thermodynamics. When equation (3.16) is applied throughout a closed cycle of processes then the final state is the same as the initial state, i.e. $\oint dU = 0$ and

$$\oint dW = -\oint dq \quad \frac{J}{kg} \tag{3.37}$$

where $\oint$ indicates integration around a closed cycle.

However, our everyday experience indicates that the first law, by itself, is incapable of explaining many phenomena. All mechanical work can be converted to a numerically equivalent amount of heat through frictional processes, impact, compression or other means such as electrical devices. However, when we convert heat into mechanical energy, we invariably find that the conversion is possible only to a limited extent, the remainder of the heat having to be rejected. An internal combustion engine is supplied with heat from burning fuel. Some of that heat produces a mechanical work output but, unfortunately, the majority is rejected in the exhaust gases.

Although work and heat are numerically equivalent, work is a superior form of energy. There are many common examples of the limited value of heat energy. A sea-going liner cannot propel itself by utilizing any of the vast amount of thermal energy held within the ocean. Similarly, a power station cannot draw on the heat energy held within the atmosphere. It is this constraint on the usefulness of heat energy that gives rise to the second law of thermodynamics.

Perhaps the simplest statement of the second law is that heat will always pass from a higher temperature body to a lower temperature body and can never, spontaneously, pass in the opposite direction. There are many other ways of stating the second law and the numerous corollaries that result from it. The second law is, to be precise, a statement of probabilities. The molecules in a sample of fluid move with varying speeds in apparently random directions and with a mean velocity approximating that of the speed of sound in the fluid. It is extremely improbable, although not impossible, for a chance occurrence in which all the molecules move in the same direction. In the event of such a rare condition, a cup of tea could become 'hotter' when placed in cool surroundings. Such an observation has never yet been made in practice and would constitute a contravention of the second law of thermodynamics.

The question arises, just how much of a given amount of heat energy can be converted into work? This is of interest to the subsurface ventilation engineer as some of the heat energy added to the airstream may be converted into mechanical energy in order to help promote movement of the air.

To begin an answer to this question, consider a volume of hot gas within an uninsulated cylinder and piston arrangement. When placed in cooler surroundings, the gas will lose heat through the walls of the cylinder and contract. The pressure of the external atmosphere will push the piston inwards and provide mechanical work. This will continue until the temperature of the gas is the same as that of the surroundings. Although the gas still 'contains' heat energy, it is incapable of causing further work to be done in those surroundings. However, if the cylinder were then placed in a refrigerator, the gas would contract further and more work would be generated—again until the temperatures inside and outside the cylinder were equal. From this imaginary experiment it would seem that heat can produce mechanical work only while a temperature difference exists.

In 1824, while still in his twenties, Sadi Carnot produced a text *Reflections on the Motive Power of Fire*, in which he devised an ideal heat engine operating between a supply temperature T_1 and a lower reject temperature, T_2. He showed that for a given amount of heat, q, supplied to the engine, the maximum amount of work that could be produced is

$$W_{ideal} = \frac{T_1 - T_2}{T_1} q \quad J \tag{3.38}$$

In real situations, no process is ideal and the real work output will be less than the Carnot work because of friction or other irreversible effects. However, from equation (3.38) a maximum 'Carnot efficiency', η_c, can be devised:

$$\eta_c = \frac{W_{ideal}}{q} = \frac{T_1 - T_2}{T_1} \tag{3.39}$$

Example Air enters a mine at 5 °C and leaves at 35 °C. Determine the theoretical maximum fraction of the added heat that can be used to promote airflow.

Solution

$$\eta_c = \frac{35 - 5}{273.15 + 35} = 0.097$$

Hence, 9.7% of the heat added is theoretically available to produce a natural ventilating effect. In practice, it will be much less than this because of friction and additional thermal losses in the exhaust air.

3.4 FRICTIONAL FLOW

3.4.1 The effects of friction in flow processes

In the literature of thermodynamics, a great deal of attention is given to frictionless processes, sometimes called ideal or reversible. The latter term arises from the concept that a reversible process is one that having taken place can be reversed to leave no net change in the state of either the system or surroundings.

Frictionless processes can never occur in practice. They are, however, convenient yardsticks against which to measure the performance or efficiencies of real heat engines or other devices that operate through exchanges of work and/or heat. In subsurface ventilation systems, work is added to the airflow by means of fans and heat is added from the strata or other sources. Some of that heat may be utilized in helping to promote airflow in systems that involve differences in vertical elevation. An ideal system would be one in which an airflow, having been initiated, would continue indefinitely without further degradation of energy. Although they cannot exist in practice, the concept of ideal processes assists in gaining an understanding of the behaviour of actual airflow systems.

Let us consider what we mean by 'friction' with respect to a fluid flow. The most common everyday experience of friction is concerned with the contact of two surfaces—a brake on a wheel or a tyre on the road. In fluid flow, the term 'friction' or 'frictional resistance' refers to the effects of viscous forces that resist the motion of one layer of fluid over another, or with respect to a solid boundary (section 2.3.3). In turbulent flow, such forces exist not only at boundaries and between laminae of fluid, but also between and within the very large number of vortices that characterize turbulent flow. Hence, the effect of fluid friction is much greater in turbulent flow.

In the expanded version of the steady flow energy equation, the term denoting work done against frictional effects was F_{12}. To examine how this affects the other parameters, let us restate that equation.

$$\frac{u_1^2 - u_2^2}{2} + (Z_1 - Z_2)g + W_{12} = \int_1^2 V\,dP + F_{12} = H_2 - H_1 - q_{12} \quad \frac{J}{kg}$$

from equation (3.25). However, from equation (3.33) for a perfect gas

$$H_2 - H_1 = C_p(T_2 - T_1) \quad J/kg \tag{3.40}$$

giving

$$\frac{u_1^2 - u_2^2}{2} + (Z_1 - Z_2)g + W_{12} = \int_1^2 V\,dP + F_{12} = C_p(T_2 - T_1) - q_{12} \quad J/kg \tag{3.41}$$

We observe that the friction term, F_{12}, appears in only the middle section of this three-part equation. This leads to two conclusions for any given net change in kinetic, potential and fan input energy, i.e. each part of the equation remaining at the same total value. First, frictional effects, F_{12}, appear at the expense of $\int_1^2 V\,dP$. As F_{12} increases then the flow work must decrease for the total energy to change by the same amount. However, the conversion of mechanical energy into heat through frictional effects will result in the specific volume, V, expanding to a value that is higher than would be the case in a corresponding frictionless system. It follows that the appearance of friction must result in a loss of pressure in a flow system. A real (frictional) process from station 1 to station 2 will result in the pressure at station 2 being less than would be the case of a corresponding ideal process. The difference is the 'frictional pressure drop'.

Secondly, and again for a fixed net change in kinetic, potential and fan energies, the lack of a friction term in the right-hand part of equation (3.41) shows that the change in temperature $T_2 - T_1$ is independent of frictional effects. In other words, the actual change in temperature along an airway is the same as it would be in a corresponding frictionless process. This might seem to contradict everyday experience where we expect friction to result in a rise in temperature. In the case of the steady flow of perfect gases, the frictional conversion of mechanical work to heat through viscous shear produces a higher final specific volume and a lower pressure than for the ideal process, but exactly the same temperature.

Returning to the concept of frictional pressure drop, the steady flow energy equation can also be used to illustrate the real meaning of this term. If we accept the usual situation in which the variation in density, $\rho(= 1/V)$, along the airway is near linear, we can multiply equation (3.25) by a mean value of density, ρ_m to give

$$\rho_m \frac{u_1^2 - u_2^2}{2} + \rho_m (Z_1 - Z_2)g + \rho_m W_{12} = P_2 - P_1 + \rho_m F_{12}$$

$$= \rho_m (H_2 - H_1) - \rho_m q_{12} \quad \frac{J}{kg} \frac{kg}{m^3} = \frac{J}{m^3} \tag{3.42}$$

Note the new units of the terms in this equation. They still express variations in energy levels (J), but now with reference to a unit volume (m^3) rather than a unit mass (kg). As it is the total mass flow that remains constant the steady flow energy equation (in J/kg) is the preferred version of the equation (see also section 2.1.2). However, most of the terms in equation (3.42) do have a physical meaning. We can re-express the units as

$$\frac{J}{m^3} = \frac{Nm}{m^3} = \frac{N}{m^2} = Pa \tag{3.43}$$

As this is the unit of pressure, equation (3.42) is sometimes called the **steady flow pressure equation**. Furthermore,

$$\rho_m \frac{u_2^2 - u_2^2}{2}$$

is the change in velocity pressure (see equation (2.17)),

$$\rho_m (Z_1 - Z_2)g$$

is the change in static pressure due to the column of air between Z_1 and Z_2 (equation (2.18)),

$$\rho_m W_{12}$$

is the increase in pressure across the fan,

$$P_2 - P_1$$

is the change in barometric pressure and

$$\rho_m F_{12}$$

is the 'frictional pressure drop', p (equation (2.46)).

The frictional pressure drop may now be recognized as the work done against friction per cubic metre of air. As the mass in a cubic metre varies as a result of density changes, the disadvantage of a relationship based on volume becomes clear.

3.4.2 Entropy

In section 3.3.5 we discussed work and heat as 'first- and second-class' energy terms respectively. All work can be transferred into heat but not all heat can be transferred into work. Why does this preferential direction exist? It is, of course, not the only example of 'one-wayness' in nature. Two liquids of the same density but different colours will mix readily to a uniform shade but cannot easily be separated back to their original condition. A rubber ball dropped on to the floor will bounce, but not quite to the height from which it originated—and on each succeeding bounce it will lose more height. Eventually, the ball will come to rest on the floor. All of its original potential energy has been converted to heat through impact on the floor, but that heat cannot be used to raise the ball to its initial height. Each time we engage in any non-ideal process, we finish up with a lower quality, or less organized, state of the system. Another way of putting it is that the 'disorder' or 'randomness' of the system has increased. It is a quantification of this disorder that we call **entropy**.

Suppose we build a symmetric tower out of toy building bricks. The system is well ordered and has a low level of entropy. Now imagine your favourite infant taking a wild swipe at it. The bricks scatter all over the floor. Their position is obviously now in a much greater degree of disorder. Energy has been expended on the system and the entropy (disorder) has increased. This is entropy of position. Let us carry out another imaginary experiment. Suppose we have a tray on which rest some marbles. We vibrate the tray gently and the marbles move about in a apparently random manner. Now let us vibrate the tray violently. The marbles becomes much more agitated or disordered in their movement. We have done work on the system and, again, the entropy level has increased. This is entropy of motion.

What has all this to do with heat tranfer? Imagine that we have a perfect crystal at a temperature of absolute zero. The molecules are arranged in a symmetric lattice and are quite motionless. We can state that this is a system of perfect order or zero entropy. If we now add heat to the crystal, that energy will be utilized in causing the molecules to vibrate. The more heat we add, the more agitated the molecular vibration and the greater the entropy level. Can you see the analogy with the marbles?

The loss of order is often visible. For example, ice is an 'ordered' form of water. Adding heat will cause it to melt to the obviously less organized form of liquid water. Further heat will produce the even less ordered form of water vapour.

How can we quantify this property that we call entropy? William J. M. Rankine (1820–1872), a Scots professor of engineering, showed in 1851 that during a reversible (frictionless) process, the ratio of heat exchanged to the current value of temperature, $\delta q/T$, remained constant. Clausius arrived at the same conclusion independently in Germany during the following year. He also recognized that particular ratio to be a thermodynamic function of state and coined the name **entropy**, s (after the Greek work for 'evolution'). Clausius further realized that although entropy remained constant for ideal processes, the total entropy must increase for all real processes.

$$\mathrm{d}s = \frac{\delta q}{T} \geqslant 0$$

When testing Clausius' conclusion, it is important to take all parts of the system and surroundings into account. A subsystem may be observed to experience a decrease in entropy if viewed in isolation. For example, water in a container that is placed in sufficiently colder surroundings can be seen to freeze—the ordered form of ice crystals growing, apparently spontaneously, on the surface of the less ordered liquid. The entropy of the water is visibly decreasing. Suppose the temperature of the water is T_w and that of the cooler air is T_a, where $T_a < T_w$. Then for a transfer of heat δq from the water to the air, the corresponding changes in entropy for the air, subscript a, and the water, subscript w, are

$$ds_a = \frac{\delta q}{T_a}$$

and

$$ds_w = -\frac{\delta q}{T_w} \quad \text{(negative as heat is leaving the water)}$$

Then the total change in entropy for this system is

$$ds_a + ds_w = \delta q \left(\frac{1}{T_a} - \frac{1}{T_w} \right) > 0$$

Using the thermodynamic property entropy, we can now express the total heat increase of a system, whether by heat transfer or frictional effects, as dq_c:

$$dq_c = T\,ds \quad \text{J/kg} \tag{3.44}$$

(see also equation (3.3)). The symbol q_c is used to denote the **combination** of added heat and the internally generated frictional heat.

Now to make the concept really useful, we must be able to relate entropy to other thermodynamic properties. Let us take the differential form of the steady flow energy equation (3.25):

$$V\,dP + dF = dH - dq \quad \text{J/kg} \tag{3.45}$$

Then

$$dF + dq = dH - V\,dP$$

However, the combined effect of friction and added heat is

$$dq_c = dF + dq$$

Therefore, equations (3.44) and (3.45) give

$$T\,ds = dH - V\,dP \quad \text{J/kg} \tag{3.46}$$

This is another equation that will be important to us in the analysis of subsurface ventilation circuits.

In order to derive an expression that will allow us to calculate a change in entropy directly from measurements of pressure and temperature, we can continue our analysis from equation (3.46).

$$ds = \frac{dH}{T} - \frac{V\,dP}{T} \quad \frac{\text{J}}{\text{kg K}}$$

However,

$$dH = C_p \, dT$$

(equation (3.32)) and

$$\frac{V}{T} = \frac{R}{P}$$

(from the general gas law $PV = RT$) giving

$$ds = C_p \frac{dT}{T} - R \frac{dP}{P} \qquad (3.47)$$

Integrating between end stations 1 and 2 gives

$$s_2 - s_1 = C_p \ln(T_2/T_1) - R \ln(P_2/P_1) \quad J/(kg\,K) \qquad (3.48)$$

This is known as the steady flow entropy equation.

3.4.3 The adiabatic and isentropic processes

An important thermodynamic process with which the ventilation engineer must deal is one in which there is no heat transfer between the air and the strata or any other potential source. This can be approached closely in practice, particularly in older return airways that contain no equipment and where the temperatures of the air and the surrounding rock have reached near equilibrium.

The steady flow energy equation for adiabatic flow is given by setting $q_{12} = 0$. Then

$$\int_1^2 V \, dP + F_{12} = H_2 - H_1 \quad J/kg \qquad (3.49)$$

The ideal, frictionless, or reversible adiabatic process is a particularly useful concept against which to compare real adiabatic processes. This is given simply by eliminating the friction term giving

$$\int_1^2 V \, dP = H_2 - H_1 \quad \frac{J}{kg} \qquad (3.50)$$

or

$$V \, dP = dH$$

In section 3.4.2, we defined a change in entropy, ds, by equation (3.44), i.e.

$$dq_c = T \, ds \quad J/kg \qquad (3.51)$$

where q_c is the combined effect of added heat and heat that is generated internally by frictional effects. It follows that during a frictionless adiabatic process where both q_{12} and F_{12} are zero, then dq_c is also zero and the entropy remains constant, i.e. an **isentropic process**.

The governing equations for an isentropic process follow from setting $s_2 - s_1 = 0$ in equation (3.48). Then

$$C_p \ln(T_2/T_1) = R \ln(P_2/P_1)$$

Taking antilogarithms gives

$$\frac{T_2}{T_1} = \left(\frac{P_2}{P_1}\right)^{R/C_p} \tag{3.52}$$

The index R/C_p was shown by equation (3.36) to be related to the ratio of specific heats $C_p/C_v = \gamma$ by the equation

$$\frac{R}{C_p} = \frac{\gamma - 1}{\gamma}$$

Then

$$\frac{T_2}{T_1} = \left(\frac{P_2}{P_1}\right)^{(\gamma - 1)/\gamma} \tag{3.53}$$

Values of the constants are given in Table 3.1. For dry air, $\gamma = 1.400$, giving the isentropic relationship between temperature and pressure as

$$\frac{T_2}{T_1} = \left(\frac{P_2}{P_1}\right)^{0.2856} \tag{3.54}$$

This is a particularly useful equation to relate pressures and temperatures in dry mine shafts where adiabatic conditions may be approached.

The isentropic relationship between pressure and specific volume follows from equation (3.53):

$$\frac{T_2}{T_1} = \left(\frac{P_2}{P_1}\right)^{1 - 1/\gamma} \tag{3.55}$$

However, from the general gas law (equation (3.12)),

$$\frac{T_2}{T_1} = \frac{P_2 V_2}{P_1 V_1} \tag{3.56}$$

Combining equations (3.55) and (3.56) gives

$$\frac{P_2 V_2}{P_1 V_1} = \left(\frac{P_2}{P_1}\right)^{1 - 1/\gamma}$$

$$\frac{V_2}{V_1} = \left(\frac{P_2}{P_1}\right)^{-1/\gamma} = \left(\frac{P_1}{P_2}\right)^{1/\gamma}$$

or

$$P_1 V_1^\gamma = P_1 V_2^\gamma = \text{constant} \tag{3.57}$$

3.4.4 Availability

In the context of conventional ventilation engineering, the energy content of a given airstream is useful only if it can be employed in causing the air to move, i.e. if it can be converted to kinetic energy. A more general concept is that of available energy. This is defined as the maximum amount of work that can be done by a system until it comes to complete physical and chemical equilibrium with the surroundings.

Suppose we have an airflow of total energy (equation (3.20))

$$\frac{u^2}{2} + Zg + H \quad \text{J/kg}$$

The kinetic and potential energy terms both represent mechanical energy and are fully available to produce mechanical effects, i.e. to do work. This is not true for the enthalpy. Remember that enthalpy is composed of PV and internal energy terms, and that the second law allows only a fraction of thermal energy to be converted into work. Suppose that the free atmosphere at the surface of a mine has a specific enthalpy H_0. Then when the mine air is rejected at temperature T to the surface, it will cool at constant pressure until it reaches the temperature of the ambient atmosphere T_0. Consequently, its enthalpy will decrease from H to H_0:

$$H - H_0 = C_p(T - T_0) \quad \text{J/kg} \tag{3.58}$$

The process is shown on a temperature–entropy diagram in Fig. 3.3. The heat

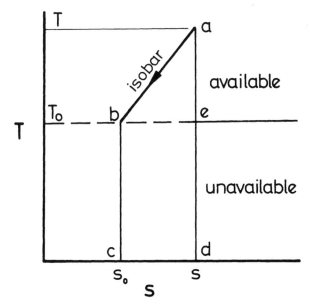

Figure 3.3 Temperature–entropy diagram for constant pressure cooling.

rejected to the free atmosphere is represented by the area under the process line

$$\int_a^b T \, ds = \text{area abcd}$$

However, for an isobar (constant pressure),

$$\int_a^b T \, ds = \int_a^b dH = H - H_0$$

(see equation (3.46) with $dP = 0$). Hence, the enthalpy change $H - H_0$ gives the heat lost to the atmosphere. Equation (3.58) actually illustrates the same fact, as heat lost is

$$\text{mass (1 kg)} \times \text{specific heat } (C_p) \times \text{change in temperature } (T - T_0)$$

for a constant pressure process.

Now the second law insists that only a part of this heat can be used to do work. Furthermore, we have illustrated earlier that when a parcel of any gas reaches the temperature of the surroundings then it is no longer capable of doing further work within those surroundings. Hence, the part of the heat energy that remains unavailable to do any work is represented by the $\int_e^b T \, ds$ area under the ambient temperature line T_0, i.e. $T_0(s - s_0)$ or area bcde on Fig. 3.3. The only part of the total heat that remains available to do work is represented by area abe or $H - H_0 - T_0(s - s_0)$.

The available energy, ψ, in any given airflow, with respect to a specified datum (subscript 0) may now be written as

$$\psi = \frac{u^2}{2} + Zg + H - H_0 - T_0(s - s_0) \quad \text{J/kg} \tag{3.59}$$

It should be made clear that the available energy represents the maximum amount of energy that is theoretically capable of producing useful work. How much of this is actually used depends on the ensuing processes.

Let us now try to show that any real airway suffers from a loss of available energy because of frictional effects. If we rewrite equation (3.59) in differential form for an adiabatic airway, then the increase in available energy along a short length of the airway is

$$d\psi = u \, du + g \, dz + dH - T_0 \, ds \quad \text{J/kg}$$

(H_0 and s_0 are constant for any given datum conditions). This equation assumes that we do not provide any added heat or work. From the differential form of the steady flow energy equation (3.25) for adiabatic conditions and no fan work ($dq = dW = 0$),

$$u \, du + g \, dz + dH = 0$$

leaving

$$d\psi = - T_0 \, ds \quad \text{J/kg} \tag{3.60}$$

Now, from equation (3.51),

$$ds = \frac{dq_c}{T}$$

where q_c is the combined effect of friction, F, and added heat, q. However, in this case, q is zero giving

$$ds = \frac{dF}{T}$$

and

$$d\psi = -\frac{T_0}{T}dF \quad \frac{J}{kg} \tag{3.61}$$

As T_0, T and dF are positive, the change in available energy must always be negative in the absence of any added work or heat. This equation also shows that the loss of available energy is a direct consequence of frictional effects.

Available energy is a 'consumable' item, unlike total energy which remains constant for adiabatic flow with no work input. During any real airflow process, the available energy is continuously eroded by the effects of viscous resistance in the laminar sublayer

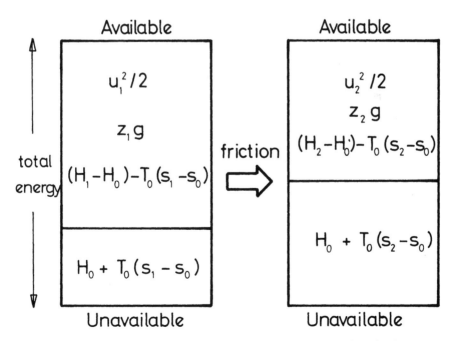

Figure 3.4 Available energy decreases and unavailable energy increases both by the amount $T_0(s_2 - s_1)$.

and within the turbulent eddies. That energy reappears as 'low grade' heat or un-available energy and is irretrievably lost in its capacity to do useful work. The process is illustrated in Fig. 3.4.

3.5 THERMODYNAMIC DIAGRAMS

During any thermodynamic process, there will be variations in the values of the fluid properties. The equations derived in the preceding sections of this chapter may be used to quantify some of those changes. However, plotting the variation of any one property against another provides a powerful visual aid to understanding the behaviour of any process path and can also give a graphical means of quantifying work or heat transfers where the complexity of the process path precludes an analytical treatment. These graphical plots are known as thermodynamic diagrams.

   The two most useful diagrams in steady flow thermodynamics are those of pressure against specific volume and temperature against specific entropy. These diagrams are particularly valuable, as areas on the PV diagram represent **work** and areas on the Ts diagram represent **heat**. Remembering that simple fact will greatly facilitate our understanding of the diagrams. In this section, we shall introdue the use of these diagrams through three compression processes. In each case, the air will be compressed from pressure P_1 to a higher pressure P_2. This might occur through a fan or compressor or by air dropping through a downcast shaft. The processes we shall consider are isothermal, isentropic and polytropic compression. As these are impor-tant processes for the ventilation engineer, the opportunity is taken to discuss the essential features of each, in addition to giving illustrations of the visual power of thermodynamic diagrams.

3.5.1   Ideal isothermal (constant temperature) compression

Suppose air is passed through a compressor so that its pressure is raised from P_1 to P_2. As work is done on the air, the first law of thermodynamics tells us that the internal energy and, hence, the temperature of tha air will increase (equation (3.16)). However, in this particular compressor, we have provided a water jacket through which flows a continuous supply of cooling water. Two processes then occur simul-taneously. First, the air is compressed and, secondly, it is cooled at just the correct rate to maintain its temperature constant. This is **isothermal** compression.

   Figure 3.5 shows the process on a pressure–specific volume (PV) and a temperature–entropy (Ts) diagram. The first stage in constructing these diagrams is to draw the isobars representing P_1 and P_2. On the PV diagram these are simply horizontal lines. Lines on the Ts diagram may be plotted using equation (3.48). Isobars curve slightly. However, over the range of pressures and temperatures of interest to the ventilation engineer, the curvature is negligible.

   The process path for the isothermal (constant temperature) compression is shown as line AB on both diagrams. On the PV diagram, it follows the slightly curved

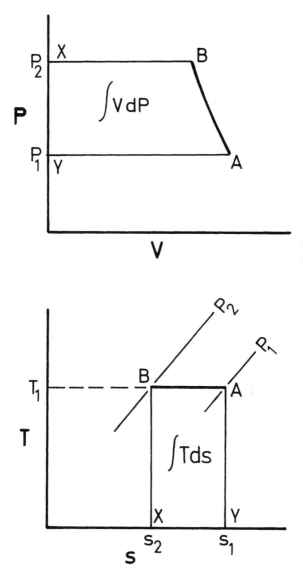

Figure 3.5 Isothermal compression.

path described by the equation $PV = RT_1 = C$ where C is a constant, or

$$P = \frac{C}{V} \quad \text{Pa} \tag{3.62}$$

On the Ts diagram, the isotherm is, of course, simply a horizontal line at temperature T_1.

Let us first concentrate on the PV diagram. From the steady flow energy equation for a frictionless process, we have

$$\frac{u_1^2 - u_2^2}{2} + (Z_1 - Z_2)g + W_{12} = \int_1^2 V\,dP$$

In this case we may assume that the flow through the compressor is horizontal $(Z_1 = Z_2)$ and that the change in kinetic energy is negligible, leaving

$$W_{12} = \int_1^2 V\,dP \quad \frac{J}{kg} \tag{3.63}$$

The integral $\int_1^2 V\,dP$ is the area to the left of the process line on the PV diagram, i.e. area ABXY for the ideal compressor is equal to the work, W_{12}, that has been done by the compressor in raising the air pressure from P_1 to P_2. We can evaluate the integral by substituting for V from equation (3.62):

$$W_{12} = \int_1^2 V\,dP = \int_1^2 \frac{C}{P}\,dP$$

$$= C\ln(P_2/P_1)$$

or

$$W_{12} = RT_1\ln(P_2/P_1) \quad J/kg \tag{3.64}$$

This is positive as work is done on the air.

Now let us turn to the Ts diagram. Remember that the integral $\int T\,ds$ represents the combined heat from actual heat transfer and also that generated by internal friction. However, in this ideal case, there is no friction. Hence the $\int T\,ds$ area under the process line AB represents the heat that is removed from the air by the cooling water during the compression process. The heat area is the rectangle ABXY or $-T_1(s_1 - s_2)$ on the Ts diagram. In order to quantify this heat, recall the entropy equation (3.48) and apply the condition of constant temperature. This gives

$$T_1(s_1 - s_2) = -RT_1\ln(P_2/P_1) \quad J/kg \tag{3.65}$$

The sign is negative as heat is removed from the air.

Now compare the work input (equation (3.64)) with the heat removed (equation (3.65)). Apart from the sign, they are numerically identical. This means that, as work energy is done on the system, exactly the same quantity of energy is removed as heat. This is also shown directly by the first law (equation (3.16)) with $dU = C_v\,dT = 0$ giving $\delta W = -\delta q$ for an isothermal process. Despite the zero net increase in energy, the air has been pressurized and is certainly capable of doing further useful work through a compressed air motor. How can that be? The answer lies in the discussion on availability given in section 3.4.4. All of the work input is available energy capable of producing mechanical effects. However, all of the heat removed is unavailable energy that already existed in the ambient air and which could not be used to do useful work. The fact that this heat is, indeed, completely unavailable is illustrated on the Ts diagram by the corresponding heat area $-T_1(s_1 - s_2)$ lying completely below the ambient temperature line.

This process should be studied carefully until it is clearly understood that there is a very real distinction between available energy and unavailable energy. The high pressure air that leaves the compressor retains the work input as available energy but has suffered a loss of unavailable energy relative to the ambient air. On passing through a compressed air motor, the available energy is utilized in producing mechanical work output, leaving the air with only its depressed unavailable energy to be exhausted back to the atmosphere. This explains why the air emitted from the exhaust ports of a compressed air motor is very cold and may give rise to problems of freezing.

Maintaining the temperature constant during isothermal compression minimizes the work that must be done on the air for any given increase in pressure. True isothermal compression cannot be attained in practice, as a temperature difference must exist between the air and the cooling medium for heat transfer to occur. In large multistage compressors, the actual process path on the Ts diagram proceeds from A to B along a zig-zag line with stages of adiabatic compression alternating with isobaric cooling attained through interstage water coolers.

3.5.2 Isentropic (constant entropy) compression

During this process we shall, again, compress the air from P_1 to P_2 through a fan or compressor, or perhaps by gravitational work input as the air falls through a downcast shaft. This time, however, we shall assume that the system is not only frictionless but is also insulated so that no heat transfer can take place. We have already introduced the frictionless adiabatic in section 3.4.2 and shown that it maintains constant entropy, i.e. an isentropic process.

The PV and Ts diagrams are shown on Fig. 3.6. The corresponding process lines for the isothermal case have been retained for comparison. The area to the left of the isentropic process line, AC, on the PV diagram is, again, the work input during compression. It can be seen that this is greater than for the isothermal case. This area, ACXY, or $\int_1^2 V\,dP$, is given directly by the steady flow energy equation (3.25) with $F_{12} = q_{12} = 0$:

$$\int_1^2 V\,dP = H_2 - H_1 = C_p(T_2 - T_1) \quad \text{J/kg} \tag{3.66}$$

On the Ts diagram, the process line AC is vertical (constant entropy) and the temperature increases from T_1 to T_2. The $\int T\,ds$ area under the line is zero. This suggests that the Ts diagram is not very useful in further evaluating an isentropic process. However, we are about to reveal a feature of Ts diagrams that enhances their usefulness very considerably. Suppose that, having completed the isentropic compression and arrived at point C on both diagrams, we now engage upon an imaginary second process during which we cool the air at constant pressure until it reattains its original ambient temperature T_1. The process path for this second process of isobaric

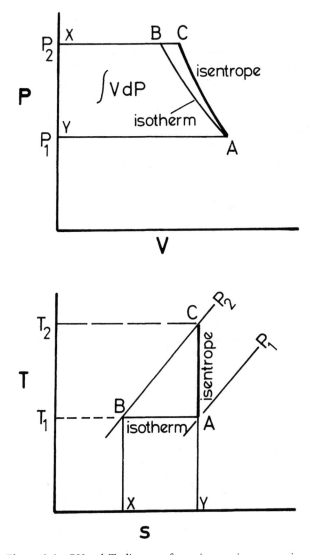

Figure 3.6   *PV* and *Ts* diagrams for an isentropic compression.

cooling is CB on both diagrams. The heat removed during the imaginary cooling is the area $\int_1^2 T\,\mathrm{d}s$ under line CB on the *Ts* diagram, i.e. area CBXY. However, from the steady flow entropy equation (3.47)

$$\mathrm{d}s = C_\mathrm{p}\frac{\mathrm{d}T}{T} - R\frac{\mathrm{d}P}{P} \qquad \frac{\mathrm{J}}{\mathrm{kg\,K}}$$

and for an isobar, $dP = 0$, giving

$$ds = C_p \frac{dT}{T} \text{ or } T\,ds = C_p\,dT$$

Then heat removed during the imaginary cooling becomes

$$\int_2^1 T\,ds = -\int_1^2 T\,ds = -\int_1^2 C_p\,dT = -C_p(T_2 - T_1) \quad \frac{J}{kg} \qquad (3.67)$$

Now compare equations (3.66) and (3.67). It can be seen that the heat removed during the imaginary cooling is numerically equivalent to the work input during the isentropic compression. Hence, the work input is not only shown as area ACBXY on the *PV* diagram but also as the same area on the *Ts* diagram. Using the device of imaginary isobaric cooling, the *Ts* diagram can be employed to illustrate work done as well as heat transfer. However, the two must never be confused—the *Ts* areas represent true heat energy and can differentiate between available and unavailable heat, while work areas shown on the *Ts* diagram are simply convenient numerical equivalents with no other physical meaning on that diagram.

The relationships between pressure, temperature and specific volume for an isentropic process are analysed in section 3.4.3.

3.5.3 Polytropic compression

The relationship between pressure and specific volume for an isentropic process has been shown to be (equation (3.57))

$$PV^\gamma = \text{constant}$$

where the isentropic index γ is the ratio of specific heats C_p/C_v. Similarly, for an isothermal process,

$$PV^1 = \text{constant}$$

These are, in fact, special cases of the more general equation

$$PV^n = \text{constant}, C \qquad (3.68)$$

where the index n remains constant for any given process but will take a different value for each separate process path. This general equation defines a **polytropic** system and is the type of process that occurs in practice within subsurface engineering. It encompasses the real situation of frictional flow and the additional increases in entropy that arise from heat transfer to the air.

Figure 3.7 shows the *PV* and *Ts* process lines for a polytropic compression. Unlike the isothermal and isentropic cases, the path line for the polytropic process is not rigidly defined but depends on the value of the polytropic index n. The polytropic curve shown on the diagrams indicates the most common situation in underground ventilation involving both friction and added heat.

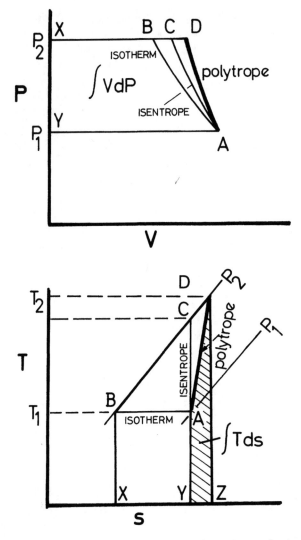

Figure 3.7 *PV* and *Ts* diagrams for a polytropic compression.

The flow work shown as area ADXY on the *PV* diagram may be evaluated by integrating

$$\int_1^2 V \, dP \quad \text{where} \quad V = \left(\frac{C}{P}\right)^{1/n} \quad \text{from equation (3.68)}$$

Then

$$\int_1^2 V \, dP = \int_1^2 \left(\frac{C}{P}\right)^{1/n} dP = C^{1/n} \frac{[P^{1-1/n}]_1^2}{1-1/n}$$

or, as $C^{1/n} = P^{1/n}V$,

$$\int_1^2 V\,dP = \frac{n}{n-1}(P_2 V_2 - P_1 V_1) \quad J/kg$$

$$= \frac{n}{n-1} R(T_2 - T_1) \tag{3.69}$$

using the general gas law $PV = RT$. This enables the flow work to be determined if the polytropic index, n, is known. It now becomes necessary to find a method of calculating n, preferably in terms of the measurable parameters, pressure and temperature.

From the polytropic law (3.68)

$$P_1 V_1^n = P_2 V_2^n$$

or

$$\frac{V_1}{V_2} = \left(\frac{P_2}{P_1}\right)^{1/n} \tag{3.70}$$

and from the general gas law (equation (3.12)),

$$\frac{V_1}{V_2} = \frac{P_2}{P_1}\frac{T_1}{T_2} \tag{3.71}$$

We have isolated V_1/V_2 in equations (3.70) and (3.71) in order to leave us with the desired parameters of pressure and temperature. Equating equations (3.70) and (3.71) gives

$$\left(\frac{P_2}{P_1}\right)^{1/n} = \frac{P_2}{P_1}\frac{T_1}{T_2}$$

or

$$\frac{T_2}{T_1} = \left(\frac{P_2}{P_1}\right)^{1-1/n}$$

giving

$$\frac{n-1}{n} = \frac{\ln(T_2/T_1)}{\ln(P_2/P_1)} \tag{3.72}$$

This enables the polytropic index, n, to be determined for known end pressures and temperatures. However, we can substitute for $n/(n-1)$, directly, into equation (3.69) to give the flow work as

$$\int_1^2 V\,dP = R(T_2 - T_1)\frac{\ln(P_2/P_1)}{\ln(T_2/T_1)} \quad \frac{J}{kg} \tag{3.73}$$

This is an important relationship that we shall use in the analysis of mine ventilation thermodynamics (Chapter 8).

Turning to the Ts diagram, the $\int T\,ds$ area under the polytrope AD, i.e. area ADZY, is the combined heat increase arising from internal friction, F_{12}, and added heat, q_{12}:

$$\int_1^2 T\,ds = F_{12} + q_{12} \quad \frac{J}{kg} \tag{3.74}$$

However, from the steady flow energy equation (3.25),

$$F_{12} + q_{12} = H_2 - H_1 - \int_1^2 V\,dP \quad \frac{J}{kg} \tag{3.75}$$

giving the area under line AD as

$$\int_1^2 T\,ds = H_2 - H_1 - \int_1^2 V\,dP \quad \frac{J}{kg} \tag{3.76}$$

We arrived at the same result in differential form (equation (3.46)) during our earlier general discussion on entropy.

As $H_2 - H_1 = C_p(T_2 - T_1)$, and using equation (3.73) for the flow work, we can rewrite equation (3.76) as

$$\int_1^2 T\,ds = C_p(T_2 - T_1) - R(T_2 - T_1)\frac{\ln(P_2/P_1)}{\ln(T_2/T_1)}$$

$$= (T_2 - T_1)\left[C_p - R\frac{\ln(P_2/P_1)}{\ln(T_2/T_1)}\right] \quad \frac{J}{kg} \tag{3.77}$$

Now, using the same logic as employed in section 3.5.2, it can be shown that the area under the isobar DB on the Ts diagram of Fig. 3.7 is equal to the change in enthalpy $H_2 - H_1$. We can now illustrate the steady flow energy equation as areas on the Ts diagram.

$$H_2 - H_1 = \underbrace{F_{12} + q_{12}} + \int_1^2 V\,dP \quad J/kg$$

$$\begin{array}{ccc} \text{area} & = & \text{area} & + & \text{area} \\ \text{DBXZ} & & \text{ADZY} & & \text{ADBXY} \end{array} \tag{3.78}$$

Once again, this shows the power of the Ts diagram.

Example A dipping airway drops through a vertical elevation of 250 m between stations 1 and 2. The following observations are made.

	Velocity u (m/s)	Pressure P (kPa)	Temperature t (°C)	Airflow Q (m/s)
Station 1	2.0	93.40	28.20	43.0
Station 2	3.5	95.80	29.68	

Assuming that the airway is dry and that the airflow follows a polytropic law, determine

1. the polytropic index, n,
2. the flow work,
3. the work done against friction and the frictional pressure drop,
4. the change in enthalpy,
5. the change in entropy, and
6. the rate and direction of heat transfer with the strata, assuming no other sources of heat.

Solution   It is convenient to commence the solution by calculating the end air densities and the mass flow of air.

$$\rho_1 = \frac{P_1}{RT_1} \quad \text{(equation (3.11))}$$

$$= \frac{93\,400}{287.04 \times (273.15 + 28.20)} = 1.0798 \text{ kg/m}^3$$

$$\rho_2 = \frac{P_2}{RT_2}$$

$$= \frac{95\,800}{287.04 \times (273.15 + 29.68)} = 1.1021 \text{ kg/m}^3$$

mass flow $\qquad M = Q_1\rho_1 \quad \dfrac{\text{m}^3}{\text{s}}\dfrac{\text{kg}}{\text{m}^3}$

$$= 43.0 \times 1.0798 = 46.43 \text{ kg/s}$$

1. *Polytropic index, n.* From equation (3.72),

$$\frac{n-1}{n} = \frac{\ln(T_2/T_1)}{\ln(P_2/P_1)}$$

where $T_1 = 273.15 + 28.20 = 301.35$ K and $T_2 = 273.15 + 29.68 = 302.83$ K.

$$\frac{n-1}{n} = \frac{\ln(302.83/301.35)}{\ln(95.80/93.40)}$$

$$= 0.1931$$

giving

$$n = 1.239$$

This polytropic index is less than the isentropic index for dry air, 1.4, indicating that heat is being lost from the air to the surroundings.

2. *Flow work.* From equation (3.73),

$$\int_1^2 V \, dP = R(T_2 - T_1) \frac{\ln(P_2/P_1)}{\ln(T_2/T_1)}$$

$$= 287.04(29.68 - 28.20) \frac{\ln(95.80/93.40)}{\ln(302.83/301.35)}$$

$$= 2200.0 \quad \text{J/kg}$$

Degrees Celsius can be used for a difference $T_2 - T_1$ but remember to employ kelvins in all other circumstances.

3. *Friction.* From the steady flow energy equation with no fan

$$F_{12} = \frac{u_1^2 - u_2^2}{2} + (Z_1 - Z_2)g - \int_1^2 V \, dP$$

$$= \frac{2^2 - 3.5^2}{2} + 250 \times 9.81 - 2200.0$$

$$= -4.1 + 2452.5 - 2200.0$$

$$= 248.4 \quad \text{J/kg}$$

Note how small is the change in kinetic energy compared with the potential energy and flow work

In order to determine the frictional pressure drop, we use equation (2.46):

$$p = \rho F_{12}$$

For this to be meaningful, we must specify the value of density to which it is referred. At a mean density (subscript m) of

$$\rho_m = \frac{\rho_1 + \rho_2}{2} = \frac{1.0798 + 1.1021}{2} = 1.0909 \, \text{kg/m}^3$$

$$p_m = 1.0909 \times 248.4 = 271 \quad \text{Pa}$$

or, for comparison with other pressure drops, we may choose to quote our frictional pressure drop referred to a standard air density of $1.2 \, \text{kg/m}^3$ (subscript st), giving

$$p_{st} = 1.2 \times 248.4 = 298 \quad \text{Pa}$$

4. *Change in enthalpy.* From equation (3.33),

$$H_2 - H_1 = C_p(T_2 - T_1)$$

where $C_p = 1005$ for dry air (Table 3.1)

$$= 1005(29.68 - 28.20)$$

$$= 1487.4 \, \text{J/kg}$$

5. *Change in entropy.* From equation (3.51),

$$(s_2 - s_1) = C_p \ln(T_2/T_1) - R \ln(P_2/P_1)$$

where $R = 287.04$ J/(kg K) from equation (3.15)

$$= 1005 \ln(302.83/301.35) - 287.04 \ln(95.8/93.4)$$

$$= \qquad 4.924 \qquad\qquad -7.283$$

$$= -2.359 \quad \text{J/(kg K)}$$

The decrease in entropy confirms that heat is lost to the strata.

6. *Rate of heat transfer.* Again, from the steady flow energy equation (3.25),

$$q_{12} = H_2 - H_1 - \frac{u_1^2 - u_2^2}{2} - (Z_1 - Z_2)g$$

Each term has already been determined, giving

$$q_{12} = 1487.4 - 4.1 - 2452.5$$

$$= -969.2 \text{ J/kg}$$

To convert this to kilowatts, multiply by mass flow

$$q_{12} = \frac{-969.2 \times 46.43}{1000} = -45 \text{ kW}$$

The negative sign shows again that heat is transferred from the air to the strata and at a rate of 45 kW.

FURTHER READING

Hinsley, F. B. (1943) Airflow in mines: a thermodynamic analysis. *Proc. South Wales Inst. Eng.* **LIX** (2).

Look, D. C. and Sauer, H. J. (1982) *Thermodynamics*, Brooks-Cole.

Rogers, G. F. C. and Mayhew, Y. R. (1957) *Engineering Thermodynamics, Work and Heat Transfer*, Longmans, Green and Co.

Van Wylen, G. J. (1959) *Thermodynamics*, Wiley.

PART TWO

Subsurface Ventilation Engineering

4

Subsurface ventilation systems

4.1 INTRODUCTION

Practically every underground opening is unique in its geometry, extent, geological surroundings, environmental pollutants and reasons for its formation—natural or man made. The corresponding patterns of airflow through those openings are also highly variable. There are, however, certain features that are sufficiently common to permit classifications of structured ventilation systems and subsystems to be identified.

In this chapter, we shall discuss the essential characteristics of subsurface ventilation systems, first on the basis of complete mines and primary airflow routes. The opportunity will be taken to introduce some of the technical terms used by ventilation engineers. The terms chosen are those that are in common use throughout the English-speaking mining countries. Secondly, we shall look at district systems for more localized areas of a mine. These, in particular, vary considerably depending on the geometry of the geologic deposit being mined. Although reference will be made to given mining methods, the treatment here will concentrate on principles rather than detailed layouts. In most countries, state or national mining law impacts upon the ventilation layout. System designers must, as a prerequisite, become familiar with the relevant legislation.

Thirdly, auxiliary ventilation systems are examined, these dealing with the ventilation of blind headings. The chapter also includes the principles of controlled partial recirculation and the ventilation of underground repositories for nuclear waste or other stored material.

4.2 MINE SYSTEMS

4.2.1 General principles

Figure 4.1 depicts the essential elements of a ventilation system in an underground mine or other subsurface facility. Fresh air enters the system through one or more

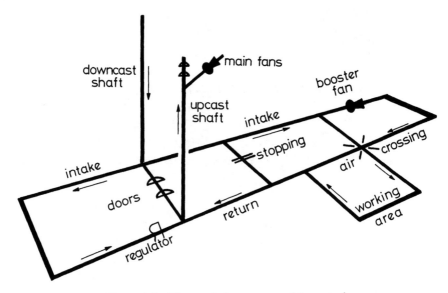

Figure 4.1   The ventilation system and its control.

downcast shafts, or other connections to surface. The air flows along **intake** airways to the working areas or places where the majority of pollutants are added to the air. These include dust and a combination of many other potential hazards including toxic or flammable gases, heat, humidity, and radiation. The contaminated air passes back through the system along **return** airways. In most cases, the concentration of contaminants is not allowed to exceed mandatory threshold limits imposed by law and safe for the entry of personnel into all parts of the ventilation system including return airways. The intake and return airways are often referred to simply as **intakes** and **returns** respectively. The return air eventually passes back to the surface via one or more **upcast** shafts, or through inclined or level drifts.

Fans

The primary means of producing and controlling the airflow are also illustrated on Fig. 4.1. Main fans, either singly or in combination, handle all of the air that passes through the entire system. These are usually, but not necessarily, located on surface, either **exhausting** air through the system as shown on Fig. 4.1 or, alternatively, connected to downcast shafts or main intakes and **forcing** air into and through the system. Because of the additional hazards of gases and dust that may both be explosive, legislation governing the ventilation of coal mines is stricter than for most other underground facilities. In many countries, the main ventilation fans for coal mines are required, by law, to be placed on the surface and may also be subject to other restrictions such as being located out of line with the connected shaft or drift and equipped with 'blow-out' panels to help protect the fan in case of a mine explosion.

Stoppings and seals

In developing a mine, connections are necessarily made between intakes and returns. When these are no longer required for access or ventilation, they should be blocked by **stoppings** in order to prevent short-circuiting of the airflow. Stoppings can be constructed from masonry, concrete blocks or fire-proofed timber blocks. Pre-fabricated steel stoppings may also be employed. Stoppings should be well keyed into the roof, floor and sides, particularly if the strata are weak or in coal mines liable to spontaneous combustion, Leakage can be reduced by coating the high pressure face of the stopping with a sealant material and particular attention paid to the perimeter. Here again, in weak or chemically active strata, such coatings may be extended to the rock surfaces for a few metres back from the stopping. In cases where the airways are liable to convergence, precautions should be taken to protect stop-pings against premature failure or cracking. These measures can vary from 'crush pads' located at the top of the stopping to sliding or deformable panels on pre-fabricated stoppings. In all cases, components of stoppings should be fireproof and should not produce toxic fumes when heated.

As a short term measure, fire-resistant **brattice curtains** may be tacked to roof, sides and floor to provide temporary stoppings where pressure differentials are low such as in locations close to the working areas.

Where abandoned areas of a mine are to be isolated from the current ventilation infrastructure, **seals** may be constructed at the entrances of the connecting airways. These consist of two or more stoppings, 5 to 10 metres apart, with the intervening space occupied by sand, stonedust, compacted non-flammable rock waste, cement-based fill or other manufactured material. Steel girders, laced between roof and floor, add structural strength. Grouting the surrounding strata adds to the integrity of the seal in weak ground. In coal mines, mining law or prudent regard for safety may require seals to be explosion proof.

Doors and airlocks

Where access must remain available between an intake and a return airway, a stopping may be fitted with a **ventilation door**. In its simplest form, this is merely a wooden or steel door hinged such that it opens towards the higher air pressure. This self-closing feature is supplemented by angling the hinges so that the door lifts slightly when opened and closes under its own weight. It is also advisable to fit doors with latches to prevent their opening in cases of emergency when the direction of pressure differentials may be reversed. Contoured rubber strips attached along the bottom of the door assist greatly in reducing leakage, particularly when the airway is fitted with rail track.

Ventilation doors located between main intakes and returns are usually built as a set of two or more to form an **airlock**. This prevents short-circuiting when one door is opened for passage of vehicles or personnel. The distance between doors should be capable of accommodating the longest train of vehicles required to pass through the airlock. For higher pressure differentials, multiple doors also allow the pressure break to be shared between doors.

Mechanized doors, opened by pneumatic or electrical means, are particularly convenient for the passage of vehicular traffic or where the size of the door or applied pressure would make manual operation difficult. Mechanically operated doors may, again, be side-hinged or take the form of roll-up or concertina devices. They may be activated manually by a pull-rope or automatic sensing of an approaching vehicle or person. Large doors may be fitted with smaller hinged openings for access by personnel. Man–doors exposed to the higher pressure differentials may be difficult to open manually. In such cases, a sliding panel may be fitted in order to reduce that pressure differential temporarily while the door is opened. Interlock devices may also be employed on an airlock to prevent all doors from being opened simultaneously.

Regulators

A **passive regulator** is simply a door fitted with one or more adjustable orifices. Its purpose is to reduce the airflow to a desired value in a given airway or section of the mine. The most elementary passive regulator is a rectangular orifice cut in the door and partially closed by a sliding panel. The airflow may be modified by adjusting the position of the sliding panel manually. Louvre regulators can also be employed. Another form of regulator is a rigid duct passing through an airlock. This may be fitted with a damper, louvres or butterfly valve to provide a passive regulator, or a fan may be located within the duct to produce an active regulator. Passive regulators may be actuated by motors, either to facilitate their manual adjustment or to react automatically to monitored changes in the quantity or quality of any given airflow.

When the airflow in a section of the mine must be adjusted to a magnitude beyond that obtainable from the open system then a means of achieving this is by **active regulation**. This implies the use of a **booster fan** to enhance the airflow through a part of the mine. Section 9.6 deals with the subject of booster fans in more detail. Where booster fans are employed, they should be designed into the system in order to help control leakage without causing undesired recirculation in either normal or emergency situations. In some countries, coal mine legislation inhibits the use of booster fans.

Air crossings

Where intake and return airways are required to cross over each other then leakage between the two must be controlled by the use of an **air crossing**. The sturdiest form is a **natural** air crossing in which the horizon of one of the airways is elevated above the other to leave a sill of strata between the two, perhaps reinforced by roof bolts, girders or timber boards. A more usual method is to intersect the two airways during construction, then to heighten the roof of one of them and/or excavate additional material from the floor of the other. The two airstreams can then be separated by masonry or concrete blocks, or a steel structure with metal or timber shuttering,

Sealants may be applied on the high pressure side. Control of the airway gradients approaching the air crossing reduces the shock losses caused by any sudden change of airflow direction. Man-doors can be fitted into the air crossing for access.

Completely fabricated air crossings may be purchased or manufactured locally. These take the form of a stiffened metal tunnel. Such devices may offer high resistance to airflow and should be sized for the flow they are required to pass. They are often employed for conveyor crossings. Another type of air crossing used mainly for lower airflows and that requires no additional excavation is to course one of the airstreams through one or more ducts that intersect a stopping on either side of the crossing. An advantage of this technique is that the ducted airflow may be further restricted by passive regulators, or uprated by fans in the ducts.

In all cases, the materials used in the construction of air crossings should be fireproof and capable of maintaining their integrity in case of fire. Neither aluminium nor any other low melting point material should be employed in an air crossing.

4.2.2 Location of main fans

In the majority of the world's mines, main fans are sited on the surface. In the case of coal mines, this may be a mandatory requirement. A surface location facilitates installation, testing, access and maintenance while allowing better protection of the fan during an emergency situation. Siting main fans underground may be considered where fan noise is to be avoided on surface or when shafts must be made available for hoisting and free of airlocks. A problem associated with underground main fans arises from the additional doors, airlocks and leakage paths that then exist in the subsurface.

In designing the main ventilation infrastructure of a mine, a primary decision is whether to connect the main fans to the upcast shafts, i.e. an **exhausting** system or, alternatively, to connect the main fans to the downcast shaft in order to provide a **forcing** or **blowing** system. These choices are illustrated in Fig. 4.2(a) and 4.2(b).

From the time of the shaft bottom furnaces of the nineteenth century, the upcast shaft has, traditionally, been regarded as associated with the means of producing ventilation. Most mines are ventilated using the exhaust system. An examination of the alternatives continues to favour a primary exhaust system in the majority of cases. The choice may be based on the following four concerns.

Gas control

Figure 4.2 shows that air pressure in the subsurface is depressed by the operation of an exhausting main fan but is increased by a forcing fan. The difference is seldom more than a few kilopascals. As strata gases are, typically, held within the rock matrix at gauge pressures of 1000 kPa or more, it is evident that the choice of an exhausting or forcing system producing a few kilopascals will have little effect on the rate of gas production from the strata.

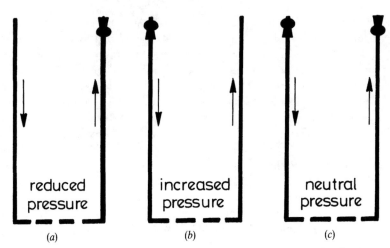

reduced pressure

increased pressure

neutral pressure

(a) (b) (c)

Figure 4.2 Possible locations of main fans: (a) exhausting system; (b) forcing system; (c) push–pull system.

Unfortunately, much of the gas is not emitted directly into the ventilating airstream but collects in worked-out areas, relaxed strata or voidage that is connected to, but not part of, the main ventilation system. Such accumulations of gas are at near equilibrium pressure with the adjacent airways. Hence, any reduction in barometric pressure within the ventilation system will result in an isothermal expansion of the accumulated gases and produce a transient emission of those gases into the ventilation system. This occurs naturally during periods of falling barometric pressure on the surface. In some countries, a mandatory record of surface barometric readings is updated at the beginning of each shift at coal mines. During a period of falling barometric pressure, unusually heavy emissions of methane or de-oxygenated air may be expected. At such times, it is not wise to sit down for lunch against a seal or stopping beyond which are old workings.

Steady state operation of either exhausting or forcing fans will produce no changes of air pressure in the subsurface. However, consider the situation of the stoppage of a forcing fan. The barometric pressure throughout the ventilation system will fall rapidly, Accumulations of voidage gas will expand and flood into the workings at the worst possible time, i.e. when the airflow is considerably diminished, causing a peak concentration of gas. Conversely, when a main exhausting fan stops, air pressure in the system increases, compressing accumulations of gas. Hence, no peak of general body gas concentration occurs within the airstream. It is true that when the main exhaust fan restarts, the sudden reduction in barometric pressure will then cause expansion and emission of accumulated gas. However, this occurs at a time of full ventilation and the peak of gas concentration will be much less than that caused by stoppage of a forcing fan. A consideration of strata gas control favours a primary exhausting system.

Transportation

The choice between main forcing and exhausting systems should take into account the preferred routes for the transportation of mineral, personnel and materials. Ideally, conveyors, locomotives or other modes of moving broken rock should not be required to pass through airlocks. Hence, a mine design that has mineral or rock transportation routes in main intakes and rock hoisting in a downcast shaft will favour an exhausting system. Alternatively, if there are good reasons to transport mineral in the returns and upcast shaft then a forcing system may be preferred. This could be necessary, for example, in evaporite mines producing potash or halite where the hygroscopic nature of the mineral could give ore-handling problems if transported within the variable humidity of intake airways.

In American coal mines, conveyors are normally required to be located in 'neutral' airways, ventilated by air that neither will pass on to working faces nor is returning from work areas. This system has the advantage that smoke and gases produced by any conveyor fire will not pollute the working faces. The major disadvantages are the additional potential for leakage and difficulties in controlling the air quantity along the conveyor routes.

Fan maintenance

An exhausting fan will pass air that carries dust, water vapour and perhaps liquid water droplets, and is usually at a higher temperature than that of the air entering the mine. The combined effects of impact and corrosion on impeller blades is much greater on main exhaust fans. Forcing fans handle relatively clean air and require less maintenance for any given duty. On the other hand, corrosive air passing up through the headgear of an upcast shaft can cause much damage. This can be prevented by drawing the air out of the shaft into a near surface fan drift by means of a low pressure–high volume fan.

Fan performance

A forcing fan normally handles air that is cooler and denser than that passing through an exhausting fan. For any given mass flow, the forcing fan will pass a lower volume flow at a reduced pressure. The corresponding power requirement is, therefore, also lower for a forcing fan. However, the effect is not great and unlikely to be of major significance.

A counteracting influence is that forcing fans must be fitted with inlet grilles to prevent the ingress of birds or other solid objects. These grilles necessarily absorb available energy and result in an additional frictional pressure drop. Furthermore, the expanding evasee fitted to a main exhaust fan recovers some of the kinetic energy that would otherwise be lost to the surface atmosphere.

When air is compressed through a fan its temperature is increased. If the air contains no liquid droplets and there is insignificant heat transfer through the fan casing then the temperature rise is given as (see derivation in section 10.6.1)

$$\Delta T = \frac{0.286}{\eta}\frac{T_1}{P_1}\Delta P \quad °C \tag{4.1}$$

where η = fan isentropic efficiency (fractional), T_1 = absolute temperature at inlet (K), P_1 = barometric pressure at inlet (Pa) and ΔP = increase in absolute pressure across the fan (Pa). The rise in temperature through a forcing fan will be reflected by an increase in average temperature in the intake airways. However, heat exchange with the strata is likely to dampen the effect before the air reaches the work areas (section 15.2.2).

Figure 4.2(c) shows a combination of main forcing and exhausting fans, known descriptively as a **push–pull** system. The primary application of a main push–pull system is in metal mines practicing caving techniques and where the zone of fragmented rock has penetrated through to the surface. Maintaining a neutral pressure underground with respect to the surface minimizes the degree of air leakage between the workings and the surface. This is particularly important if the rubbelized rock is subject to spontaneous combustion. In cold climates, drawing air through fragmented strata intentionally can help to smooth out extremes of temperature of the intake air entering the workings (section 18.4.6).

In the more general case of multishaft mines, the use of multiple main fans (whether exhausting, forcing or push–pull) offers the potential for an improved distribution of airflow, better control of both air pressures and leakage, greater flexibility and reduced operating costs. On the other hand, these advantages may not always be realized as a multifan system requires particularly skilled adjustment, balancing and planning.

4.2.3 Infrastructure of main ventilation routes

Although the simplified sketch of Fig. 4.1 depicts the main or **trunk** intakes and returns as single airways, this is seldom the case in practice other than for small mines. In designing or examining the underground layout that constitutes a subsurface ventilation system, the following matters should be addressed.

Mine resistance

For any given total airflow requirement, the operational cost of ventilation is proportional to the resistance offered to the passage of air (section 9.5.5). This resistance, in turn, depends on the size and number of the openings and the manner in which they are interconnected. Problems of ground stability, air velocity and economics limit the sizes of airways. Hence, multiple main intakes and returns are widely employed.

The mine resistance is greatly reduced and environmental conditions improved by providing a separate **split** of air to each working panel. The advantages of parallel circuits over series ventilation were realized early in the nineteenth century (section 1.2).

Leakage control

The **volumetric efficiency** of a mine is defined as

$$VE = \frac{\text{airflow usefully employed}}{\text{total airflow through main fans}} \times 100\% \qquad (4.2)$$

where the 'airflow usefully employed' is the sum of the airflows reaching the working faces and those used to ventilate equipment such as electrical gear, pumps or battery charging stations, the volumetric efficiency of mines may vary from 75% down to less than 10%. The latter value indicates the large and, often, expensive amount of air leakage that can occur in a mine. It is, therefore, important to design a subsurface ventilation system to minimize leakage potential and to maintain the system in order to control that leakage. Whenever possible, intake and return airways, or groups of airways, should be separated geographically or by barrier pillars with a minimum of interconnections.

A prerequisite is that all doors, stoppings, seals and air crossings should be constructed and maintained to a good standard. A stopping between a main intake and return that has been carelessly holed in order to insert a pipe or cable, or one that has been subject to roadway convergence without the necessary repairs, may be a source of excessive leakage. Unfortunately, if a large number of stoppings exist between an intake and adjacent return then the leakage may become untenable even when each individual stopping is of good quality. This can occur in workings that have been developed by room and pillar methods. The reason for this is the dramatic decrease in effective resistance to airflow when the flowpaths are connected in parallel. For n stoppings constructed between two adjacent airways, their combined (effective) resistance becomes (see section 7.3.1)

$$R_{\text{eff}} = \frac{R}{n^2}$$

where R is the resistance of a single stopping. Figure 4.3 shows the dramatic reduction in effective resistance that occurs as the number of stoppings increases. In such cases, it becomes important not only to maintain good quality stoppings but also to design the system such that pressure differentials between the airways are minimized.

Air pressure management is a powerful tool in controlling leakage and, hence, the effectiveness, volumetric efficiency and costs of a ventilation system. It is particularly important for mines that are liable to spontaneous combustion. Ideally, resistance to airflow should be distributed equitably between intakes and returns. In practice, one often observes return airways of smaller cross-section than intakes and that have been allowed to deteriorate because they are less frequently used for travelling or transportation. This will increase the pressure differentials between intakes and returns.

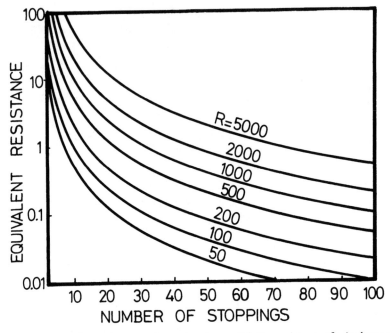

Figure 4.3 Equivalent resistances of stoppings in parallel. R = resistance of a single stopping.

Similarly, local obstructions caused by falls of roof, stacked materials, equipment or parked vehicles will affect the pressure distribution and may exacerbate leakage. The positions and settings of booster fans or regulators also have a marked influence on leakage patterns and should be investigated thoroughly by network analysis (Chapter 7) during design procedures.

Direction of airflow

There are two considerations regarding the direction of the airstream—first with respect to the transportation of the mined material. An **antitropal** system is one in which the airflow and transported rock move in opposite directions. Other than the 'neutral' airways of American conveyor roads, this implies mineral transportation in intake airways. Conversely, a **homotropal** system is one in which the airflow and the fragmented rock move in the same direction. This implies mineral transportation in the return airways and is often associated with a main forcing system. The homotropal system ensures that any pollution generated from the fragmented rock along the transportation route passes directly out of the mine without affecting working faces. Such pollution may include dust, heat, humidity and gases issuing from the broken rock or equipment. The higher relative velocity between conveyed material and the airflow in an antitropal system can result in a greater entrainment of dust particles within the airstream. Furthermore, a homotropal system is preferable in the event of a fire occurring along the mineral transportation route. On the other hand, siting electrical or other equipment capable of igniting a methane –air mixture in a return airway may be inadvisable or, indeed, illegal for gassy mines.

The second concern in the matter of airflow direction is the inclination of the airway. An **ascentional** ventilation system implies that the airflow moves upwards-through inclined workings. This takes advantage of the natural ventilating effects caused by the addition of heat to the air (section 8.3.1). In open stoping or mining layouts that involve multiple connections in inclined workings, ascentional ventilation may be the only technique capable of either controlling or utilizing natural ventilating effects. **Descentional** ventilation may be employed on more compact mining systems such as longwall faces and normally then becomes also a homotropal system with both air and conveyed mineral moving downhill. However, this may cause difficulties in controlling the natural buoyancy effects of methane in waste areas. The advantage claimed for descentional ventilation is that because the air enters the workings at a higher elevation it is then cooler and drier than if it were first coursed to the lower end of the workings.

Escapeways

Except for blind headings, there should always be at least two means of egress from each working place in an underground mine or facility. Preferably, there should be two separate intake routes designated as escapeways in case of a fire or other emergency. Within this context the term 'separate' is taken to imply that those airways have different and identifiable sources of intake air such that a source of pollution in one of them will not affect the other—either through leakage or series ventilation. Nevertheless, at least one return air route must always remain open and travellable without undue discomfort, to allow for an emergency situation where the working face itself becomes impossible to traverse.

Escapeways should be marked clearly on maps and by signs underground. Personnel should be made familiar with those routes through regular travel or organized escape drills. Mining legislation may dictate minimum sizes for escapeways and the frequency of their inspection.

Airflow travel distance and use of old workings

The routes utilized for main intake and return airflows should be reviewed from the viewpoint of travel distance and corresponding time taken for a complete traverse by the air. For high strata temperatures, it is advantageous for intake air to reach the workings as quickly as possible in order to minimize the gain of heat and humidity. However, this is tempered by air velocity constraints and ventilation operating costs.

In mines located in cold climates, it may be preferable to encourage natural heating of the intake air by allowing it to take a circuitous and slow route in order to maximize its exposure to rock surfaces. Another situation arises when variations in air humidity or temperature cause problems of slaking of strata from the roof or sides of the airways or workings. Here again, a case may be made for the natural air conditioning gained by passing the intake air through a network of older airways prior to reaching the current work areas (section 18.4.6).

The employment of old workings as an integral part of a ventilation system can result in significant reductions in mine resistance and, hence, the operating costs of ventilation. Furthermore, return air passing through abandoned areas will help to prevent build-up of toxic, asphyxiating or flammable gases. However, using old workings in this way must be treated with caution. It is inadvisable to rely on such routes as they may be subject to sudden closure from falls of roof. Secondly, travellable intake and return airways must always be maintained for reasons of safety and, third, old workings liable to spontaneous combustion must be sealed off and the pressure differentials across them reduced to a minimum.

From a practical viewpoint, where old workings can be employed safely for airflow then it is sensible to use them. However, during system design exercises they should not be relied on to provide continuous airflow routes but, rather, as a bonus in reducing the cost of ventilation. In any event, as a mine develops, it becomes advisable to seal off old areas that are remote from current workings. Unless this is done, then overall management and control of the airflow distribution will become increasingly difficult.

4.3 DISTRICT SYSTEMS

4.3.1 Basics of district system design

Underground ventilation layouts serving one or more districts of a mine may be divided into two broad classifications, **U-tube** and **through-flow** ventilation. Each of these takes on a diversity of physical configurations depending on the type of mine and disposition of the local geology.

As illustrated on Fig. 4.4 the basic feature of U-tube ventilation is that air flows towards and through the workings, then returns along adjacent airways separated from the intakes by stoppings and doors. Room and pillar layouts, and advancing longwalls, tend to be of this type.

Figure 4.5 illustrates the alternative through-flow ventilation system. In this layout, intakes and returns are separated geographically. Adjacent airways are either all intakes or all returns. There are far fewer stoppings and air crossings but additional regulation (regulators or booster fans) is required to control the flow of air through the work area. Practical examples of through-flow ventilation are the parallel flows from downcast to upcast shafts across the multilevels of a metal mine, or the back-bleeder system of a retreating long wall.

The simplest possible application of the U-tube system is for a set of twin development headings. Indeed, the U-tube method is the only one capable of ventilating pilot workings that are advancing into an unmined area. Through ventilation requires the prior establishment of one or more connections between main intake and return airways. Once that has been accomplished then through ventilation has

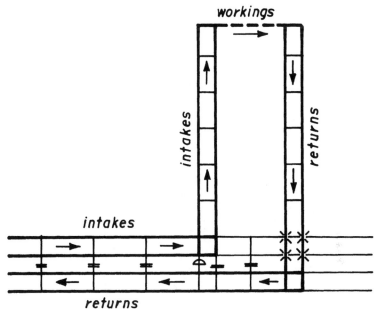

Figure 4.4 U-tube ventilation.

several significant advantages. First, leakage of air from intake to return is virtually eliminated. Hence, lower total airflows are required to provide any required ventilation at the working face. Secondly, the parallel airways and, often, shorter total travel distance of the airstream give a lower district resistance—particularly for workings distant from the main shafts. This permits reduced ventilating pressures. The combination of lower total airflows and lower ventilating pressures leads to large reductions in ventilation operational costs. Furthermore, the fan duties will remain much more stable in a through-flow system than the escalating demands of an advancing U-tube layout.

4.3.2 Stratified deposits

The vast majority of underground mines extracting coal, potash or other tabular forms of mineral deposits normally do so by one of two techniques, longwall or room and pillar (bord and pillar) mining. While the actual layouts can vary quite significantly from country to country and according to geological conditions, this section highlights the corresponding modes of airflow distribution that may be employed.

Longwall systems

The two major features of longwall mining that have influenced the design of their ventilation systems are, first, the control of methane or other gases that accumulate

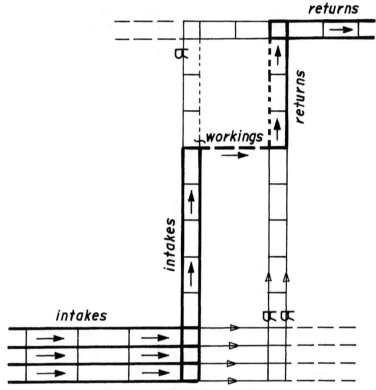

Figure 4.5 Through-flow ventilation.

in the waste (gob) areas and, second, the high rate of rock breakage on heavily mechanized longwalls that has exacerbated the production of dust, gas, heat and humidity.

Figure 4.6 illustrates some of the ventilation layouts used on longwall districts. Single-entry systems are employed primarily in European coal mines. Figures 4.6(a) and 4.6(b) show the application of the U-tube principle to advancing and retreating longwalls respectively. With the advancing system, leakage of some of the intake air occurs through the waste area, controlled by the resistance offered by the roadside packing material and the distribution of resistance and, hence, air pressure around the district. This can give rise to problems of gob fires in mines liable to spontaneous combustion. Gases from the waste may also flush onto the face leading to unacceptable concentrations toward the return end. The same difficulty may arise to a lesser extent when the U-tube principle is applied to a retreating face, the abandoned airways being stopped off as the face retreats.

Figure 4.6(c) shows a single-entry longwall with the back (or bleeder) return held open in order to constrain the gas fringe safely back in the waste area and, hence, to prevent flushes of waste gas onto the face. The system illustrated in Fig. 4.6(c) is a combination of U-tube and through-flow ventilation.

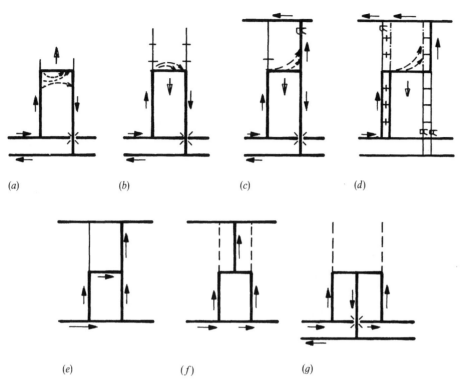

Figure 4.6 Classifications of longwall district ventilation systems: (a) single-entry advancing; (b) single-entry retreating; (c) single-entry retreating with back bleeder; (d) double-entry retreating with back bleeder; (e) Y-system; (f) double-Z system; (g) W-system.

Figure 4.6(d) illustrates the longwall system more often used in coal mining countries that have a tradition of room and pillar mining such as the United States, Australia or South Africa. Two or more entries are driven initially using room and pillar mining, these serving as the lateral boundaries of retreat longwall panels. Again, back bleeders are used to control waste gas.

Figures 4.6(e), 4.6(f) and 4.6(g) illustrate a classification of systems for longwall faces where the make of gas from the face itself is particularly heavy. The Y system provides an additional feed of fresh air at the return end of the face. This helps to maintain gas concentrations at safe levels along the back return airway(s). Figure 4.5(d) is, in fact, a double-entry through-flow Y system. The double-Z layout is also a through-flow system and effectively halves the length of face ventilated by each airstream. The W system accomplishes the same end but is based on the U-tube principle. Both the double-Z and W systems may be applied to advancing or retreating faces, depending on the ability of the centre return to withstand front abutment and waste area strata stresses. Again, in both the double-Z and W systems, the directions of airflow may be reversed to give a single intake and two returns (or two sets of multiple returns). This may be preferred if heavy emissions of gas are experienced from solid rib sides.

Room (bord) and pillar systems

Figure 4.7 shows two methods of ventilating a room and pillar development panel: (a) a bidirectional or W system in which intake air passes through one or more central airways with return airways on both sides, and (b) a unidirectional or U-tube system with intakes and returns on opposite sides of the panel. In both cases the conveyor is shown to occupy the central roadway with a brattice curtain to regulate the airflow through it. It is still common practice in room and pillar mines to course air around the face ends by means of line brattices pinned to roof and floor but hung loosely in the cross-cuts to allow the passage of equipment. An advantage of the bidirectional system is that the air splits at the end of the panel with each airstream ventilating the operational rooms sequentially over one half of the panel only. Conversely, in the unidirectional or U-tube system the air flows in series around all of the faces in turn. A second advantage of the bidirectional system arises from the fact that rib-side gas emission is likely to be heavier in the outer airways. This can become the dominant factor in gassy coal seams of relatively high permeability necessitating that the outer

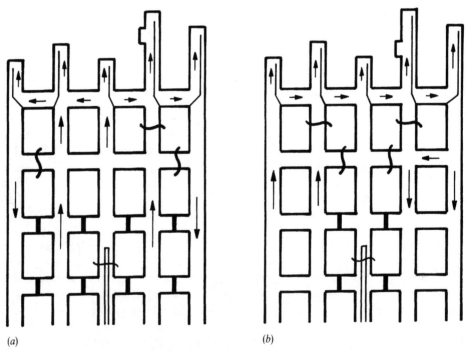

(a)   (b)

Figure 4.7 Room and pillar development with line brattices (high face resistance): (a) bidirectional (W) system; (b) unidirectional (U-tube) system.

airways be returns. In most coal mining countries, legislation requires that gas concentration in intake airways be maintained at very low levels.

Unfortunately, the bidirectional system suffers from one very great disadvantage. The number of stoppings required to be built and the number of leakage paths created between intakes and returns are both doubled. In long development panels, the amount of leakage can become excessive, allowing insufficient air to reach the last open cross-cuts (section 4.2.3). In such circumstances, attempts to increase the pressure differential across the outbye ends of the panel exacerbate the leakage and give a disappointing effect at the faces.

The unidirectional system has a higher volumetric efficiency because of the reduced number of leakage paths. However, in both cases, the line brattices in the rooms offer a high resistance to airflow compared with an open cross-cut. This is particularly so in the case of the unidirectional system where the useful airflow is required to pass around all of these high resistance line brattices in series.

The imposition of line brattice resistance at the most inbye areas of a mine ventilation system forces more air to be lost to return airways at all leakage points throughout the entire system. An analogy may be drawn with a leaky hosepipe. If the end of the pipe is unobstructed then water will flow out of it fairly freely and dribble from the leakage points. If, however, the end of the pipe is partially covered then the flow from it will decrease but water will now spurt out of the leakage points.

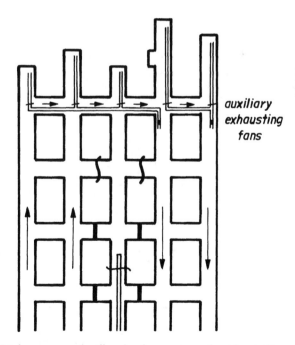

Figure 4.8 U-tube room and pillar development panel with auxiliary fans (zero face resistance).

The problem can be overcome by employing auxiliary fans and ducts either to force air into the rooms or exhaust air from them (section 4.4). Figure 4.8 illustrates a room and pillar panel equipped with exhausting auxiliary ventilation. With such a system, the fans provide the energy to overcome frictional resistance in the ducts. The effective resistance of the whole face area becomes zero. Smaller pressure differentials are required between intakes and returns for any given face airflows and, hence, there is a greatly reduced loss of air through leakage. The electrical power taken by the auxiliary fans is more than offset by the savings in main fan duties.

A further advantage of employing auxiliary fans is that each room is supplied with its own separate and controllable supply of air. However, the fans must be sized or ducts regulated such that no undesired recirculation occurs.

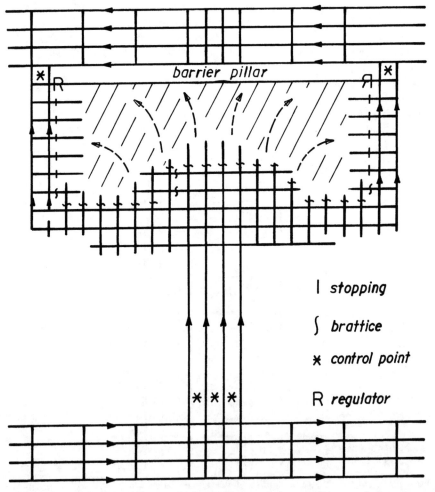

Figure 4.9   A retreating room and pillar district using a through-flow ventilation system.

The choice between auxiliary fan and duct systems and line brattices in room and pillar workings should also take into account the height and width of the airways, the size and required mobility of equipment, the placement of ducts or brattices, the extent of pollution from dust, gas and heat, fan noise, and visibility within the workings.

The systems shown on Fig. 4.6 for longwalls each have their counterparts in room and pillar mining. An example of a retreating double-Z (through-flow) system applied to a room and pillar section is shown on Fig. 4.9. There are, however, significant differences in the ventilation strategy between the two mining methods. The larger number of interconnected airways and higher leakage result in room and pillar layouts having lower resistance to airflow than longwall mines. It follows that room and pillar mines tend to require higher volume flows at lower fan pressures than longwall systems. Similarly, because of the increased number of airways and leakage paths it is particularly important to maintain control of airflow distribution paths as a room and pillar mine develops. It is vital that barrier pillars be left between adjacent panels and to separate the panels from trunk airway routes. Such barriers are important not only to protect the integrity of the mine in case of pillar failure but also to provide ventilation control points and to allow sealing of the panel in cases of emergency or when mining has been completed.

4.3.3 Orebody deposits

Metalliferous orebodies rarely occur in deposits of regular geometry. Zones of mineralization appear naturally in forms varying from tortuous veins to massive irregularly shaped deposits of finely disseminated metal and highly variable concentration. The mining layouts necessarily appear less ordered than those for stratified deposits. Furthermore, the combination of grade variation and fluctuating market prices results in mine development that often seems to be chaotic. The same factors may also necessitate many more stopes or working places than would be usual in a modern coal mine, with perhaps only a fraction of them operating in any one shift. Hence, the ventilation system must be sufficiently flexible to allow airflow to be directed wherever it is needed on a day-by-day basis.

Ventilation networks for metal mines, therefore, tend to be more complex than for stratified deposits and are usually also three dimensional. Figure 4.10 illustrates the ventilation strategy of many metal mines although, again, the actual geometry will vary widely. Air moves in a through-flow manner from a downcast shaft or ramp, across the levels, sublevels and stopes towards return raises, ramps or upcast shaft. Airflow across each of the levels is controlled by regulators or booster fans. Movement of air from level to level, whether through stopes, or by leakage through ore passes or old workings, tends to be ascentional in order to utilize natural ventilating effects and to avoid thermally induced and uncontrolled recirculation.

Airflow distribution systems for individual stopes are also subject to great variability depending on the geometry and grade variations of the orebody. There are, however, certain guiding principles. These are illustrated in Figures 4.11–4.13 for

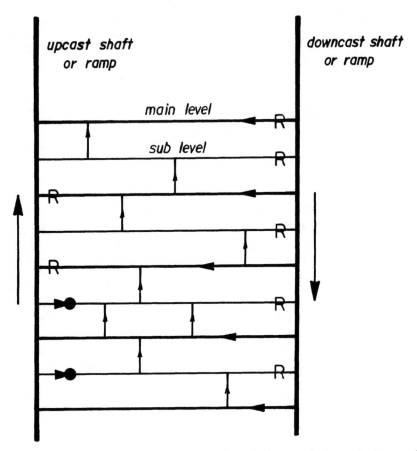

Figure 4.10 Section showing the principle of through-flow ventilation applied across the levels of a metal mine.

three stoping methods. In the majority of cases, where controlled vertical movement of the air is required, stope airflow systems employ ascentional through-flow ventilation. Although auxiliary fans and ducts may be necessary at individual drawpoints, every effort should be made to utilize the mine ventilation system to maintain continuous airflow through the main infrastructure of the stope. Series ventilation between stopes should be minimized in order that blasting fumes may be cleared quickly and efficiently.

Leakage through ore passes creates a problem in metal mines as the ore passes may often be emptied, allowing a direct connection between levels. Airflows emerging from ore passes can also produce unacceptable dust concentrations. Closed ore chutes and instructions to maintain some rock within the passes at all times are both beneficial but are difficult to enforce in the necessarily production orientated activities of an operating mine. The design of the ventilating system and operation of regulators and

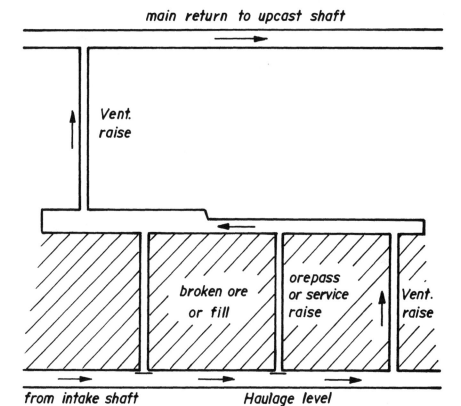

Figure 4.11 Simple ventilation system for shrinkage or cut-and-fill stopes.

booster fans should attempt to avoid significant pressure differences across ore passes. Attrition on the sides of ore passes often enlarges their cross-section and may produce fairly smooth surfaces. When no longer required for rock transportation, such openings may usefully be employed as low resistance ventilation raises.

Figure 4.13 for a block caving operation illustrates another guideline. Wherever practicable, each level or sublevel of a stope should be provided with its own through-flow of air between shafts, ventilation raises or ramps. While vertical leakage paths must be taken into account during planning exercises, maintaining an identifiable circuit on each level facilitates system design, ventilation management and control in case of emergency.

4.4 AUXILIARY SYSTEMS

Auxiliary ventilation refers to the systems that are used to supply air to the working faces of blind headings. Auxiliary ventilation may be classified into three basic types,

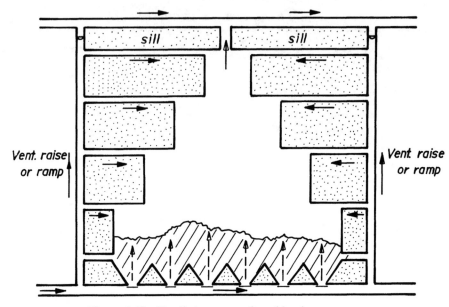

Figure 4.12   Ventilation system for sublevel open stopes.

line brattices, fan and duct systems, and 'ductless' air movers. Ideally, auxiliary systems should have no impact on the distribution of airflows around the main ventilation infrastructure, allowing auxiliary ventilation to be planned independently from the full mine ventilation network. Unfortunately, this ideal is not always attained, particularly when line brattices are employed.

4.4.1   Line brattices and duct systems

The use of line brattices was introduced in section 4.3.2 (Figure 4.7) in relation to room and pillar workings where they are most commonly employed. It was shown that a major disadvantage of line brattices is the resistance they add to the mine ventilation network at the most sensitive (inbye) points, resulting in increased leakage throughout the system. This resistance depends primarily on the off set distance of the line brattice from the nearest side of the airway, and the condition of the flow-path behind the brattice. This is sometimes obstructed by debris from sloughed sides, indented brattices or, even, items of equipment put out of sight and out of mind, despite legislative prohibitions of such obstructions. In this section we shall examine the further advantages and disadvantages of line brattices.

Figure 4.14 shows line brattices used in the forcing (Fig. 4.14(a)) and exhausting (Fig. 4.14(b)) modes. The flame-resistant brattice is pinned between roof and floor, and supported by a framework at a position some one-quarter to one-third of the airway width from the nearest side. This allows access by continuous miners and

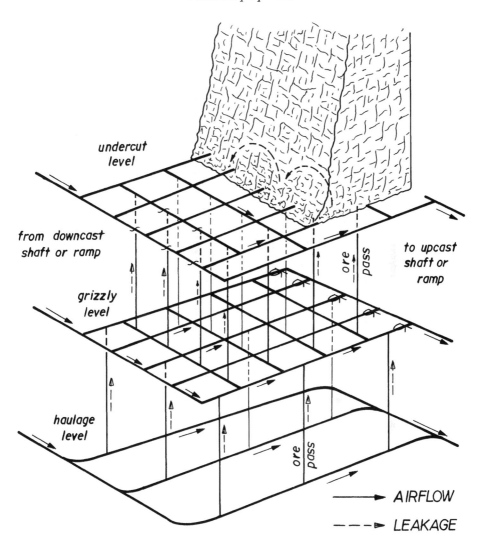

Figure 4.13 Ventilation system for a block caving operation.

other equipment. Even with carefully erected line brattices, leakage is high with often less than a third of the air that is available at the last open cross-cut actually reaching the face. This limits the length of heading that can be ventilated by a line brattice. The need for line brattices to be extended across the last 'open' cross-cut inhibits visibility, creating a hazard where moving vehicles are involved. The advantages of line brattices are that the capital costs are low in the short term, they require no power and produce no noise.

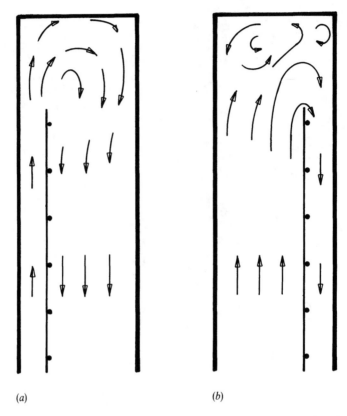

(a)   (b)

Figure 4.14   Line brattices used for auxiliary ventilation: (a) forcing; (b) exhausting.

Figure 4.15 shows the corresponding forcing and exhausting systems using auxiliary fans and ducting. In most cases, in-line axial fans are used although centrifugal fans are quieter and give higher pressures for the longer headings. The advantages of an auxiliary fan and duct are that they provide a more positive and controlled ventilating effect at the face, they cause no additional resistance to the mine ventilation system or any consequential leakage throughout the network, and they are much less liable to leakage in the heading itself. For headings longer than some 30 m, auxiliary fans are the only practicable means of producing the required airflows. An exhausting duct also allows the air to be filtered, an advantage for dust control where series ventilation is practiced. The disadvantages involve the initial capital cost, the need for electrical power at the fans, the space required for ducts and the noise produced by the fans.

Care should be taken to ensure that the pressure—volume characteristics of the fan are commensurate with the resistance offered by the duct and the airflow to be passed. The later is determined on the basis of the type and magnitude of pollutants to be removed (Chapter 9). The duct resistance is established as a combination of the

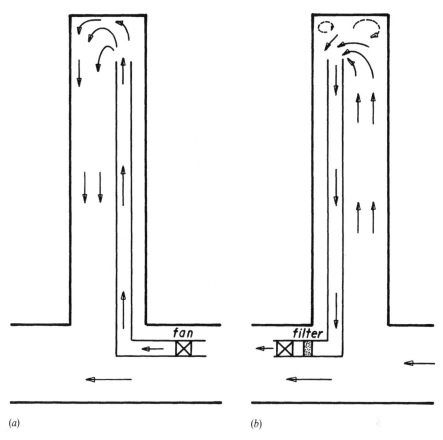

(a)   (b)

Figure 4.15 Fan and duct systems of auxiliary ventilation: (a) forcing; (b) exhausting.

wall losses within the duct, shock losses at any bend or change of cross-section and at discharge. The equations employed are those derived for airway resistance in section 5.4.

Example An airflow of 15 m³/s is to be passed through a 0.9 m diameter fibreglass duct, 200 m long, with one sharp right-angled bend. From manufacturer's literature, the friction factor for the duct is 0.0032 kg/m³. Calculate the total pressure to be developed by the fan and the fan power, assuming a fan efficiency of 60% and an air density of 1.2 kg/m³.

Solution

$$\text{duct area} = \pi \times 0.9^2/4 = 0.636\,\text{m}^2$$
$$\text{perimeter} = \pi \times 0.9 \quad = 2.827\,\text{m}$$

Let us first determine the shock loss (X) factors for the system from section A5

Entry (section A5.4). In the absence of any inlet fitting, the shock loss factor is given as $X_{in} = 1.0$. This is caused by turbulence as the air enters the duct and should not be confused with the conversion of static pressure to velocity pressure at entry.

Bend (Fig. 5A.1). For a sharp right-angled bend, $X_b = 1.2$.

Exit (section A5.4). This is not really a shock loss but represents the kinetic energy of the air provided by the fan and lost to the receiving atmosphere. $X_{ex} = 1.0$.

The total shock loss factor is

$$X_{sh} = 1.0 + 1.2 + 1.0 = 3.2$$

Equivalent resistance of shock losses (see equation (5.18))

$$R_{sh} = \frac{X_{sh}\rho}{2A^2} \qquad \frac{Ns^2}{m^8}$$

$$= \frac{3.2 \times 1.2}{2 \times (0.636)^2} = 4.744 \quad Ns^2/m^8$$

Duct resistance (see equation (5.4))

$$R_d = \frac{kL \text{ per}}{A^3} \cdot \frac{Ns^2}{m^8}$$

$$= \frac{0.0032 \times 200 \times 2.827}{(0.636)^3} = 7.028 \quad Ns^2/m^8$$

Total resistance:

$$R_{tot} = R_{sh} + R_d$$

$$= 4.744 + 7.028 = 11.772 \quad Ns^2/m^8$$

Required fan total pressure (see equation (5.5)):

$$P_t = R_{tot}Q^2$$

$$= 11.772 \times 15^2 = 2649 \, Pa$$

The required fan power is

$$\frac{P_t \times Q}{\eta} \qquad \text{where } \eta = \text{fan efficiency}$$

or

$$\frac{2649 \times 15}{1000 \times 0.6} = 66.2 \, kW$$

4.4.2 Forcing, exhausting, and overlap systems

Figures 4.14 and 4.15 illustrate forcing and exhausting systems of auxiliary ventilation for line brattices and fan–duct systems respectively. The choice between forcing and exhausting arrangements depends mainly upon the pollutants of greatest concern, dust, strata gas or heat.

The higher velocity airstream emerging from the face-end of a forcing duct or, to a lesser extent, a forcing brattice gives a scouring effect as the air sweeps across the face. This assists in the turbulent mixing of any methane that may be emitted from fragmented rock or newly exposed surfaces. It also helps to prevent the formation of methane layers at roof level (Chapter 12). In hot mines, the forcing system provides cooler air at the face, even having taken the energy added by the fan into account. Furthermore, as the system is under positive gauge pressure, the cheaper type of flexible ducting may be used. This is also easier to transport and enable leaks to be detected more readily.

The major disadvantage of a forcing system is that pollutants added to the air at the face affect the full length of the heading as the air passes back, relatively slowly, along it.

Where dust is the main hazard, an exhausting system is preferred. The polluted air is drawn directly into the duct at the face-end allowing fresh air to flow through the length of the heading. However, the lack of a jet effect results in poor mixing of the air. Indeed, unless the end of the duct or brattice line is maintained close to the face then local pockets of sluggish and uncontrolled recirculation may occur. In all cases, it is important that the ducting or brattice line be extended regularly so that it remains within some three metres of the face. This distance may be prescribed by legislation.

A further advantage of a ducted exhaust system is that a dust filter may be included within the system. In this case, the additional pressure drop across the filter must be taken into account in choosing the fan, and the filter serviced regularly in order that its resistance does not become excessive. Exhaust ducts must necessarily employ the more expensive rigid ducting or reinforced flexible ducting.

For long headings, the resistance of the duct may become so great that multiple fans connected in series must be employed. If these are grouped as a cluster at the outbye end of the ducting then the high (positive or negative) gauge pressure will exacerbate leakage. It is preferable to space the fans along the length of the ducting in order to avoid excessive gauge pressures. The use of hydraulic gradient diagrams assists in the optimum location of fans and to prevent uncontrolled recirculation of leakage air. Multiple fans must be interlinked electrically and airflow or pressure monitors employed to detect accidental severing or blockage of the duct, in which case all fans inbye the point of damage must be switched off—again, to prevent uncontrolled recirculation.

It is clear that forcing and exhausting systems both have their advantages and disadvantages. Two-way systems have been devised that can be switched from forcing mode to exhausting for cyclic mining operations. These may employ a reversible axial fan or, alternatively, both a forcing and an exhausting fan, only one of which is operated at any one time with an appropriate adjustment of valves or shutter doors within the duct arrangment.

The more common methods of combining the advantages of forcing and exhausting ducts are overlap systems. Examples are shown on Figure 4.16. The direction and mean velocity of the air in the heading within the overlap zone clearly depend on the airflows in each of the ducts. These should be designed such that the general body airflow in this region does not become unacceptably low. Where permitted by law, controlled recirculation may be used to advantage in overlap systems (section 4.5.2). Where continuous miners or tunnelling machines are employed, the overlap ventilator may be mounted on the machine. In all cases, it is important that the fans are interlinked so that the overlap system cannot operate when the primary duct fan is switched off.

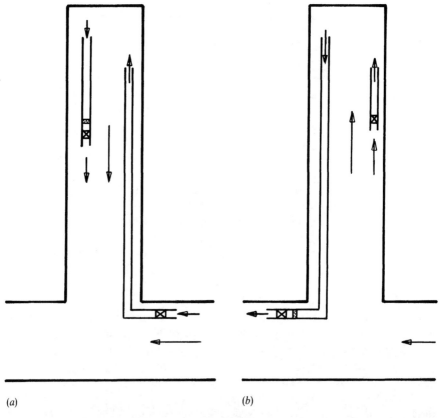

(a)   (b)

Figure 4.16 Overlap systems of auxiliary ventilation: (a) forcing system with exhaust overlap; (b) exhausting system with force overlap.

4.4.3 Air movers

In addition to conventional ducted systems of auxiliary ventilation, a number of other techniques may be employed to enhance or control the movement of air within localized areas of a mine or tunnel.

Jet fans, sometimes known as **ductless**, **vortex** or **induction** fans, are free-standing units that produce a relatively high velocity outlet airstream. The jet of air produces two effects. First, the reach and integrity of the air vortex depends on the velocity at the fan outlet, the size of the heading and whether the airway is a blind heading or part of a through-flow system. Satisfactory ventilation and turbulent mixing at the face of a large heading can be obtained from a jet fan sited 100 m outbye. Secondly, an induction effect occurs at the outer boundaries of the expanding cone of projected air. This entrains additional air from the surroundings producing a forward-moving flow that is greater than the airflow through the fan itself. The conversion of velocity pressure into static pressure as the plume of air decelerates also generates a true ventilating pressure. This is seldom greater than some 20 Pa but is sufficient to create significant airflows in large, low resistance airways. The induction effect inhibits excessive recirculation provided that incoming intake air is provided at the fan inlet. Jet fans have particular application in large room and pillar operations and may also be used in series to promote airflow through vehicular tunnels.

An airflow can also be generated by a spray of water giving rise to **spray fans**. Inertia from the motion of the water droplets is transmitted to the air by viscous drag and turbulent induction. Spray fans may be used very effectively to control the local movement of air around rock-winning machines such as continuous miners or longwall shearers. This assists in the rapid dilution of methane and in diverting dust-laden air away from operator positions. The effect depends on the shape, velocity and fineness of the spray. Although dust suppression sprays also cause air induction it is usually necessary to add additional sprays if these are to be used for local airflow control. Provided that the service water is chilled, spray fans are also an efficient means of cooling the air in a work area.

Compressed air injectors are also induction devices. The compressed air is supplied through one or more forward-pointing jets within a cylindrical or shaped tube. The best effect is obtained when the compressed air is supplied at the throat of a venturi. These devices are noisy and of much lower efficiency than fans. However, they have a role in areas where electrical power is unavailable. Mining law may proscribe the use of compressed air for the promotion of airflow in gassy mines as the high velocity flow of air through the jets can cause the build-up of an electrostatic charge at the nozzle. This produces the possibility of sparks that could ignite a methane–air mixture.

4.5 CONTROLLED PARTIAL RECIRCULATION

4.5.1 Background and principles of controlled partial recirculation

The idea of recirculating air in any part of a gassy mine has, traditionally, been an anathema to many mining engineers. Most legislation governing coal mines prohibits any ventilation system or device that causes air to recirculate. The background to such legislation is the intuitive fear that recirculation will cause concentrations of pollutants to rise to dangerous levels.

A rational examination of controlled recirculation was carried out by Leach, Slack and Bakke during the 1960s at the Safety in Mines Research Establishment in England. Those investigators made a very simple and obvious statement but one that had, to that time, apparently been denied or ignored within the context of air recirculation. They argued that the general body gas concentration, C, leaving any ventilated region of a mine is given by

$$C = \frac{\text{flow of gas into the region}}{\text{flow of fresh air passing through the region}} \qquad (4.3)$$

The value of C is quite independent of the flowpaths of the air within the region, including recirculation. It is true, of course, that if the through-flow of fresh air falls while the gas emission remains constant then the concentration of gas will rise. This would happen, for example, if the air duct serving a long gassy heading was dislocated while an inbye fan continued to run. This example illustrates a case of **uncontrolled** recirculation. The definition of a system of **controlled partial recirculation** is one in which a controlled fraction of the air returning from a work area is passed back into the intake while, at the same time, the volume flow of air passing through the region is monitored to ensure that it remains greater than a predetermined minimum value.

The advantages of controlled partial recirculation lie in the improved environmental conditions it can provide with respect to gases, dust and heat, as well as allowing mining to proceed in areas of a mine that are too distant from surface connections to be ventilated economically by conventional means.

As illustrated in section 4.5.2, general body gas concentrations may actually be reduced by controlled recirculation. Furthermore, the higher air velocities that occur within a recirculation zone assist in the turbulent mixing of gas emissions, reducing the tendency to methane layering and diminishing the probability of accumulations of explosive methane–air mixtures.

As with gas, concentrations of respirable dust reach predictable maximum levels in a system of controlled recirculation and may be reduced significantly by the use of filters. The greater volume of air being filtered results in more dust being removed. The effect of controlled recirculation on climatic conditions is more difficult to predict. However, both simulation programming and practical observations have indicated the improvements in the cooling power of partially recirculated air for any given value of throughflow ventilation.

As workings proceed further away from shaft bottoms, the cost of passing air along the lengthening primary intakes and returns necessarily increases. Where new surface connections closer to the workings are impractical, perhaps because the workings lie beneath the sea or because of great depth then the cost of conventional ventilation will eventually become prohibitive—even when booster fans are employed. Using the air more efficiently through controlled partial recirculation then becomes an attractive proposition. If, for example, the methane concentration returning from a conventionally ventilated face is 0.3%, and the safe mandatory limit is 1.0%, then the through-flow provided from the main airways might be reduced to one-half, giving a methane concentration of 0.6% while maintaining or increasing the face velocities by controlled recirculation.

During the 1970s the concept of controlled partial recirculation gained respectability and is now practiced by several of the world's mining industries operating, in some cases, by authorized exemptions from existing legislation.

The greatest disincentive against the introduction of controlled recirculation has been the risk of combustion gases from a fire being returned to working areas. Further potential problems arise from a consideration of transient phenomena such as blasting, or rapid changes in barometric pressure caused by the operation of doors or fans, and the possible resulting peak emissions of gases. The introduction of fail-safe monitoring systems with continuous computer surveillance has revolutionized the situation (section 9.6.3). These self-checking systems involve monitoring the concentration of gases, air pressure differentials and airflows at strategic locations as well as the operating conditions of fans and other plant. Fans can be interlinked electrically to obviate the possibility of uncontrolled recirculation. Should any monitored parameter fall outside prescribed limits then the system will automatically revert to a conventional non-recirculating circuit. The introduction of reliable monitoring technology has allowed the advantages of controlled partial recirculation to be realized safely.

4.5.2 Controlled recirculation in headings

The most widespread application of controlled recirculation has been in headings. One of the disadvantages of the conventional overlap systems shown in Fig. 4.16 is the reduction in general body air velocity within the overlap zone. This can be overcome completely by arranging for the overlap fan to pass an airflow that is greater than that available within the heading, i.e. a system of controlled recirculation, accompanied by the corresponding monitoring system and electrical interlocks. This is particularly advantageous when applied to the scheme depicted in Fig. 4.16(a) as filtered air is then available to machine operators as well as throughout the length of the heading.

Figure 4.17 shows two examples of a primary exhaust system configured for controlled recirculation. In both cases, an airflow Q_t (m^3/s) is available at the last open cross-cut and contains a gas flow of G_i (m^3/s). An airflow of Q_h passes up the heading where a gas emission of G_h is added.

Let us try to find the maximum general body gas concentrations that will occur in the systems. Referring to Figure 4.17(a) and using the locations A, B, C and D as identifying subscripts, the fractional gas concentration, C_g, at position D (leaving the

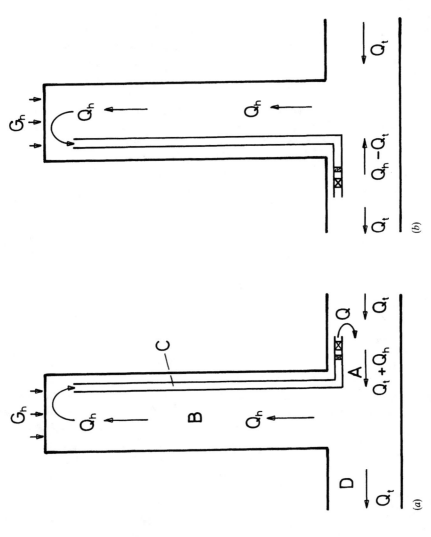

Figure 4.17 Controlled recirculation exhaust systems for headings: (a) exhausting back into the intake; (b) exhausting back into the return.

system) must be (see equation (4.3))

$$Cg_D = \frac{G_i + G_h}{Q_t}$$

(In these relationships, it is assumed that the gas flow is much smaller than the airflow.) However, inspection of the figure shows that this must also be the gas concentration at locations A and B. In particular,

$$Cg_B = \frac{G_i + G_h}{Q_t}$$

but gas flow G = gas concentration $\times$ airflow:

$$G_B = \frac{G_i + G_h}{Q_t} Q_h$$

The gas flow in the duct, $G_C = G_B + G_H$, must then be

$$G_C = \frac{G_i + G_h}{Q_t} Q_h + G_h$$

Hence, the gas concentration in the duct is

$$Cg_C = \frac{G_C}{Q_h}$$

$$= \frac{G_i + G_h}{Q_t} + \frac{G_h}{Q_h} \tag{4.4}$$

This is the highest general body gas concentration that can occur anywhere within the system shown on Fig. 4.17(a).

Examination of equation (4.4) shows that if the gas flows, G_i and G_h, are fixed and the fresh air supply, Q_t, remains unchanged then the maximum general body gas concentration, Cg_C, must fall as Q_h is increased—that is, as the degree of recirculation rises. In the limit, at very high Q_h, the gas concentration in the duct tends towards that leaving the system at position D.

In a conventional non-recirculating system, the airflow taken into a heading is often limited to no more than half of that available at the last open cross-cut, i.e. $Q_h = 0.5 Q_t$. Applying these conditions to equation (4.4) gives

$$Cg_C \text{ (conventional)} = \frac{G_i + G_h}{Q_t} + \frac{G_h}{0.5 Q_t}$$

$$= \frac{G_i + 3G_h}{Q_t} \tag{4.5}$$

Recirculation commences when $Q_h = Q_t$, giving

$$Cg_C \text{(maximum, recirculating)} = \frac{G_i + G_h}{Q_t} + \frac{G_h}{Q_t}$$

$$= \frac{G_i + 2G_h}{Q_t} \qquad (4.6)$$

Cg_C must be less than this at all greater values of Q_h, i.e. higher degrees of recirculation. Comparing equations (4.5) and (4.6) shows that in this configuration the maximum general body gas concentration is always less using controlled recirculation than with a conventional system.

Turning to Fig. 4.17(b), the analysis is even simpler. In this case, the maximum gas concentration (in the duct) must be the same as that leaving the system $(G_i + G_h)/Q_t$ provided that Q_h is equal to or greater than Q_t, i.e. controlled recirculation must exist. Here again, this is always less than would be attainable with a conventional non-recirculating system.

A similar analysis for dust concentration, Cd, on the system shown on Fig. 4.17 but with no dust filter gives an analogous expression to that for gas

$$Cd_C = \frac{D_i + D_h}{Q_t} + \frac{D_h}{Q_h} \quad \frac{mg}{m^3} \qquad (4.7)$$

where $D_i =$ dust flow in intake air (mg/s) and $D_h =$ dust make in heading (mg/s).

Again it can be seen that the maximum concentration falls as Q_h is increased. For dust it is more pertinent to state the concentration at position B, i.e. in the main length of the heading: this becomes

$$Cd_B = \frac{D_i + D_h}{Q_t} \quad \frac{mg}{m^3} \qquad (4.8)$$

and is completely independent of the degree of recirculation.

If the filters shown in Fig. 4.17(a) remove a fraction, η, of the dust in the duct then it can be shown that the corresponding concentrations become

$$Cd_C = \frac{Q_h(D_i + D_h) + D_h Q_t}{Q_h(Q_t + \eta Q_h)} \quad \frac{mg}{m^3} \qquad (4.9)$$

and

$$Cd_B = \frac{D_i + (1 - \eta) D_h}{Q_t + \eta Q_h} \quad \frac{mg}{m^3} \qquad (4.10)$$

These equations show that, when filters are used, dust concentrations fall throughout the system. It has been assumed in these analyses that there is no settlement of dust.

Similar relationships can be derived for other configurations of controlled recirculation in headings.

4.5.3 District systems

The extension of controlled partial recirculation to complete areas of a mine has particular benefits in decreasing the costs of heating or cooling the air and for workings distant from the surface connections.

Positions of fans

Figure 4.18 shows simplified schematics illustrating three configurations of fan locations in a district recirculation system. In each case, the through-flow ventilation in the mains is shown as Q_m with Q_c passing from return to intake in the recirculation cross-cut, to give an enhanced airflow of $Q_m + Q_c$ in the workings. The ratio $F = Q_c / (Q_m + Q_c)$ is known as the recirculation fraction. The fan that creates the recirculation develops a pressure of p_r while the pressure differentials applied across the outbye ends for the three systems shown are p_{o1}, p_{o2} and p_{o3} respectively.

The simplest configuration is shown in Fig. 4.18(a) with the recirculating fan sited in the cross-cut. This maintains the intakes and returns free for travel and un-

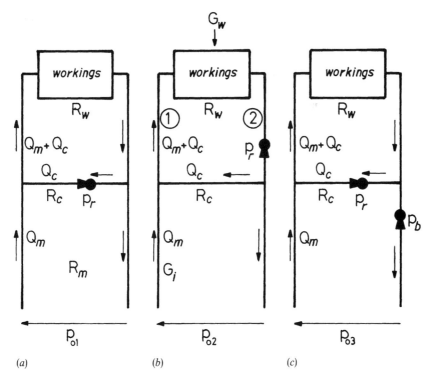

Figure 4.18 Schematics of district recirculation systems: (a) cross-cut fan; (b) in-line fan; (c) combined cross-cut and booster fans. Q = airflow, G = gasflow, R = resistance, p = ventilating pressure.

obstructed by airlocks. Locating the fan in this position will tend to decrease the through-flow, Q_m. Hence, if the total flow is to be maintained, the applied pressure differential must be increased from

$$p_{o1} = R_m Q_m^2 + R_w Q_m^2 \quad \text{(with no recirculation)}$$

where R_m is the combined resistance of the intake and return mains, to

$$p_{o1} = R_m Q_m^2 + R_w (Q_m + Q_c)^2 \tag{4.11}$$

with the recirculation shown on Fig. 4.18(a). These equations are derived by summing the frictional pressure drops, p, around the path of the mains (subscript m) and workings (subscript w), and by applying the square law $p = RQ^2$, where R = airway resistance (section 5.2).

Similarly, in all three cases in Fig. 4.18 the pressure required of the recirculating fan is given by summing the frictional pressure drops around the workings and cross–cut.

$$p_r = R_c Q_c^2 + R_w (Q_m + Q_c)^2 \tag{4.12}$$

In the system of Fig. 4.18(b), the fan is located within either the intake or return inbye the recirculation cross–cut. In this position, the fan acts as a district booster fan as well as creating the controlled recirculation. Hence, the through–flow, Q_m, will tend to increase. Alternatively, if the through–flow is to remain constant then either the outbye pressure differential may be reduced to

$$p_{o2} = R_m Q_m^2 - R_c Q_c^2 \tag{4.13}$$

or a regulator can be introduced into either of the mains.

The system in Fig. 4.18(c) combines a booster fan with a cross–cut fan and is the preferred configuration where recirculation is employed due to the workings being distant from the surface connections. In this system, the ventilating pressure applied across the district may be reduced by the magnitude of the booster fan pressure to maintain a constant Q_m, i.e.

$$p_{o3} = R_m Q_m^2 + R_w (Q_m + Q_c)^2 - p_b \tag{4.14}$$

The total airpower consumed within the systems is given as the sum of the pQ products for all airways. If corresponding airflows and the resistances are the same in each of the three systems then the required total airpowers must also be equal, irrespective of the locations of the fans. In practice, differences in the efficiencies of the fans will cause variations in the required total electrical input to the fan motors. This may, however, be of minor significance.

A major consideration in all designs of controlled partial recirculation is that in conditions of an emergency or plant stoppage, the system must fail safe and revert to a conventional non-recirculating configuration. The airflows must then remain sufficient to allow personnel to evacuate the area safely and for the necessary ameliorative measures to be taken. The detection of such conditions is provided by monitoring the environmental parameters and the operation of the fans.

If the recirculating cross-cut fan in the system of Fig. 4.18(a) fails, then doors in the cross-cut must close automatically. The through-flow ventilation will increase, reducing the general body gas consideration in the return airways. However, the reduced air velocity in the working area will increase the probability of local accumulations of gas to that of a conventional non-recirculating system.

Stoppage of the in-line fan of the system in Fig. 4.18(b) is more serious. Again, doors in the cross-cut must be closed but the through-flow of air will decrease resulting both in diminished airflows in the workings and also higher general body gas concentrations in the returns. However, this system is capable of better airflow control than the cross-cut fan. The degree of recirculation may be varied by modifying the duty of the fan installation (fan speed, vane settings or number of fans operating), regulating the airflow in either the intake or return inbye the cross-cut or by adjustment of a bypass path around the fan.

The system of Fig. 4.18(c) gives the greatest degree of flexibility. Stoppage of the cross-cut fan and closure of the corresponding doors will increase the through-flow, Q_m. However, should the booster fan fail, then electrical interlocks should close down the cross-cut fan. Reduced airflows throughout the system are maintained by the outbye pressure differential. Adjustment of the two fans allows a much greater degree of independent control of the airflow distribution than either of the systems of Figs. 4.18(a) or 4.18(b).

Pollution levels

Although the general body gas concentration leaving any zone is independent of airflow distribution within the zone—recirculating or otherwise (section 4.5.1), any airflow passed from a return to an intake airway may affect the quality as well as the quantity of the air in that intake.

Referring to Fig. 4.18(b), suppose that the incoming intake air contains a gas flow of G_i (m^3/s) and a constant gas emission of G_w occurs in the workings. Let us derive expressions for the general body gas concentrations in the face return (position 2) and intake (position 1).

The return concentration must be the same as that leaving the complete district and in the cross-cut, assuming no other sources of gas emission.

$$Cg_2 = Cg_c = Cg \text{ (main return)} = \frac{G_i + G_w}{Q_m} \qquad (4.15)$$

(See equation (4.3)). This is independent of the degree of recirculation.

To determine the gas concentration in the intake at position 1, consider first the gas flow passing through the cross-cut (subscript c):

$$G_c = Q_c \times Cg_c$$

$$= Q_c \frac{G_i + G_w}{Q_m}$$

from equation (4.15). Now the gas flow at position 1 is

$$G_1 = G_i + G_c$$

$$= G_i + \frac{Q_c}{Q_m}(G_i + G_w)$$

The corresponding concentration is given by dividing by the airflow, $Q_m + Q_c$, giving

$$Cg_1 = \frac{G_1}{Q_m + Q_c} = \frac{1}{Q_m}\left[\frac{G_i Q_m}{Q_m + Q_c} + \frac{Q_c}{Q_m + Q_c}(G_i + G_w)\right] \qquad (4.16)$$

However, if we define the recirculation fraction as

$$F = \frac{Q_c}{Q_m + Q_c}$$

where, also,

$$1 - F = \frac{Q_m}{Q_m + Q_c}$$

then equation (4.16) becomes

$$Cg_1 = \frac{1}{Q_m}\left[(1 - F)\,G_i + F(G_i + G_w)\right]$$

$$= \frac{1}{Q_m}\left(G_i + FG_w\right) \qquad (4.17)$$

This verifies the intuitive expectation that as the degree of recirculation, F, increases then the gas concentration in the intake also increases, However, as F is never greater than 1, and comparing with equation (4.15), we can see that the intake, or face, gas concentration can never be greater than the return concentration. Hence, in a district recirculation system the general body gas concentration at no place is greater than the return general body concentration with or without recirculation. The maximum allowable methane concentrations in coal mine intakes may be prescribed by law at a low value such as 0.25%. The value of F should be chosen such that this limit is not exceeded.

Similar analyses may be carried out for dust concentrations. In this case, drop-out and the use of filters can result in significant reductions in the concentrations of dust in a system of controlled partial recirculation. However, the enhanced air velocities in the work area should not exceed some 4 m/s as the re-entrainment of settled particles within the airstream accelerates rapidly at greater velocities.

The climatic conditions within a system of controlled partial recirculation depend not only on the airflows and positions–duties of the fans but also on the highly interactive nature of heat transfer between the strata and the ventilating airstreams

(Chapter 15). The locations, types and powers of other mechanized equipment, and the presence of free water, also have significant effects. The only practicable means of handling the large numbers of variables is through a computer program to simulate the interacting physical processes (Chapter 16). Such analyses, together with practical observations, indicate that wet and dry bulb temperatures at any point may either increase or decrease when controlled partial recirculation is initiated without air cooling. The increased air velocities within the recirculation zone enhance the cooling power of the air on the human body for any given temperature and humidity. However, when controlled recirculation is practiced in hot mines, it is normally accompanied by cooling of the recirculated air. This is less expensive and more effective than bulk cooling the intake air in a conventional ventilation system and significant improvements in climatic conditions may be realized. Again, practical experience has shown that the higher airflows within a recirculation zone improves the effectiveness of existing refrigeration capacity.

In closing this section the reader is reminded, once again, that air recirculation may be prohibited by the governing legislation. The relevant statutes should be read, and/or enforcement agencies consulted, before instituting a system of controlled partial recirculation.

4.6 UNDERGROUND REPOSITORIES

4.6.1 Types of repository

Underground space is increasingly being utilized for purposes other than the extraction of minerals or transportation. The high cost of land, overcrowding and aesthetic considerations within urban areas encourage use of the subsurface for office accommodation, manufacturing, warehousing, entertainment facilities and many other purposes. The safety and stability of a well-chosen geologic formation makes underground space particularly suitable for the storage of materials, varying from foodstuffs and liquid or gaseous fuels to toxic wastes. The design and operation of environmental systems in such repositories require the combined skills of mine ventilation engineers and HVAC (heating, ventilating and air-conditioning) personnel. The repositories must be constructed and operated in a manner that preserves the integrity of the stored material and also protects the public from hazardous emissions or effluents.

Perhaps the most demanding designs arise out of the perceived need to store radioactive waste in deep underground repositories. There are basically two types of this waste. First, there is the transuranic, or low level, radioactive waste such as contaminated clothing, cleaning materials or other consumable items that are produced routinely by establishments that handle radioactive materials. Such waste may be compressed into containers which may be stacked within excavated chambers underground. Secondly, there is the concentrated and highly radioactive waste produced from some defence establishments and as the plutonium-rich spent fuel rods from nuclear power stations. This waste may be packed into heavily shielded and corrosion resistant cylinders and emplaced within boreholes, about one metre in diameter, drilled from underground airways into the surrounding rock.

4.6.2   Ventilation circuits in repositories for nuclear waste

Figure 4.19 depicts the primary ventilation structure of a high level nuclear waste repository. As in the figures illustrating mining circuits shown earlier in the chapter, this sketch is conceptual in nature and is not intended to represent all airways.

During the operation of an underground repository two activities must proceed in phase with each other. One is the mining of the rooms or drifts where the material is to be placed, together with the excavation of transportation routes, ventilation airways and all of the other infrastructure required in an underground facility. This is referred to simply as the mining activity. Secondly, the hazardous waste material must be transported through the relevant shafts and airways to the selected rooms for emplacement. Accordingly, this is known as the emplacement activity.

For reasons of environmental safety, the ventilation circuits for mining and emplacement activities in a nuclear waste repository must be kept separate. Furthermore, any leakage of air through doors or bulkheads between the two systems must always leave the mining zone and flow into the emplacement zone—even in the event of the failure of any fan. Figure 4.19 shows how this is achieved. The mining circuit operates as a through-flow **forcing** system with the main fan(s) sited at the top of the mining downcast shafts(s). On the other hand, the emplacement circuit operates as a through flow **exhaust** system with the main fans located at the top of the waste upcast shafts. It should be remembered that, within the nomenclature of underground repositories, the term **waste** refers to the hazardous waste to be emplaced, and not waste rock produced by mining activities. With this design, any accidental release of radionucleides into the underground atmosphere is contained completely within the emplacement circuit and will not contaminate the mining zones.

The shaft or surface-connecting ramp used for transporting the nuclear waste underground is not shown on Fig. 4.19. This shaft will normally not form part of the main ventilation system but will have a limited downcasting airflow which passes directly into a waste main return. Similar arrangements may be made for the waste transportation routes underground, thus limiting the potential dispersion of radioactive contamination in the event of a waste container being damaged during transportation. Separate maintenance and repair shops are provided in the mining and emplacement circuits.

When emplacement activities have been completed in any given room then the ends of that room may be sealed. In the case of high level nuclear waste, this may result in the envelope of rock surrounding the airway reaching temperatures in excess of 150 °C, depending on the rate of heat emission from the waste, the distance of the canisters from the airway, the thermal properties of the rock, and thermal induction of water and vapour migration within the strata. If the drift is to be reopened for retrieval or inspection of any canister then a considerable period of

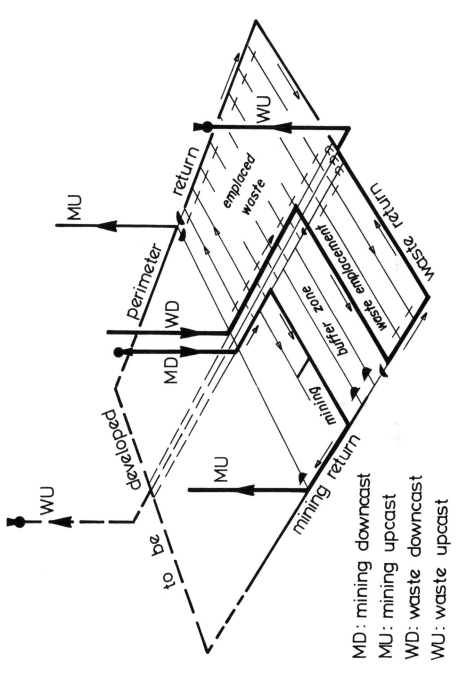

MD : mining downcast
MU : mining upcast
WD: waste downcast
WU: waste upcast

Figure 4.19 Example of primary ventilation circuits for a nuclear waste repository.

cooldown by refrigerated air may be required before unprotected personnel can re-enter. To reduce the time and expense of the cooldown period, the emplaced room may not be completely sealed but allowed to pass a regulated airflow sufficient to maintain the rock surface temperature at a controlled level.

4.6.3 Additional safety features

Before any repository for hazardous waste is commissioned it must conform to the strictest standards of safety and quality assurance in order to protect both the workers and the general public from chemical or radioactive contamination. In the case of an underground repository, design safeguards commence with an extensive examination of the suitability of the geologic formation to act as a natural containment medium. This will involve the physical and chemical properties of the rock, the presence and natural migration rate of groundwater and the probability of seismic activity. Other factors that influence the choice of site include population density and public acceptance of the surface transportation of hazardous waste to the site.

In addition to continuous electronic surveillance of the quality of the atmosphere throughout the main ventilation routes of a nuclear waste repository, the fans, bulkheads and regulators must be monitored to ensure that they operate within design limits and that pressure differentials are maintained in the correct direction at all times. Regulators and doors may be fitted with electrical or pneumatic actuators suitable for both local and remote operation. Should airborne radioactivity be detected at any time and any location in the circuit then the air emerging from the top of the waste upcast shafts is diverted automatically through banks of high efficiency particulate (HEPA) filters. Separate or additional precautions should also be taken to protect surface buildings and shaft tops from tornadoes, floods or fall-out from volcanic activity.

FURTHER READING

Burton, R. C., *et al.* (1984) Recirculation of air in the ventilation and cooling of deep gold mines. *Trans. 3rd Int. Congr. on Mine Ventilation, Harrogate.*

Dhar, N. and McPherson, M. J. (1987) The effect of controlled recirculation on mine climate. *Proc. 5th Annu. Workshop, Generic Mineral Technical Center, Mine Systems Design and Ground Control,* AL, 139–51.

Hardcastle, S. G. (1985) Computer predicted airborne respirable dust concentrations in mine air recirculation systems. *Proc. 2nd US Mine Ventilation Symp., Reno,* 239–48.

Leach, S. J. and Slack, A. (1969) Recirculation of mine ventilation systems. *Min. Eng. London,* (100).

Lee, R. D. and Longson, I. (1987) Controlled recirculation of mine air in working districts. *J. Mine Vent. Soc. S. Afr.* **40**(2).

Longson, I., Lee, R. D. and Lowndes, I. S. (1985) The feasibility of controlled air recirculation around operating longwall coal faces. *Proc. 2nd US Mine Ventilation Symp. Reno,* 227–37.

Robinson, R. (1989) Use of booster fans and recirculation systems for environmental control in British coal mines. *Proc. 4th US Mine Ventilation Symp., Berkeley, CA*, 235–42.

Wallace, K. G. (1988) Ventilation strategy for a prospective US nuclear waste repository. *Proc. 4th Int. Mine Ventilation Congr., Brisbane*, 65–72.

Wallace, K. G., *et al.* (1986) Impact of using auxiliary fans on mine ventilation efficiency and cost. *AIME Meet., New Orleans, March 1986.*

5

Incompressible flow relationships

5.1 INTRODUCTION

In Chapter 2 we introduced some of the basic relationships of incompressible fluid flow. With the exception of shafts greater than 500 m in vertical extent, changes in air density along individual airways may be ignored for most practical ventilation planning. Furthermore, in areas that are actively ventilated, airflows are turbulent in nature, other than in very large openings.

In this chapter, we shall confine ourselves to incompressible turbulent flow in order to develop and illustrate the equations and concepts that are most commonly employed in the practice of subsurface ventilation engineering.

Although the basic relationships that were derived in Chapter 2 may be employed directly for ventilation planning, John J. Atkinson introduced certain simplifications in his classical paper of 1854. These facilitate practical application but were achieved at the expense of precision. As the resulting 'laws of airflow' remain in common use, they are introduced and discussed in this chapter. The important concept of airway resistance is further expanded by an examination of the factors that influence it.

5.2 THE ATKINSON EQUATION AND THE SQUARE LAW

In seeking to quantify the relationships that govern the behaviour of airflows in mines, Atkinson utilized earlier work of the French hydraulic engineers and, in particular, the Chézy–Darcy relationship of the form expressed in equation (2.49),

$$p = fL \frac{\text{per}}{A} \rho \frac{u^2}{2} \quad \text{Pa}$$

It must be remembered that Atkinson's work was conducted in the middle of the nineteenth century, some 30 years before Reynolds' experiments and long before Stanton, Prandtl and Nikuradse had investigated the variable nature of the coefficient of friction, f. As far as Atkinson was aware, f was a true constant for any given airway. Furthermore, the mines of the time were relatively shallow, allowing the air density, ρ, to be regarded also as constant. Atkinson was then able to collate the 'constants' in the equation into a single factor

$$k = \frac{f\rho}{2} \quad \text{kg/m}^3 \tag{5.1}$$

giving

$$p = kL\frac{\text{per}}{A}u^2 \quad \text{Pa} \tag{5.2}$$

This has become known as **Atkinson's equation** and k as the Atkinson friction factor. Notice that unlike the dimensionless Chézy–Darcy coefficient, f, Atkinson's friction factor, is a function of air density and, indeed, has the dimensions of density.

Atkinson's equation may be written in terms of airflow, $Q = uA$, giving

$$p = kL\frac{\text{per}}{A^3}Q^2 \quad \text{Pa} \tag{5.3}$$

Now for any given airway, the length, L, perimeter, per, and cross-sectional area, A, are all known. Ignoring its dependence on density, the friction factor varies only with the roughness of the airway lining for fully developed turbulence. Hence we may collect all of those variables into a single characteristic number, R, for that airway.

$$R = kL\frac{\text{per}}{A^3} \quad \frac{\text{N s}^2}{\text{m}^8} \quad \text{or} \quad \frac{\text{kg}}{\text{m}^7} \tag{5.4}$$

giving

$$p = RQ^2 \quad \text{Pa} \tag{5.5}$$

This is known as the **square law** of mine ventilation and is probably the single most widely used relationship in subsurface ventilation engineering.

The parameter R is called the Atkinson resistance of the airway and, as shown by the square law, is the factor that governs the amount of airflow, Q, that will pass when an airway is subjected to any given pressure differential, p.

The simplicity of the square law has been achieved at the expense of precision and clarity. For example, the resistance of an airway should, ideally, vary only with the geometry and roughness of that airway. However, equations (5.1) and (5.4) show that R depends also on the density of the air. Hence, any variations in the temperature and/or pressure of the air in an airway will produce a change in the Atkinson resistance of that airway.

Secondly, the frictional pressure drop, p, depends on the air density as well as the geometry of the airway for any given airflow Q. This fact is not explicit in the conventional statement of the square law (5.5) as the density term is hidden within the definition of Atkinson resistance, R.

A clearer and more rational version of the square law was, in fact, derived as equation (2.50) in Chapter 2, namely

$$p = R_t\rho Q^2 \quad \text{Pa}$$

where R_t was termed the rational turbulent resistance, dependent only upon geometric factors and having units of m^{-4}:

$$R_t = \frac{fL \text{ per}}{2A^3} \quad m^{-4}$$

(from equation (2.51)).

It is interesting to reflect upon the influence of historical development and tradition within engineering disciplines. The k factor and Atkinson resistance, R, introduced in 1854 have remained in practical use to the present time, despite their weakness of being functions of air density. With our increased understanding of the true coefficient of friction, f, and with many mines now subject to significant variations in air density, it would be circumspect to abandon the Atkinson factors k and R, and to continue with the more fundamental coefficient of friction f, and rational resistance, R_t. However, the relinquishment of concepts so deeply rooted in tradition and practice does not come about readily, no matter how convincing the case for change. For these reasons, we will continue to utilize the Atkinson friction factor and resistance in this book in addition to their more rational equivalents. Fortunately, the relationship between the two is straightforward. Values of k are usually quoted on the basis of standard density, $1.2\,kg/m^3$. Then equation (5.1) gives

$$k_{1.2} = 0.6f \quad kg/m^3 \tag{5.6}$$

and equations (5.5) and (2.50) give

$$R_{1.2} = 1.2R_t \quad N\,s^2/m^8 \tag{5.7}$$

Again, on the premise that listed values of k and R are quoted at standard density (subscript 1.2), equations (5.3)–(5.5) may be utilized to give the frictional pressure drop and resistance at any other density, ρ:

$$p = k_{1.2} L \frac{\text{per}}{A^3} Q^2 \frac{\rho}{1.2} \quad Pa \tag{5.8}$$

$$R = k_{1.2} L \frac{\text{per}}{A^3} \frac{\rho}{1.2} \quad \frac{N\,s^2}{m^8} \tag{5.9}$$

and

$$p = RQ^2 \quad Pa \tag{5.10}$$

5.3   DETERMINATION OF FRICTION FACTOR

The degree of roughness of an underground opening has an important influence on its resistance and, hence, the cost of passing any given airflow. The surface roughness also has a direct bearing on the rate of heat transfer between the rock and the airstream (Chapter 15).

The coefficients of friction, f, shown on the Moody diagram, Fig. 2.7, remain based on the concept of sand grain roughness. Furthermore, as shown by equation (5.6), the Atkinson friction factor, k, is directly related to f. However, the k factor must be tolerant to wide deviations in the size and distribution of asperities on any given surface.

The primary purpose of a coefficient of friction, f, or friction factor, k, is to be able to predict the resistance of a planned but yet unconstructed airway. There are three main methods of determining an appropriate value of the friction factor.

By analogy with similar airways

During ventilation surveys, measurements of frictional pressure drops, p, and corresponding airflows, Q, are made in a series of selected airways (Chapter 6). During major surveys, it is pertinent to choose a few airways representative of, say, intakes, returns, conveyor roadways or particular support systems, and to conduct additional tests in which the airway geometry and air density are also measured. The corresponding values of the friction factor may then be calculated, and referred to standard density from equations (5.9) and (5.10) as

$$k_{1.2} = \frac{p}{Q^2} \frac{A^3}{L} \frac{1.2}{\text{per } \rho} \quad \frac{\text{kg}}{\text{m}^3} \tag{5.11}$$

Those values of k may subsequently be employed to predict the resistances of similar planned airways and, if necessary, at different air densities. Additionally, where a large number of similar airways exist, representative values of the friction factor can be employed to reduce the number or lengths of airways to be surveyed. In this case, care must be taken that unrepresentative obstructions or blockages in those airways are not overlooked.

In practice, where the k factor is to be used to calculate the resistances of airways at similar depths and climatic conditions, the density correction $1.2/\rho$ is usually ignored.

Experience has shown that local determinations of friction factor lead to more accurate planning predictions than those given in published tables (see the following subsection). Mines may vary considerably in their mechanized or drill-and-blast techniques of roadway development, as well as in methods of support. Furthermore, many modes of roadway drivage, or the influence of rock cleavage, leave roughenings on the surface that have a directional bias. In such cases, the value of the friction factor will depend also on the direction of airflow.

From design tables

Since the 1920s, measurements of the type discussed in the previous subsection have been conducted in a wide variety of mines, countries and airway conditions. Table 5.1 has been compiled from a combination of reported tests and the results of numerous observations made during the conduct of unpublished ventilation surveys. It should be mentioned, again, that empirical design data of this type should be used only as a guide and when locally determined friction factors are unavailable.

Table 5.1 Friction factors (referred to air density of $1.2\,kg/m^3$) and coefficients of friction (independent of air density)

	Friction factor, k $(kg/m^3)^a$	Coefficient of friction, f (dimensionless)
Rectangular airways		
Smooth concrete lined	0.004	0.0067
Shotcrete	0.0055	0.0092
Unlined with minor irregularities only	0.009	0.015
Girders on masonry or concrete walls	0.0095	0.0158
Unlined, typical conditions, no major irregularities	0.012	0.020
Unlined, irregular sides	0.014	0.023
Unlined, rough or irregular conditions	0.016	0.027
Girders on side props	0.019	0.032
Drift with rough sides, stepped floor, handrails	0.04	0.067
Steel arched airways		
Smooth concrete all round	0.004	0.0067
Bricked between arches all round	0.006	0.010
Concrete slabs or timber lagging between flanges all round	0.0075	0.0125
Slabs or timber lagging between flanges to spring	0.009	0.015
Lagged behind arches	0.012	0.020
Arches poorly aligned, rough conditions	0.016	0.027
Shafts[b]		
Smooth lined, unobstructed	0.003	0.005
Brick lined, unobstructed	0.004	0.0067
Concrete lined, rope guides, pipe fittings	0.0065	0.0108
Brick lined, rope guides, pipe fittings	0.0075	0.0125
Unlined, well trimmed surface	0.01	0.0167
Unlined, major irregularities removed	0.012	0.020
Unlined, mesh bolted	0.014	0.023
Tubbing lined, no fittings	0.007–0.014	0.0012–0.023
Brick lined, two side buntons	0.018	0.030
Two side buntons, each with a tie girder	0.022	0.037
Longwall faceline with steel conveyor and powered supports[c]		
Good conditions, smooth wall	0.035	0.058
Typical conditions, coal on conveyor	0.05	0.083
Rough conditions, uneven faceline	0.065	0.108
Ventilation ducting[d]		
Collapsible fabric ducting (forcing systems only)	0.0037	0.0062

Table 5.1 *(Contd.)*

	Friction factor, k $(kg/m^3)^a$	Coefficient of friction, f *(dimensionless)*
Flexible ducting with fully stretched spiral spring reinforcement	0.011	0.018
Fibreglass	0.0024	0.0040
Spiral-wound galvanized steel	0.0021	0.0035

[a] To convert friction, factor k (kg/m^3), to imperial units (lb min^2/ft^4), multiply by 5.39×10^{-7}.
[b] See section 5.4.6 for more accurate assessment of shaft resistance.
[c] k factors in excess of 0.015 kg/m^3 are likely to be caused by the aerodynamic drag of free-standing obstructions in addition to wall drag.
[d] These are typical values for new ducting. Manufacturer's test data should be consulted for specific ducting. It is prudent to add about 20% to allow for wear and tear.

From geometric data

The coefficient of friction, f, and, hence, the Atkinson friction factor, k, can be expressed as a function of the ratio e/d, where e is the height of the roughenings or asperities and d is the hydraulic mean diameter of the airway or duct ($d = 4A/\text{per}$). The functional relationships are given in section 2.3.6 and are illustrated graphically in Fig. 5.1 (see, also, Fig. 2.7).

For fully developed turbulent flow, the Von Kármán equation gives (see equation (2.55))

$$f = \frac{k_{1.2}}{0.6} = \frac{1}{4[2\log_{10}(d/e) + 1.14]^2}$$

Here again, the equation was developed for uniformly sized and dispersed asperities on the surface (sand grain roughness). For friction factors that are determined empirically for non-uniform surfaces, this equation can be transposed, or Fig. 5.1 can be used, to find the equivalent e/d value. For example, a k factor of 0.012 kg/m^3 gives an equivalent e/d value of 0.063. In an airway of hydraulic mean diameter 3.5 m, this gives the effective height of asperities to be 0.22 m.

The direct application of the e/d method is limited to those cases where the height of the asperities can be measured or predicted. The technique is applicable for supports that project a known distance into the airway.

Example The projection of the flanges, e, in a tubbed shaft is 0.152 m. The wall to wall diameter is 5.84 m. Calculate the coefficient of friction and Atkinson friction factor.

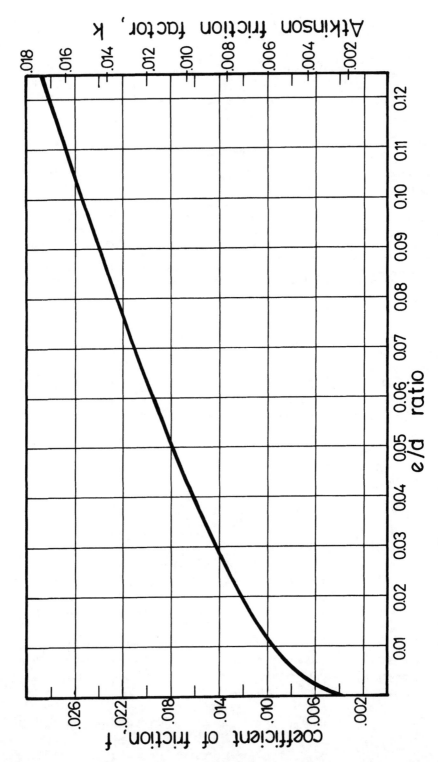

Figure 5.1 The coefficient of friction varies with the height of roughenings to airway diameter ratio, *e/d*.

Solution   The e/d ratio is

$$\frac{0.152}{5.84} = 0.026$$

From equation (2.55) or Fig. 5.1, this gives

$$f = 0.0135 \quad \text{and} \quad k_{1.2} = 0.0081\,\text{kg/m}^3$$

It should be noted that, for regularly spaced projections such as steel rings, the effective friction factor becomes a function of the spacing between those projections. Immediately downstream from each support, wakes of turbulent eddies are produced. If the projections are sufficiently far apart for those vortices to have died out before reaching the next consecutive support, then the projections act in isolation and independently of each other. As the spacing is reduced, the number of projections per unit length of shaft increases. So, also, does the near-wall turbulence, the coefficient of friction, f, and, hence, the friction factor, k. At that specific spacing where the vortices just reach the next projection, the coefficient of friction reaches a maximum. It is this value that is given by the e/d method. For wider spacings, the method will overestimate the coefficient of friction and, hence, errs on the side of safe design.

The maximum coefficient of friction is reached at a spacing/diameter ratio of 1/8. Decreasing the spacing further will result in 'wake interference'—the total degree of turbulence will reduce and so, also, will the coefficient of friction and friction factor. However, in this condition, the diameter available for effective flow is also being reduced towards the inner dimensions of the projections.

5.4   AIRWAY RESISTANCE

The concept of airway resistance is of major importance in subsurface ventilation engineering. The simple form of the square law (see equation (5.5))

$$p = RQ^2$$

shows the resistance to be a constant of proportionality between frictional pressure drop, p, in a given airway and the square of the airflow, Q, passing through it at a specified value of air density. The parabolic form of the square law on a p, Q, plot is known as the airway resistance curve. Examples are shown in Fig. 5.2.

The cost of passing any given airflow through an airway varies directly with the resistance of that airway. Hence, as the total operating cost of a complete network is the sum of the individual airway costs, it is important that we become familiar with the factors that influence airway resistance. Those factors are examined in the following subsections.

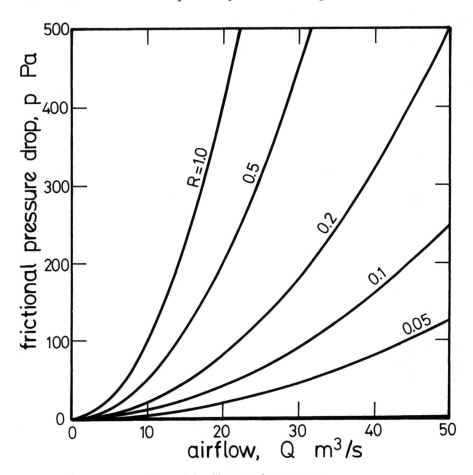

Figure 5.2 Airway resistance curves.

5.4.1 Size of airway

Equation (5.4) showed that for a given length of airway, L, and friction factor, k,

$$R \propto \frac{\text{per}}{A^3} \tag{5.12}$$

where $\propto$ means 'proportional to'. However, for any given shape or cross-section,

$$\text{per} \propto \sqrt{A} \tag{5.13}$$

Substituting for per in proportionality (5.12) gives

$$R \propto \frac{1}{A^{2.5}} \tag{5.14}$$

or, for a circular airway,

$$R \propto \frac{1}{d^5} \tag{5.15}$$

These two latter proportionalities show the tremendous effect of airway size on resistance. Indeed, the cross-sectional area open for flow is the dominant factor in governing airway resistance. Driving an airway at only half its design diameter will result in the resistance being 2^5 or 32 times greater. Hence, for any required airflow, the cost of passing that ventilation through the airway will also increase by a factor of 32. It is clear that, when sizing an underground opening that will form part of a main ventilation route, the resistance and ventilation operating costs must be taken into account. Airway sizing is considered further in Chapter 9.

5.4.2 Shape of airway

Again, for any given length of airway, L, and friction factor, k, the proportionality (5.12) can be rewritten as

$$R \propto \frac{\text{per}}{A^{1/2}} \frac{1}{A^{5/2}} \tag{5.16}$$

However, for any given shape of cross-section (see proportionality (5.13)),

$$\frac{\text{per}}{A^{1/2}} \text{ is a constant}$$

We term this parameter the **shape factor**, SF, for the airway. Then if all other parameters remain constant, including cross-sectional area, A, the resistance of an airway varies with respect to its shape factor.

The planar figure having the minimum possible shape factor is a circle

$$SF(\text{circle}) = \frac{\text{per}}{A^{1/2}} = \frac{\pi d}{\sqrt{\pi/4}\,d}$$

$$= 3.5449$$

All other shapes have a greater shape factor than this value. For this reason, shape factors are usually normalized with respect to a circle by dividing by 3.5449 and are then quoted as **relative shape factor** (RSF) as shown in Table 5.2. The further we depart from the ideal circular shape then the greater will be the RSF and, hence, the airway resistance.

The purpose of relative shape factors in ventilation planning is limited to little more than comparing the effect of shape for proposed airways of given cross-sectional area. It may also be useful as a correction factor for older nomograms relating airway resistance, area and k factor and which were produced on the basis of a circular cross-section.

Table 5.2 Relative shape factors

Shape of airway	Relative shape factor
Circular	1.00
Arched, straight legs	1.08
Arched, splayed legs	1.09
Square	1.13
Rectangular	
width:height = 1.5:1	1.15
2:1	1.20
3:1	1.30
4:1	1.41

5.4.3 Airway lining

Equation (5.4) shows that airway resistance is proportional to the Atkinson friction factor, k, and, hence, is also directly proportional to the more fundamental coefficient of friction, f. The latter depends only on the roughness of the airway lining for fully developed turbulent flow.

5.4.4 Air density

The Atkinson resistance, R, as used in the square law, $p = R Q^2$, depends on the friction factor, k. However, equation (5.1) shows that k, itself, depends on the density of the air. It follows that the Atkinson resistance also varies with the density of the air.

On the other hand, the rational resistance, R_t, as used in the rational expression of the square law, $p = R_t \rho \, Q^2$ (equation (2.50)) is a function of the geometry and lining of the airway only and is independent of air density.

5.4.5 Shock losses

Whenever the airflow is required to change direction, additional vortices will be initiated. The propagation of those large-scale eddies consumes mechanical energy and the resistance of the airway may increase significantly. This occurs at bends, junctions, changes in cross-section, obstructions, regulators and at points of entry or exit from the system.

The effects of shock losses remain the most uncertain of all the factors that affect airway resistance. This is because fairly minor modifications in geometry can cause significant changes in the initiation of vortices and, hence, the airway resistance. Analytical techniques may be employed for simple and well-defined geometries. For the more complex situations that arise in practice, scale models may be employed to investigate the flow patterns and shock losses.

There are two methods that may be used to assess the additional resistance caused by shock losses.

Shock loss factor

In textbooks on fluid mechanics, shock losses are often referred to in terms of the head loss or drop in total pressure, p_{shock}, caused by the shock loss. This, in turn, is expressed in terms of 'velocity heads':

$$p_{shock} = X\rho \frac{u^2}{2} \quad \text{Pa} \tag{5.17}$$

where ρ = air density (kg/m³), u = mean velocity of air (m/s) and X = shock loss factor (dimensionless).

The shock loss factor can be converted into an Atkinson type of resistance, R_{shock}, by rewriting equation (5.17) as a square law:

$$p_{shock} = \frac{X\rho}{2} \frac{Q^2}{A^2}$$

$$= R_{shock} Q^2$$

where

$$R_{shock} = \frac{X\rho}{2A^2} \quad \frac{\text{N s}^2}{\text{m}^8} \tag{5.18}$$

If rational resistances are employed then the density term is eliminated and the corresponding shock resistance becomes simply

$$R_{shock} = \frac{X}{2A^2} \quad \text{m}^{-4} \tag{5.19}$$

The major cause of the additional resistance is the propagation of vortices downstream from the cause of the shock loss. Accordingly, in most cases, it is the downstream branch to which the shock resistance should be allocated. However, for junctions, the cross-sectional area used in equations (5.18) and (5.19) is usually that of the **main** or **common** branch through which all of the local airflow passes.

One of the most comprehensive guides to the selection of X factors is contained within the *Fundamentals Handbook* of the American Society of Heating, Refrigerating and Air Conditioning Engineers (ASHRAE). Similar design information is produced by corresponding professional societies in other countries.

An appendix is given at the end of this chapter which contains graphs and formulae relating to shock loss factors commonly required in subsurface ventilation engineering.

Equivalent length

Suppose that, in a subsurface airway of length L, there is a bend or other cause of a shock loss. The resistance of the airway will be greater than if that same airway contained no shock loss. We can express that additional resistance, R_{shock}, in terms of the length of corresponding straight airway which would have that same value of shock resistance. This 'equivalent length' of shock loss, L_{eq}, may be incorporated into equation (5.9) to give an Atkinson resistance of

$$R = k(L + L_{eq}) \frac{\text{per}}{A^3} \frac{\rho}{1.2} \quad \frac{\text{N s}^2}{\text{m}^8} \tag{5.20}$$

The resistance due to the shock loss is

$$R_{shock} = kL_{eq} \frac{\text{per}}{A^3} \frac{\rho}{1.2} \quad \frac{\text{N s}^2}{\text{m}^8} \tag{5.21}$$

Equation (5.20) gives a very convenient and rapid method of incorporating shock losses directly into the calculation of airway resistance.

The relationship between shock loss factor, X, and equivalent length, L_{eq}, is obtained by comparing equations (5.18) and (5.21). This gives

$$R_{shock} = \frac{X\rho}{2A^2} = kL_{eq} \frac{\text{per}}{A^3} \frac{\rho}{1.2} \quad \frac{\text{N s}^2}{\text{m}^8}$$

or

$$L_{eq} = \frac{1.2X}{2k} \frac{A}{\text{per}} \quad \text{m}$$

This can be expressed in terms of hydraulic mean diameters, $d = 4A/\text{per}$, giving

$$L_{eq} = \frac{1.2X}{8k} d \quad \text{m} \tag{5.22}$$

leading to a very convenient expression for equivalent length,

$$L_{eq} = 0.15 \frac{X}{k} \text{ hydraulic mean diameters} \tag{5.23}$$

Reference to the appendix for X factors, together with a knowledge of the expected friction factor and geometry of the planned airway, enables the equivalent length of shock losses to be included in airway resistance calculation sheets, or during data preparation for computer exercises in ventilation planning.

When working on particular projects, the ventilation engineer will soon acquire a knowledge of equivalent lengths for commonly occurring shock losses. For example, a well-used rule of thumb is to allocate an equivalent length of 20 hydraulic mean diameters for a sharp right-angled bend.

Example A 4 m by 3 m rectangular tunnel is 450 m long and contains one right-angled bend of centre-line radius of curvature 2.5 m. The airway is unlined but is in good condition with major irregularities trimmed from the sides. If the tunnel is to pass 60 m³/s of air at a mean density 1.1 kg/m³, calculate the Atkinson and rational resistances at that density and the frictional pressure drop.

Solution   Let us first state the geometric factors for the airway: hydraulic mean diameter,

$$d = \frac{4A}{\text{per}} = \frac{4 \times 12}{14} = 3.429 \text{ m}$$

height/width ratio,

$$\frac{H}{W} = \frac{3}{4} = 0.75$$

radius of curvature/width ratio

$$\frac{r}{W} = \frac{2.5}{4} = 0.625$$

From Table 5.1, we estimate that the friction factor for this airway will be $k = 0.012 \text{ kg/m}^3$ at standard density $(f = 0.02)$. At $r/W = 0.625$ and $H/W = 0.75$, Fig. 5A.2 (at the end of this chapter) gives the shock loss factor for the bend to be 0.75.

We may now calculate the Atkinson resistance at the prevailing air density of 1.1 kg/m^3 by two methods.

1. *Calculate the resistance produced by the shock loss separately.* For the airway, equation (5.9) gives

$$
\begin{aligned}
R_{\text{length}} &= kL\frac{\text{per}}{A^3}\frac{\rho}{1.2} \\
&= \frac{0.012 \times 450 \times 14 \times 1.1}{12^3 \times 1.2} \\
&= 0.040\,10 \text{ N s}^2/\text{m}^8
\end{aligned}
$$

For the bend, equation (5.18) gives

$$
\begin{aligned}
R_{\text{shock}} &= \frac{X\rho}{2A^2} \\
&= \frac{0.75 \times 1.1}{2 \times 12^2} \\
&= 0.002\,86 \text{ N s}^2/\text{m}^8
\end{aligned}
$$

Then for the full airway

$$
\begin{aligned}
R &= R_{\text{length}} + R_{\text{shock}} \\
&= 0.040\,10 + 0.002\,86 \\
&= 0.042\,96 \text{ Ns}^2/\text{m}^8
\end{aligned}
$$

2. *Using the equivalent length of the bend.* Equation (5.23) gives

$$L_{eq} = 0.15 \frac{X}{k} d$$

$$= \frac{0.15 \times 0.75 \times 3.429}{0.012}$$

$$= 32.15 \, m$$

Then, from equation (5.20)

$$R = k(L + L_{eq}) \frac{per}{A^3} \frac{\rho}{1.2}$$

$$= 0.012(450 + 32.15) \times \frac{14}{12^3} \times \frac{1.1}{1.2}$$

$$= 0.042 \, 96 \, Ns^2/m^8$$

which agrees with the result given by method 1.

If the coefficient of friction, f, and the rational resistance, R_t, are employed, the density terms become unnecessary and the equivalent length method gives the rational resistance as

$$R_t = f(L + L_{eq}) \frac{per}{2A^3}$$

$$= 0.02(450 + 32.15) \times \frac{14}{2 \times 12^3}$$

$$= 0.039 \, 06 \, m^{-4}$$

The Atkinson resistance at a density of $1.1 \, kg/m^3$ becomes

$$R = \rho R_t$$

$$= 1.1 \times 0.039 \, 06 = 0.042 \, 96 \quad Ns^2/m^8$$

At an airflow of $60 \, m^3/s$ and a density of $1.1 \, kg/m^3$, the frictional pressure drop becomes

$$p = R_{1.1} Q^2$$

$$= 0.042 \, 96 \times 60^2$$

$$= 155 \, Pa$$

or, using the rational resistance,

$$P = \rho R_t Q^2$$

$$= 1.1 \times 0.039 \, 06 \times 60^2$$

$$= 155 \, Pa$$

5.4.6 Mine shafts

In mine ventilation planning exercises, the airways that create the greatest difficulties in survey observations or in assessing predicted resistance are vertical and inclined shafts.

Shafts are quite different in their airflow characteristics to all other subsurface openings, not only because of the higher air velocities that may be involved, but also because of the aerodynamic effects of ropes, guide rails, buntons, pipes, cables, other shaft fittings, the fraction of cross-section filled by the largest conveyance (coefficient of fill, C_F) and the relatively high velocity of shaft conveyances. Despite such difficulties, it is important to achieve acceptable accuracy in the estimation of the resistance of ventilation shafts. In most cases, the total airflow supplied underground must pass through the restricted confines of the shafts. The resistance of shafts is often greater than the combined effect of the rest of the underground layout. In a deep room and pillar mine, the shafts may account for as much as 90% of the mine total resistance. Coupled with a high airflow, Q, the frictional pressure drop, p, will absorb a significant part of the fan total pressure. It follows that the operational cost of airflow (proportional to $p \times Q$) is usually greater in ventilation shafts than in any other airway.

At the stage of conceptual design, shaft resistances are normally estimated with the aid of published lists of friction (k) factors (Table 5.1) that make allowance for shaft fittings. For an advanced design, a more detailed and accurate analysis that in-corporates both theoretical and empirical factors may be employed. An outline of one such method is given in this section.

The resistance to airflow, R, offered by a mine shaft is composed of the effects of four identifiable components:

1. the shaft walls,
2. the shaft fittings (buntons, pipes etc.),
3. the conveyances (skips or cages) and
4. insets, loading and unloading points.

Shaft walls

The component of resistance offered by the shaft walls, R_{tw}, may be determined from equation (5.9) or (2.51) and using a friction factor, k, or coefficient of friction, f, determined by one of the methods described in section 5.3. Experience has shown that the relevant values in Table 5.1 are preferred either for smooth-lined walls or where the surface irregularities are randomly dispersed. On the other hand, where projections from the walls are of known size as, for example, in the case of tubbed lining, then the e/d method gives satisfactory results.

Shaft fittings

It is the permanent equipment in a shaft, particularly cross-members (buntons) that account, more than any other factor, for the large variation in k values reported for mine shafts.

Longitudinal fittings

These include ropes, guide rails, pipes and cables, situated longitudinally in the shaft and parallel to the direction of airflow. Such fittings add very little to the coefficient of friction and, indeed, may help to reduce swirl. If no correction is made for longitudinal fittings in the calculation of cross-sectional area then their effect may be approximated during preliminary design by increasing the k factor (see Table 5.1). However, measurements on model shafts have actually shown a decrease in the true coefficient of friction when guide rails and pipes are added. It is, therefore, better to account for longitudinal fittings simply by subtracting their cross-sectional area from the full shaft area to give the 'free area' available for airflow. Similarly, the rubbing surface is calculated from the sum of the perimeters of the shaft and the longitudinal fittings.

Buntons

The term 'bunton' is used here to mean any cross-member in the shaft located perpendicular to the direction of airflow. The usual purpose of buntons is to provide support for longitudinal fittings. The resistance offered by buntons is often dominant.

Form drag and resistance of buntons

While the resistance offered by the shaft walls and longitudinal fittings arises from skin friction (shear) drag, through boundary layers close to the surface, the kinetic energy of the air causes the pressure on the projected area of cross-members facing the airflow to be higher than that on the downstream surfaces. This produces an inertial or 'form' drag on the bunton. Furthermore, the breakaway of boundary layers from the sides of trailing edges causes a series of vortices to be propagated downstream from the bunton. The mechanical energy dissipated in the formation and maintenance of these vortices is reflected by a significant increase in the shaft resistance.

The drag force on a bunton is given by the expression

$$\text{drag} = C_D A_b \frac{\rho u^2}{2} \quad \text{N} \tag{5.24}$$

where C_D = coefficient of drag (dimensionless), depending on the shape of the bunton, A_b = frontal or projected area facing into the airflow (m^2), u = velocity of approaching airstream (m/s) and ρ = density of air (kg/m^3).

The frictional pressure drop caused by the bunton is

$$p = \frac{\text{drag}}{A} = C_D \frac{A_b}{A} \frac{u^2}{2} \rho \quad \text{N/m}^2 \tag{5.25}$$

where A = area available for free flow (m^2). However, the frictional pressure drop caused by the buntons is also given by the square law:

$$p = R_b Q^2 = R_{tb} \rho Q^2 = R_{tb} \rho u^2 A^2 \quad \text{N/m}^2 \tag{5.26}$$

where R_b = resistance offered by the bunton $(N s^2/m^8)$, R_{tb} = rational resistance of the bunton (m^{-4}) and Q = airflow (m^3/s).

Equating equations (5.25) and (5.26) gives

$$R_{tb} = C_D \frac{A_b}{2A^3} \quad m^{-4} \tag{5.27}$$

If there are n buntons in the shaft and they are sufficiently widely spaced to be independent of each other, then the combined resistance of the buntons becomes

$$R_{tn} = nR_{tb} = nC_D \frac{A_b}{2A^3} \quad m^{-4} \tag{5.28}$$

It is, however, more convenient to consider the spacing, S, between buntons:

$$S = L/n \tag{5.29}$$

where L = length of shaft (m), giving,

$$R_{tn} = \frac{L}{S} C_D \frac{A_b}{2A^3} \quad m^{-4} \tag{5.30}$$

The effect of the buntons can also be expressed as a coefficient of friction, f, or friction factor, $k = 0.6f$, since,

$$R_{tn} = \frac{f_b L \text{ per}}{2A^3} \quad m^{-4} \tag{5.31}$$

Equations (5.30) and (5.31) give

$$f_b = \frac{C_D A_b}{S \times \text{per}} \quad \text{(dimensionless)} \tag{5.32}$$

Interference factor

The preceding subsection assumed that the buntons were independent of each other, i.e. that each wake of turbulent eddies caused by a bunton dies out before reaching the next bunton. Unless the buntons are streamlined or are far apart, this is unlikely to be the case in practice. Furthermore, the drag force may be somewhat less than linear with respect to the number of buntons in a given cross-section, or to A_b at connection points between the buntons and the shaft walls or guide rails. For these reasons, an 'interference factor', F, is introduced into equation (5.30) (Bromilow) giving a reduced value of resistance for buntons:

$$R_{tn} = \frac{L}{S} C_D \frac{A_b}{2A^3} F \quad m^{-4} \tag{5.33}$$

For simple wake interference, F is a function of the spacing ratio Δ,

$$\Delta = \frac{S}{W} \quad \text{(dimensionless)} \tag{5.34}$$

where $W =$ width of the bunton. However, because of the difficulties involved in quantifying the F function analytically, and in evaluating the other factors mentioned above, the relationship has been determined empirically (Bromilow), giving

$$F = 0.0035\Delta + 0.44 \qquad (5.35)$$

for a range of Δ from 10 to 40.

Combining equations (5.33), (5.34) and (5.35) gives

$$R_{tn} = \frac{L}{S} C_D \frac{A_b}{2A^3} \left(0.0035 \frac{S}{W} + 0.44 \right) \quad \text{m}^{-4} \qquad (5.36)$$

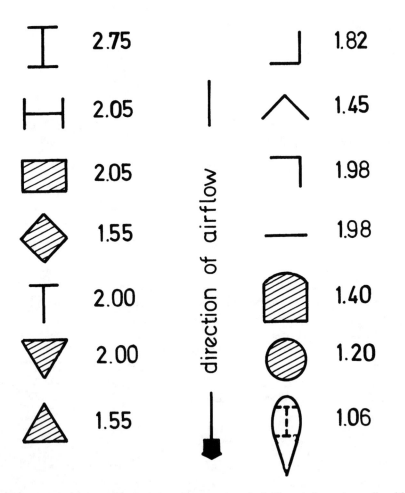

Figure 5.3 Coefficients of drag, C_D, for buntons in a shaft (from information collated by Bromilow).

If expressed as a partial coefficient of friction, this becomes

$$f_b = C_D \frac{A_b}{S \times \text{per}} \left(0.0035 \frac{S}{W} + 0.44 \right) \tag{5.37}$$

Values of the coefficient of drag, C_D, measured for buntons in shafts are invariably much higher than those reported for free-standing aerofoil sections. Figure 5.3 indicates drag coefficients for the shapes normally used for buntons.

Conveyances

Resistance of a Stationary Conveyance
The main factors that govern the resistance of a stationary cage or skip in a mine shaft are as follows.

1. The percentage of the free area in a cross-section of the shaft occupied by the conveyance. This is sometimes termed the 'coefficient of fill', C_F.
2. The area and shape of a plan view of the conveyance and, to a lesser degree, its vertical height.

The parameters of secondary practical importance include the shape of the shaft and the extent to which the cage is totally enclosed.

An analysis of the resistances offered by cages or skips commences by considering the conveyance to be stationary within the shaft with a free (approaching) air velocity of u_a (m/s). The conveyance is treated as an obstruction giving rise to a frictional pressure drop (shock loss) of X velocity heads:

$$p_c = X \frac{\rho u_a^2}{2} \quad \frac{\text{N}}{\text{m}^2} \tag{5.38}$$

where p_c = frictional pressure drop due to the conveyance (N/m^2) and u_a = velocity of the approaching airstream (m/s).

The effective rational resistance of the stationary cage, R_{tc}, can be expressed in terms of X from equation (5.19)

$$R_{tc} = \frac{X}{2A^2} \quad \text{m}^{-4} \tag{5.39}$$

where the free shaft area, A, is measured in square metres. The problem now becomes one of evaluating X.

One of the most comprehensive studies on cage resistance was carried out by A. Stevenson in 1956. This involved the construction of model cages and measuring the pressure drops across them at varying airflows in a circular wind tunnel. From the results produced in this work, Fig. 5.4 has been constructed.

All of the curves on Fig. 5.4 refer to a cage whose length, L, is 1.5 times the width, W. A multiplying correction factor may be applied to the value of X for other plan dimensions. Hence, for a cage that is square in plan, $L/W = 1$ and the correction nomogram given also on Fig. 5.4 shows that X should be increased by 12%.

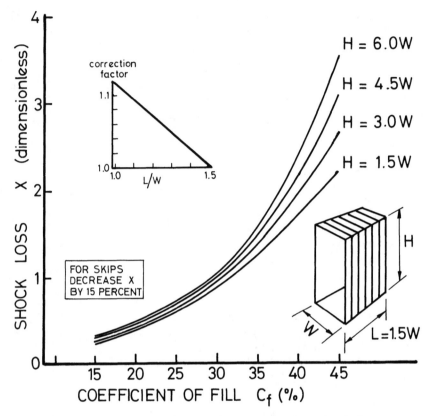

Figure 5.4 Shock loss factors for conveyance in a mine shaft.

The shape of the curves indicates the dominant effect of the coefficient of fill, increasing rapidly after the cage occupies more than 30% of the shaft cross-section. The major cause of the shock loss is form drag and the resulting turbulent wake. Skin friction effects are relatively small. Hence, so also is the influence of the vertical height of the cage.

The curves refer to cages with covered roof, floor and sides, but open ends. In general, the effect of covering the ends is to reduce the shock loss. For totally enclosed skips, the value of X read from the curves should be reduced by some 15%.

Dynamic effects of a moving conveyance

In addition to the presence of a conveyance in a shaft, its motion will also influence the frictional pressure drop and effective resistance of the shaft. Furthermore, in systems in which two conveyances pass in the shaft, their stability of motion will depend on the velocities of the airflow and the conveyances, the positions of the conveyances within the shaft cross-section, their shape, and the coefficient of fill.

When two synchronized skips or cages pass, they will normally do so at mid-shaft and at the maximum hoist velocity. The coefficient of fill will, momentarily, be increased (doubled if the two cages have the same plan area). There will be a very short-lived peak of pressure drop across the passing point. However, the inertia of the columns of air above and below that point, coupled with the compressibility of the air, dampens out that peak quickly and effectively.

More important is the effect of passing on the stability of the conveyances themselves. They pass at a relative velocity of twice maximum hoisting speed, close to each other and usually without any continuous intervening barrier. Two mechanisms then influence the stability of the cages. First, thin boundary layers of air will exist on the surfaces of the conveyances and moving at the same velocity. Hence, there will be a very steep velocity gradient in the space between passing conveyances. Shear resistance (skin friction drag) will apply a braking action on the inner sides of the conveyances and produce a tendency for the skips or cages to turn in towards each other. Secondly, the mean absolute velocity in the gap between the passing cages is unlikely to equal that surrounding the rest of the cages. There exists a variation in static pressure and, hence, applied force on the four sides of each conveyance (the venturi effect). This again, will result in a tendency for lateral movement.

Passengers in conveyances will be conscious not only of the pressure pulse when the cages pass, but also of the lateral vibrations caused by the aerodynamic effects. The sideways motion can be controlled in practice by the use of rigid guide rails or, if rope guides are employed, by using a tensioned tail rope beneath the cages passing around a shaft-bottom pulley.

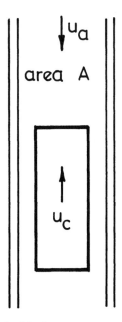

Figure 5.5 Moving conveyance in a shaft.

The aerodynamics of moving conveyances are more complex than for free bodies because of the proximity of the shaft walls. The total drag on the conveyance will reach a maximum when the vehicle is moving at its highest speed against the airflow. The apparent frictional pressure drop and resistance of the shaft will also vary and reach a maximum value at this same time. The opposite is also true. The drag, apparent frictional pressure drop, and apparent resistance of the shaft will reach a minimum when the conveyance is moving at maximum speed in the same direction as the airflow. The amplitude of the cyclic variation depends on the resistance of the stationary cage (which in turn depends on the coefficient of fill and other factors as described in the previous subsection), and the respective velocities of the airflow and the cage.

In the following analysis, expressions are derived to approximate the effective pressure drop and effective resistance of a moving conveyance.

Consider a conveyance moving at its maximum velocity u_c against an airflow of velocity u_a (Fig. 5.5).

If the rational resistance of the cage, when stationary, is R_{tc} (determined from the methodology of the previous section) then the corresponding pressure drop caused by the stationary cage will be

$$p_c = R_{tc}\rho Q^2 \tag{5.40}$$

$$= R_{tc}\rho u_a^2 A^2 \quad N/m^2$$

Now suppose that the cage were stationary in the shaft when the approaching airflow has a velocity of u_c. The corresponding pressure drop would be

$$p'_c = \rho R_{tc}u_c^2 A^2 \quad N/m^2 \tag{5.41}$$

This latter expression approximates the additional pressure drop caused by movement of the conveyance.

Adding equations (5.40) and (5.41) gives the maximum pressure drop across the moving conveyance:

$$p_{c\,max} = p_c + p'_c = \rho R_{tc}(u_a^2 + u_c^2)A^2$$

$$= \rho R_{tc}u_a^2 A^2\left(1 + \frac{u_c^2}{u_a^2}\right) \tag{5.42}$$

$$= p_c\left(1 + \frac{u_c^2}{u_a^2}\right) \quad N/m^2 \tag{5.43}$$

Similarly, the minimum pressure drop, occurring when the cage and airflow are moving in the same direction is given by

$$p_{c\,min} = p_c\left(1 - \frac{u_c^2}{u_a^2}\right) \quad N/m^2 \tag{5.44}$$

Hence, the cyclic variation in frictional pressure across the shaft is

$$\pm p_c u_c^2/u_a^2 \quad N/m^2 \tag{5.45}$$

This variation occurs much more slowly than the pressure pulse caused by passing conveyances. The pressure drop due to a conveyance will rise to $p_{c\,max}$ as the conveyance accelerates to its highest speed against the airflow, fall to p_c when the conveyance decelerates to rest, drop further to $p_{c\,min}$ as the conveyance accelerates to full speed in the same direction as the airflow, and again rise to p_c when the cage comes to rest.

If two cages of equal dimensions are travelling in synchronization within the shaft, but in opposite directions, then the effect cancels out. Otherwise, the cyclic variation in pressure may be measurable throughout the ventilation network and, particularly, at locations close to the shaft. This is most likely to occur in a single conveyance shaft with a large coefficient of fill. A further effect will be to impose a fluctuating load on any fans which influence the airflow in that shaft.

The corresponding range of effective resistance for the moving conveyance may also be determined from equation (5.42)

$$p_{c\,max} = \rho R_{tc}(u_a^2 + u_c^2)A^2$$

However, if the effective resistance of the moving conveyance is $R_{tc\,max}$, then

$$p_{c\,max} = \rho R_{tc\,max} Q^2 = \rho R_{tc\,max} u_a^2 A^2 \qquad (5.46)$$

Equating these two relationships gives

$$R_{tc\,max} = R_{tc}\left(1 + \frac{u_c^2}{u_a^2}\right) \quad m^{-4} \qquad (5.47)$$

Similarly, when the airflow and conveyance are moving the same direction,

$$R_{tc\,min} = R_{tc}\left(1 - \frac{u_c^2}{u_a^2}\right) \quad m^{-4} \qquad (5.48)$$

It may be noted from equations (5.44) and (5.48) that when the cage is moving more rapidly than the airflow ($u_c > u_a$), and in the same direction, both the effective pressure drop and effective resistance of the cage become negative. In this situation, the cage is assisting rather than impeding the ventilation.

Entry and exit losses

In badly designed installations or for shallow shafts, the shock losses that occur at shaft stations and points of air entry and exit may be greater than those due to the shaft itself. Such losses are, again, normally quoted on the basis of X velocity heads, using the velocity in the free area of the shaft.

The shock losses at shaft stations may be converted into a rational resistance:

$$R_{t,shock} = \frac{X}{2A^2} \qquad (5.49)$$

The shock loss factors, X, may be estimated from the guidelines given in the appendix to this chapter.

Total shaft resistance

Although interference will exist between components of shaft resistance, it is conservative to assume that they are additive. Hence, the rational resistance for the total shaft is

$$R_t = R_{tw} + R_{tn} + R_{tc}\left(1 \pm \frac{u_c^2}{u_a^2}\right) + R_{t,\text{shock}} \quad \text{m}^{-4} \tag{5.50}$$

This may be converted to an Atkinson resistance, R, referred to any given air density in the usual manner:

$$R = R_t \rho \quad \text{N s}^2/\text{m}^8$$

Having determined the total resistance of a shaft, the effective coefficient of friction, f, or friction factor k (at any given value of air density) may be determined

$$f = 2R_t \frac{A^3}{L \text{ per}} \tag{5.51}$$

and

$$k = \rho \frac{f}{2} \quad \text{kg/m}^3$$

Methods of reducing shaft resistance

Shaft walls

A major shaft utilized for both hoisting and ventilation must often serve for the complete life of the underground facility. Large savings in ventilation costs can be achieved by designing the shaft for low aerodynamic resistance.

Modern concrete lining of shafts closely approaches an aerodynamically smooth surface and little will be gained by giving a specially smooth finish to these walls. However, if tubbing lining is employed, the wall resistance will increase by a factor of two or more.

Buntons

A great deal can be done to reduce the resistance of buntons or other cross-members in a mine shaft. First, thought should be given to eliminating them or reducing their number in the design. Second, the shape of the buntons should be considered. An aerofoil section skin constructed around a girder is the ideal configuration. However, Fig. 5.3 shows that the coefficient of drag can be reduced considerably by such relatively simple measures as attaching a rounded cap to the upstream face of the bunton. A circular cross–section has a coefficient of drag some 60% of that for a square.

Longitudinal fittings

Ladderways and platforms are common in the shafts of metal mines. These produce high shock losses and should be avoided in main ventilation shafts. If it is necessary to include such encumbrances, then it is preferable to compartmentalize them behind a smooth wall partition. Ropes, guides, pipes and cables reduce the free area available for airflow and should be taken into account in sizing the shaft; however, they have little effect on the true coefficient of friction.

Cages and Skips

Figure 5.4 shows that the resistance of a cage or skip increases rapidly when it occupies more than 30% of the shaft cross-section. The plan area of a conveyance should be as small as possible, consistent with its required hoisting duties. Furthermore, a long narrow cage or skip offers less resistance than a square cage of the same plan area.

Table 5.3 shows the effect of attaching streamlined fairings to the upstream and downstream ends of a conveyance. This table was derived from a model of a shaft where the coefficient of fill for the cage was 43.5%. The efficiency of such fairings is reduced by proximity to the shaft wall or to another cage. There are considerable practical disadvantages to cage fairings. They interfere with cage suspension gear and balance ropes which must be readily accessible for inspection and maintenance. Furthermore, they must be removed easily for shaft inspection personnel to travel on top of the cage or for transportation of long items of equipment. Another problem is that streamlining on cages exacerbates the venturi effect when two conveyances pass. For these reasons, fairings are seldom employed in practice. However, the simple expedient of rounding the edges and corners of a conveyance to a radius of about 30 cm is beneficial.

Table 5.3 Shock loss factor (X) for a caged fitted with end fairings in a shaft (after Stevenson)

Number of cage decks	Shape of fairing	Cage without fairings	Upstream fairing only	Downstream fairing only	Fairings at both ends
1	Aerofoil	2.05	1.53	1.56	0.95
4	Aerofoil	3.16	2.46	2.71	2.11
1	Triangular with ends filled	2.13	1.58	1.59	1.13
1	Triangular without ends filled	2.13	1.65	1.79	1.48
4	Triangular with ends filled	3.29	2.96	2.92	2.70
4	Triangular without ends filled	3.29	2.86	3.18	2.80

Intersections and loading/unloading stations

For shafts that are used for both hoisting and ventilation it is preferable to employ air bypasses at the main loading and unloading stations. At shaft bottom stations, the main airstream may be diverted into one (or two) airways intersecting the shaft some 10 to 20 m above or below the loading station. Similarly, at the shaft top, the main airflow should enter or exit the shaft 10 to 20 m below the surface loading point. If a main fan is to be employed on the shaft then it will be situated in the bypass (fan drift) and an airlock becomes necessary at the shaft top. If no main fan is required at that location then a high volume, low pressure fan may be utilized simply to overcome the resistance of the fan drift and to ensure that the shaft top remains free from high air velocities.

The advantages for such air bypasses are as follows.

1. They avoid personnel being exposed to high air velocities and turbulence at loading/unloading points.
2. They reduce dust problems in rock-hoisting shafts.
3. They eliminate the high shock losses that occur when skips or conveyances are stationary at a heavily ventilated inset.

At the intersections between shafts and fan drifts or other main airways, the entrance should be rounded and sharp corners avoided. If air bypasses are not employed then the shaft should be enlarged and/or the underground inset heightened to ensure that there remains on adequate free-flow area with a conveyance stationary at that location.

5.5 AIR POWER

In chapters 2 and 3, we introduced the concept of mechanical energy within an airstream being downgraded by frictional effects to the less useful heat energy. We quantified this as the term F, the work done against friction in terms of joules per kilogram of air. In section 3.4.2 we also showed that a measurable consequence of F was a frictional pressure drop, p, where

$$F = \frac{p}{\rho} \quad \frac{\text{J}}{\text{kg}} \tag{5.52}$$

and ρ = mean density of the air.

The airpower of a moving airstream is a measure of its mechanical energy content. Airpower may be supplied to an airflow by a fan or other ventilating motivators but will diminish when the airflow suffers a reduction in mechanical energy through the frictional effects of viscous action and turbulence. The airpower loss, APL, may be quantified as

$$\text{APL} = FM \quad \frac{\text{J}}{\text{kg}} \frac{\text{kg}}{\text{s}} \tag{5.53}$$

where $M = Q\rho$ is the mass flowrate (kg/s). Then

$$APL = FQ\rho$$

However,

$$F = p/\rho$$

from equation (5.52), giving

$$APL = pQ \quad W \hspace{4cm} (5.54)$$

As both p and Q are measurable parameters, this gives a simple and very useful way of expressing how much ventilating power is dissipated in a given airway.

Substituting for p from the rational form of the square law (equation (2.50))

$$p = R_t\rho Q^2$$

gives

$$APL = R_t\rho Q^3 \quad W \hspace{3.5cm} (5.55)$$

This revealing relationship highlights the fact that the power dissipated in any given airway and, hence, the cost of ventilating that airway depend upon

1. the geometry and roughness of the airway,
2. the prevailing mean air density, and
3. the cube of the airflow.

The rate at which mechanical energy is delivered to an airstream by a fan impeller may also be written as the approximation

$$\text{air power delivered} = p_{ft}Q \quad W \hspace{2.5cm} (5.56)$$

where p_{ft} is the increase in total pressure across the fan. For fan pressures exceeding 2.5 kPa, the compressibility of the air should be taken into account. This will be examined in more detail in Chapter 10.

APPENDIX:

SHOCK LOSS FACTORS FOR AIRWAYS AND DUCTS

Shock loss (X) factors may be defined as the number of velocity heads that give the frictional pressure loss due to turbulence at any bend, variation in cross-sectional area, or any other cause of a change in the direction of airflow (equation (5.17)):

$$P_{shock} = X\rho \frac{u^2}{2} \quad Pa$$

The shock loss factor is also related to an equivalent Atkinson resistance by (equation (5.18))

$$R_{shock} = \frac{X\rho}{2A^2} \quad \frac{N\,s^2}{m^8}$$

or rational resistance by (equation (5.19))

$$R_{t,shock} = \frac{X}{2A^2} \quad m^{-4}$$

This appendix enables the X factor to be estimated for the more common configurations that occur in subsurface environmental engineering. However, it should be borne in mind that small variations in geometry can cause significant changes in the X factor. Hence, the values given in this appendix should be regarded as approximations.

For a comprehensive range of shock loss factors, reference may be made to the *Fundamentals Handbook* produced by the American Society of Heating, Refrigerating and Air Conditioning Engineers (ASHRAE).

Throughout this appendix, any subscript given to X refers to the branch in which the shock loss or equivalent resistance should be applied. However, in the case of branching flows, the conversion to an equivalent resistance $R_{sh} = X\rho/2A^2$ should employ the cross-sectional area, A, of the main or common branch. Unsubscripted X factors refer to the downstream airway.

A5.1 Bends

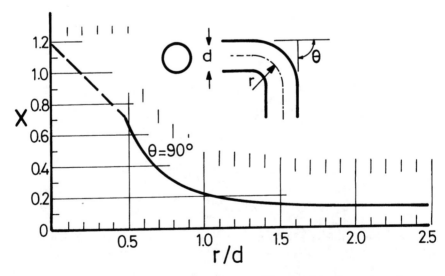

Figure 5A.1 Shock loss factor for right-angled bends of circular cross-section.

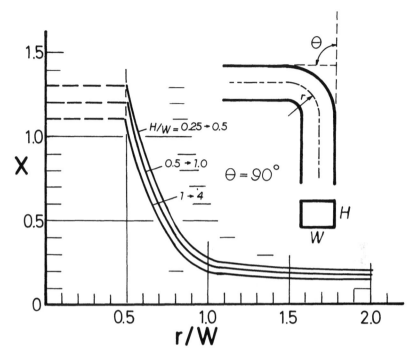

Figure 5A.2 Shock loss factor for right-angled bends of rectangular cross-section.

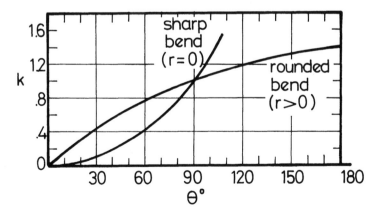

Figure 5A.3 Correction to shock loss factor for bends of angles other than 90°. $X_\theta = X_{90} \times k$. Applicable for both round and rectangular cross-sections.

A5.2 Changes in cross-section

(a) Sudden enlargement:

A = cross-sectional area

u = velocity

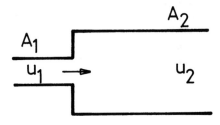

$$X_2 = \left[\frac{A_2}{A_1} - 1 \right]^2$$

where X_2 is referred to Section 2.

$$X_1 = \left[1 - \frac{A_1}{A_2} \right]^2$$

where X_1 is referred to Section 1. (Useful if A_2 is very large.)

(b) Sudden contraction:

$$X_2 = 0.5 \left(1 - \frac{A_2}{A_1} \right)^2$$

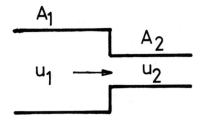

A5.3 Junctions

The formulae and graphs for shock loss factors at junctions involve the branch velocities or airflows, neither of which may be known at the early stages of subsurface ventilation planning. This requires that initial estimates of these values must be made by the planning engineer. Should the ensuing network analysis show that the estimates were grossly in error then the X values and corresponding branch resistances should be re-evaluated and the analysis run again.

(a) Rectangular main to diverging circular branch (e.g. raises, winzes)

$$X_2 = 0.5 \left[1 + 2.5 \, \frac{u_2}{u_1} \right]$$

$$R_{2,\text{shock}} = \frac{X_2 \rho}{2A_1^2} \quad \frac{\text{N s}^2}{\text{m}^8}$$

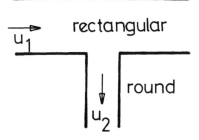

A5.4 Entry and exit

(a) Sharp-edged entry:

$$X = 0.5$$

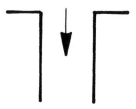

(b) Entrance to duct or pipe:

$$X = 1.0$$

(This is a real loss caused by turbulence at the inlet and should not be confused with the conversion of static pressure to velocity pressure)

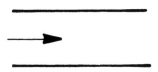

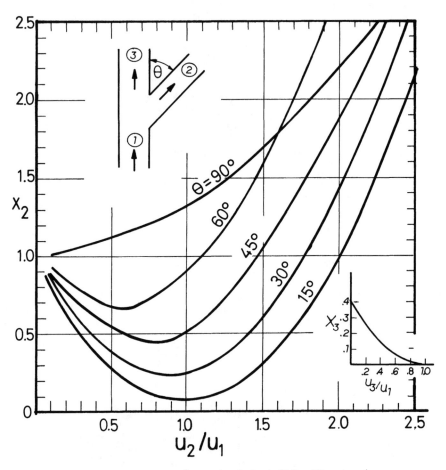

A = cross−sectional area (m²), u = air velocity (m/s) Condition: $A_1 = A_3$

$$R_2, \text{shock} = \frac{X_2\rho}{2A_1^2}\frac{\text{Ns}^2}{\text{m}^8}$$

$$R_3, \text{shock} = \frac{X_3\rho}{2A_1^2}\frac{\text{Ns}^2}{\text{m}^8}$$

Figure 5A.4 Circular main to diverging circular branch (e.g. fan drift from an exhaust shaft).

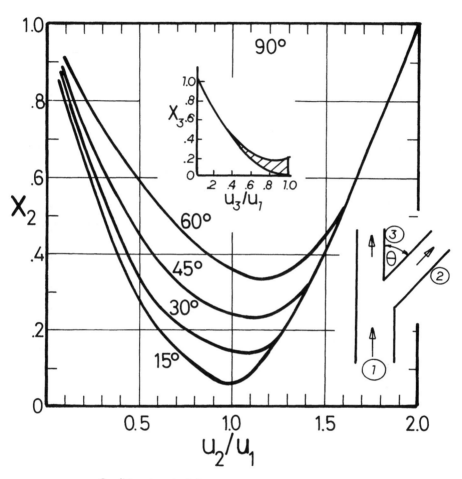

Condition: $A_1 = A_2 + A_3$

$$R_2, \text{shock} = \frac{X_2 \rho}{2A_1^2} \frac{Ns^2}{m^8}$$

$$R_3, \text{shock} = \frac{X_3 \rho}{2A_1^2} \frac{Ns^2}{m^8}$$

Figure 5A.5 Rectangular main to diverging rectangular branch.

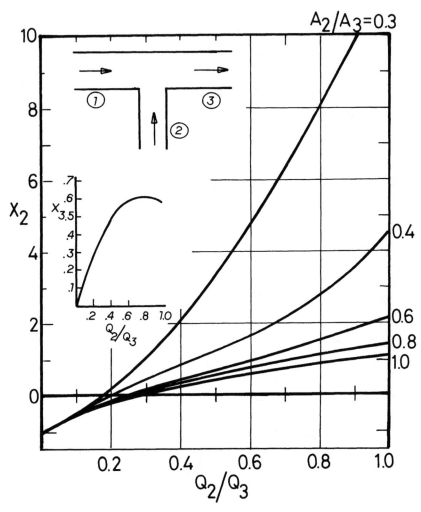

A = cross-sectional area (m^2), Q = airflow (m^3/s).

$$R_2, \text{shock} = \frac{X_2 \rho}{2A_1^2} \frac{\text{Ns}^2}{\text{m}^8}$$

$$R_3, \text{shock} = \frac{X_3 \rho}{2A_1^2} \frac{\text{Ns}^2}{\text{m}^8}$$

Figure 5A.6 Circular branch converging into circular main.

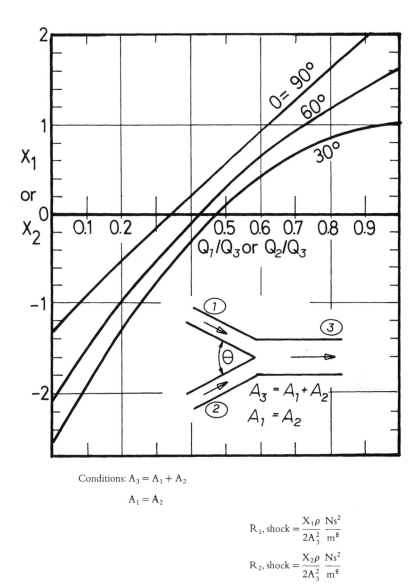

Conditions: $A_3 = A_1 + A_2$

$A_1 = A_2$

$$R_1, \text{shock} = \frac{X_1 \rho}{2A_3^2} \frac{Ns^2}{m^8}$$

$$R_2, \text{shock} = \frac{X_2 \rho}{2A_3^2} \frac{Ns^2}{m^8}$$

Figure 5A.7 Converging Y-junction, applicable to both rectangular and circular cross-sections. For a diverging Y-junction, use the branch data from Fig. 5A.5.

(c) Bell-mouth:

$$X = 0.03 \text{ for } r/D \geqslant 0.2$$

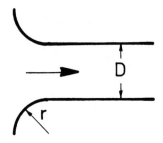

(d) Exit loss:

$$X = 1.0 \text{ (direct loss of kinetic energy)}$$

A5.5   Obstructions

(a) Obstruction away from the walls (e.g. free-standing roof support)

$$X = C_D \frac{A_b}{A}$$

where C_D = Coefficient of drag for the obstruction (see Figure 5.3), A_b = cross sectional area facing into the airflow, A = area of airway.

(b) Elongated obstruction

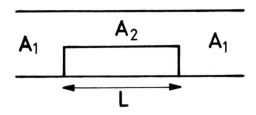

Ignoring interaction between constitutent shock losses, the effect may be approximated as the combination of the losses due to a sudden contraction, length of restricted airway and sudden enlargement. This gives

$$X_2 = 1.5 \left[1 - \frac{A_2}{A_1} \right]^2 + \frac{2kL \, per_2}{\rho A_2}$$

or, when referred to the full cross-section of the airway,

$$X_1 = X_2 \left[\frac{A_1}{A_2} \right]^2$$

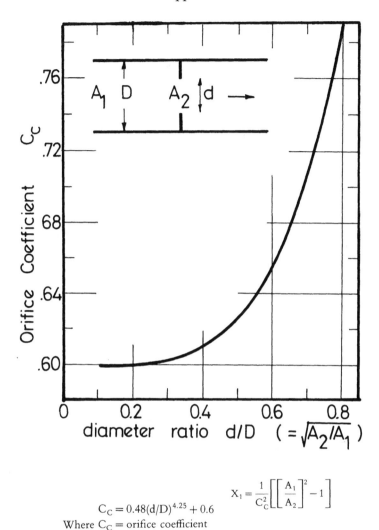

$$C_C = 0.48(d/D)^{4.25} + 0.6$$

$$X_1 = \frac{1}{C_C^2}\left[\left[\frac{A_1}{A_2}\right]^2 - 1\right]$$

Where C_C = orifice coefficient

Figure 5A.8 Orifice coefficients (of contraction) for sharp-edged circular or rectangular orifices.

where k = Atkinson friction factor for restricted length (kg/m³), per = perimeter in restricted length (m), ρ = air density (kg/m³).

A5.6 Interaction between shock losses

When two bends or other causes of shock loss in an airway are within ten hydraulic mean diameters of each other then the combined shock loss factor will usually not

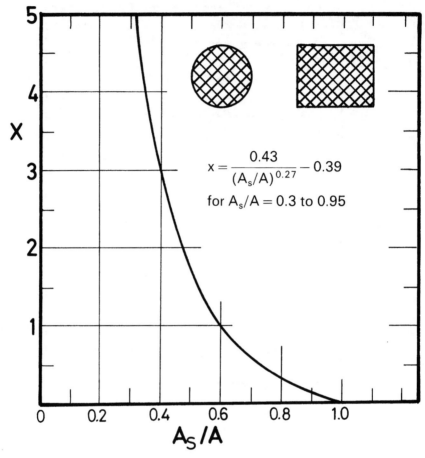

Figure 5A.9 Shock loss factor for screen in a duct. A = full area of duct, A_s = open flow area of screen.

be the simple addition of the two individual shock loss factors. Depending upon the geometrical configuration and distance between the shock losses, the combined X value may either be greater or lesser than the simple addition of the two. For example, a double bend normally has a lower combined effect than the addition of two single bends, whilst a reverse bend gives an elevated effect. Figure 5A.10 illustrates the latter situation. This figure utilizes the technique of assuming that the shock loss furthest downstream attains its normal value while any further shock within the relevant distance upstream will have its shock loss factor multiplied by an *interference factor*.

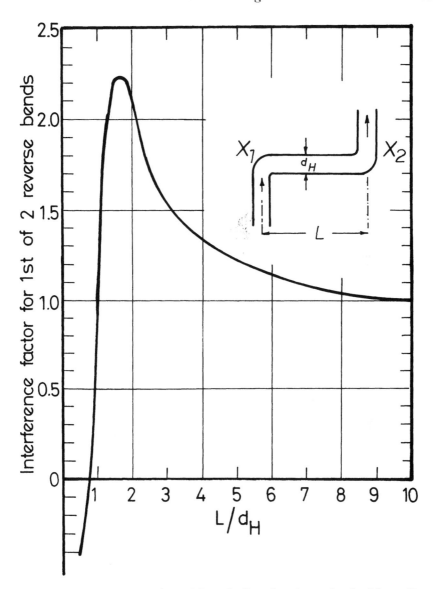

Figure 5A.10 Correction (interference) factor for first of two interacting shock losses. Combined shock loss $X_{comb} = CX_1 + X_2$, where X_1 and X_2 are obtained from the relevant graphs.

FURTHER READING

American Society of Heating, Refrigerating and Air Conditioning Engineers (1985) *Fundamentals Handbooks*, ASHRAE, 1791 Tullie Circle NE, Atlanta, GA 30329.

Atkinson, J. J. (1854) On the theory of the ventilation of mines. *Trans. N. Engl. Inst. Min. Eng.* **3**, 73–222.

Bromilow, J. G. (1960) The estimation and the reduction of the aerodynamic resistance of mine shafts. *Trans. Inst. Min. Eng.* **119**, 449–65.

McElroy, G. E. (1935) Engineering factors in the ventilation of metal mines. *US Bureau of Mines Bull.* 385.

McPherson, M. J. (1988) An analysis of the resistance and airflow characteristics of mine shafts. *4th Int. Mine Ventilation Congr., Brisbane.*

Stevenson, A. (1956) Mine ventilation investigations: (c) shaft pressure losses due to cages. *Thesis*, Royal College of Science and Technology, Glasgow.

6

Ventilation surveys

6.1 PURPOSE AND SCOPE OF VENTILATION SURVEYS

A ventilation survey is an organized procedure of acquiring data that quantify the distributions of airflow, pressure and air quality throughout the main flowpaths of a ventilation system. The required detail and precision of measurement and the rigour of the ensuing data analysis depend on the purpose of the survey. Perhaps the most elementary observation in underground ventilation is carried out by slapping one's clothing and watching the dust particles in order to ascertain the direction of a sluggish airflow.

Measurements of airflow should be taken in all underground facilities at times and places that may be prescribed by law. Even in the absence of mandatory requirements, a prudent regard for safety indicates that sufficient routine measurements of airflow be taken to ensure (a) that all working places in the mine receive their required airflows in an efficient and effective manner, (b) that ventilation plans are kept up to date and (c) to verify that the directions, quantities and separate identity of airflows throughout the ventilation infrastructure, including escapeways, are maintained. Similarly, routine measurements of pressure differentials may be made across doors, stoppings or bulkheads to ensure they are also maintained within prescribed limits and in the correct direction. The latter is particularly important in underground repositories for toxic or nuclear materials.

One of the main differences between a mine ventilation system and a network of ducts in a building is that the mine is a dynamic entity, changing continuously because of modifications to the structure of the network and resistances of individual branches. Regular measurements of airflow and pressure differentials underground are necessary as a basis for incremental adjustment of ventilation controls.

During the working life of a mine or other underground facility, there will be occasions when major modifications are required to be made to the ventilation system. These will include opening up new districts in the mine, closing off older ones, commissioning new fans or shafts, or interconnecting two main areas of the mine. The procedures of ventilation planning are detailed in chapter 9. It is important that planning the future ventilation system of any facility is based on reliable and verified data. Ventilation surveys that are carried out in order to establish a database for planning purposes must necessarily be conducted with a higher degree of organization, detail and precision than those conducted for routine control. This chapter is directed primarily towards those more accurate surveys.

A major objective of ventilation surveys is to obtain the frictional pressure drop, p, and the corresponding airflow, Q, for each of the main branches of the ventilation network. From these data, the following parameters may be calculated for the purposes of both planning and control:

1. distribution of airflows, pressure drops and leakage
2. airpower ($p \times Q$) losses and, hence, distribution of ventilation operating costs throughout the network
3. volumetric efficiency of the system
4. branch resistances ($R = p/Q^2$)
5. natural ventilating effects
6. friction factors

While observations of airflow and pressure differentials are concerned with the distribution and magnitudes of air volume flow, other measurements may be taken either separately or as an integral part of a pressure–volume survey in order to indicate the quality of the air. These measurements may include wet and dry bulb temperatures, barometric pressures, dust levels or concentrations of gaseous pollutants.

6.2   AIR QUANTITY SURVEYS

The volume of air passing any fixed point in an airway or duct every second, Q, is normally determined as the product of the mean velocity of the air, u, and the cross-sectional area of the airway or duct, A:

$$Q = u \times A \qquad \frac{m}{s} m^2 = \frac{m^3}{s}$$

Most of the techniques of observing airflow are, therefore, combinations of the methods available for measuring mean velocity and the cross-sectional area.

Prior to the invention of anemometers in the nineteenth century, the only practicable means of measuring rates of airflow in mines was to observe the velocity of visible dust or smoke particles suspended in the air. An even cruder old method was to walk steadily in the same direction as the airflow, varying one's pace, until a candle flame appeared to remain vertical. Modern instruments for the measurement of airspeed in mines divide into three groups depending on (i) mechanical effects, (ii) dynamic pressure and (iii) thermal effects.

6.2.1   Rotating vane anemometer

The vast majority of airspeed measurements made manually underground are gained from a **rotating vane** (windmill type) **anemometer**. When held in a moving air-

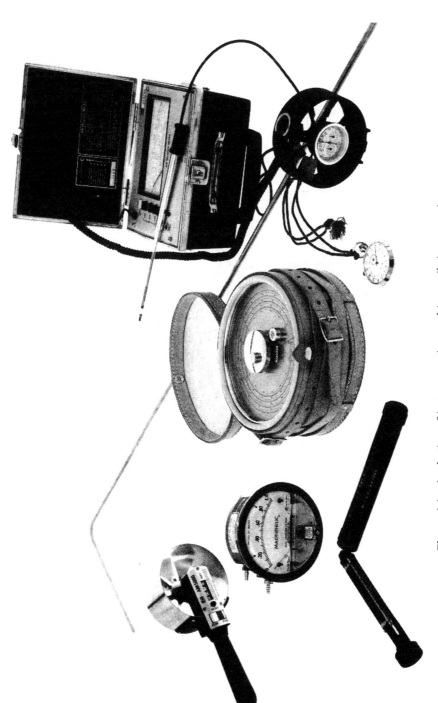

Figure 6.1 A selection of instrumentation used in ventilation surveying.

stream, the air passing through the instrument exerts a force on the angled vanes, causing them to rotate with an angular velocity that is closely proportional to the airspeed. A gearing mechanism and clutch arrangement couple the vanes to a pointer which rotates against a circular dial calibrated in metres (or feet). The instrument is used in conjunction with a stopwatch and actually indicates the number of 'metres of air' that have passed through the anemometer during a given time period. The clutch device is employed to stop and start the pointer while the vanes continue to rotate. A zero reset lever is also incorporated into the instrument. Low range vane anemometers will typically have eight vanes, jewelled bearings and give repeatable readings for velocities in the range 0.25 to 15 m/s. High range instruments may have four vanes, low friction roller or ball bearings and can be capable of measuring air velocities as high as 50 m/s. Digital vane anemometers are available that indicate directly on an odometer dial or an illuminated screen. Modern hand–held instruments may also be fitted with a microprocessor to memorize readings and to dampen out rapid variations in velocity or into which can be entered the cross–sectional area for the calculation of volume flow. Two types of vane anemometer are included in the selection of ventilation survey instruments shown on Fig. 6.1.

In order to obtain a reliable measure of the mean air velocity in an underground airway, it is important that a recommended technique of using the anemometer is employed. The following procedure has evolved from a combination of experiment and practical experience.

6.2.2 Moving Traverses

The anemometer should be attached to a rod at least 1.5 m in length, or greater for high airways. The attachment mechanism should permit the options of allowing the anemometer to hang vertically or to be fixed at a constant angle with respect to the rod. A rotating vane anemometer is fairly insensitive to yaw and will give results that do not vary by more than $\pm 5\%$ for angles deviating by up to 30° from the direction of the airstream. Hence, for most underground airways, allowing the anemometer to hang freely at the end of the rod will give acceptable results. For airways of inclination greater than 30°, the anemometer should be clamped in a fixed position relative to the rod and manipulated by turning the rod during the traverse such that the instrument remains aligned with the axis of the airway.

The traverse

The observer should face into the airflow holding the anemometer rod in front of him so that the instrument is facing him and at least 1.5 m upstream from his body. To commence the traverse, the instrument should be held in either an upper or a lower corner of the airway, with the pointer reset to zero, until the vanes have accelerated to a constant velocity. This seldom takes more than a few seconds. The observer should reach forward to touch the clutch control lever while a second observer with

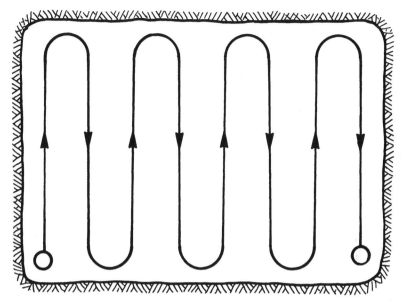

Figure 6.2 Path of an anemometer traverse.

a stopwatch counts backwards from five to zero. On zero, the anemometer clutch is activated releasing the pointer and, simultaneously, the stopwatch is started.

The path of the traverse across the airway should be similar to that shown on Fig. 6.2. The aim should be to traverse the anemometer at a constant rate not greater than about 15% of the airspeed. Ideally, equal fractions of the airway cross sectional area should be covered in equal times. This is facilitated by the stopwatch observer calling out the elapsed time at ten second intervals. The complete traverse should take not less than 60 s and may be considerably more for large or low velocity airways. The final five seconds should be counted down by the stopwatch observer during which time the traverse person stretches forward to disconnect the clutch at the end of the time period. The length indicated by the anemometer is immediately read and booked, and the instrument reset to zero.

The procedure is repeated, traversing in the opposite direction across the airway. Traverses should be repeated until three readings are obtained that agree to within $\pm 5\%$. In favourable steady-state conditions, experienced observers will often achieve repeatability to within $\pm 2\%$. Larger discrepancies may be expected in airways where there is a highly asymmetric variation in velocity across the airway, where the floor conditions are unstable, or when obstructions exist in the cross-section. Obviously, measuring stations should be chosen to avoid such difficulties wherever possible. Another highly annoying cause of discrepancy is the opening of a ventilation door during the period of measurement. At least two sets of traverses should be taken at different locations within each airway. Where cross-cuts or other leakage paths affect the airflow then a sufficient number of additional measurement points should be traversed in order to quantify the rate and direction of leakage.

Booking

The anemometer field book should be waterproof and laid out such that each double page has segments for

1. names of observers
2. the location of the measuring station, time, date
3. anemometer readings and corrections
4. dimensioned sketch of cross-section
5. calculation of area
6. calculation of air volume flow

The bookings are normally made by the stopwatch person. For each traverse, the anemometer reading is divided by the corresponding time to give the air velocity. The mean of these, ignoring any values outside the $\pm5\%$ tolerance, gives the observed mean velocity. In most cases, the time of each traverse at a station is the same, allowing the anemometer readings to be averaged before calculating the mean velocity. The observed mean velocity must then be corrected according to the calibration chart or curve for the instrument (section 6.4) to give the actual mean velocity, u.

The cross sectional area, A, is determined using one of the methods discussed in section 6.2.12. The calculation of airflow is then completed as

$$Q = u \times A \quad \frac{m}{s} m^2 = \frac{m^3}{s}$$

Anemometer traverses may also be employed at the ends of ducts. However, it is recommended that the technique should not be used for duct diameters less than six times the diameter of the anemometer.

6.2.3 Fixed point measurement

An estimate of duct airflow may be obtained by holding the anemometer at the centre of the duct and multiplying the corrected reading by a further correction factor of 0.8. A similar technique may be employed for routine check readings taken at well-established measuring stations in airways. The reading obtained from a stationary anemometer at a known location within the cross-section should, initially, be compared with that given from a series of traverses in order to obtain a 'fixed point' correction factor for that station. This is typically 0.75 to 0.8 for the fixed point located some one-half to two-thirds the height of the airway. Subsequent routine readings may be obtained simply by taking an anemometer reading at the fixed point and applying the appropriate calibration and fixed point corrections. Provided that the measuring station is well downstream of any bends or major obstructions and the airflow remains fully turbulent then the fixed point correction factor will remain near constant as the airflow varies.

6.2.4 Density correction

For precise work, anemometer readings may be further corrected for variations in air density:

$$u = u_i + C_c \sqrt{\frac{\rho_c}{\rho_m}} \tag{6.1}$$

where u = corrected velocity, u_i = indicated velocity, C_c = correction from instrument calibration curve or chart, ρ_c = air density at time of calibration and ρ_m = actual air density at time of measurement.

Equation (6.1) shows that the density adjustment is effectively applied only to the calibration correction and is ignored in most cases.

6.2.5 Swinging vane anemometer (velometer)

In its most fundamental form, the **swinging vane anemometer** (velometer) is simply a hinged vane which is displaced against a spring from its null position by a moving airstream. A connected pointer gives a direct reading of the air velocity. The air enters a port at the side of the instrument. This can be fitted with interchangeable orifices or probes to give a range of measurable velocities. Oscillations of the vane are reduced by the eddy current damping produced when an aluminium strip connected to the vane moves between two strong permanent magnets. The delicacy of the velometer together with its pronounced directional bias have limited its use in underground surveys. However, it can serve a useful purpose in giving spot readings as low as 0.15 m/s in gassy mines where hot wire probes are prohibited.

6.2.6 Vortex-shedding anemometer

For continuous monitoring systems, both rotating vane and swinging vane instruments with electrical outputs have been employed. However, they both require relatively frequent calibration checks when used in mine atmospheres. For this type of application, the **vortex-shedding anemometer** is preferred as it has no moving parts.

When any bluff object is placed in a stream of fluid, a series of oscillating vortices are formed downstream by boundary layer breakaway, first from one side of the body then the other. The propagation of the vortices is known as a Kármán street and can often be observed downstream from projecting boulders in a river. The rate of vortex production depends on the fluid velocity. In the vortex-shedding anemometer, the vortices may be sensed by the pulsations of pressure or variations in density that they produce. One apparent disadvantage noticed in practice is that when sited in a fixed location underground for monitoring purposes they require calibration for that specific location. They may also require electronic damping to eliminate large but short-lived variations in signals caused by the passage of vehicles.

6.2.7 Smoke tubes

Smoke tubes are perhaps the simplest of the mechanical techniques employed for measuring airflows and are used for very low velocities. A pulse of air forced by a rubber bulb through a glass phial containing a granulated and porous medium soaked in titanium tetrachloride or anhydrous tin will produce a dense white smoke. This is released upstream of two fixed marks in the airway. An observer with a spot-beam caplamp is located at each mark. The time taken for the cloud of smoke to travel the length of airway between the marks gives an indication of the centre-line velocity of the air. This must then be adjusted by a centre-line correction factor to give the mean velocity. The correction factor is usually taken to be 0.8 although a more accurate value can be calculated for known Reynolds' numbers. The length of airway should be chosen such that at least one minute elapses during the progression of the smoke between the two marks. The dispersion of the smoke cloud often causes the downstream observer some difficulty in deciding when to stop the stopwatch. Because of the uncertainties inherent in the technique, smoke tubes are normally employed as a last resort in slow moving airstreams.

6.2.8 Pitot-static tube

In section 2.3.2 we discussed the concepts of total, static and velocity pressures of a moving stream of fluid. A **pitot–static tube**, illustrated on Fig. 6.3, can be used to measure all three. This device consists essentially of two concentric tubes. When held facing directly into an airflow, the inner tube is subjected to the total pressure of the moving airstream, p_t. The outer tube is perforated with a ring of small holes drilled at right angles to the shorter stem of the instrument and, hence, perpendicular to the direction of air movement in the airway. This tube is, therefore, not influenced by the kinetic energy of the airstream and registers the static pressure only, p_s. A pressure gauge or manometer connected across the two tappings will indicate the difference between the total and static pressure, i.e. the velocity pressure (from equation (2.18)):

$$p_v = p_t - p_s \quad \text{Pa} \tag{6.2}$$

Furthermore, the velocity head is related to the actual velocity of the air, u (from equation (2.17)):

$$u = \sqrt{\frac{2p_v}{\rho}} \quad \text{m/s} \tag{6.3}$$

where ρ = actual density of the air (kg/m^3) (see equation (14.52) for air density).

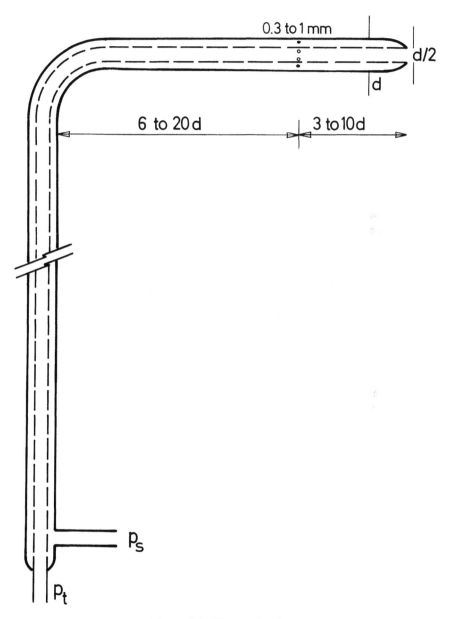

Figure 6.3 Pitot-static tube.

Pitot-static tubes vary widely in overall dimensions. For measuring air velocities in mine airways or at main fans, the longer stem may be some 1.5 m in length. Much smaller versions are available for use in ducts or pipes.

Modern pitot-static tubes reflect the total, static and velocity pressures of the airflow to an excellent degree of accuracy. Unfortunately, the precision of the measurement depends also on the manometer or pressure gauge connected to the tappings. This imposes a practical restriction on the lower limit of air velocity that can be measured by a pitot-static tube in the turbulent airflows of an underground system.

Example   If a diaphragm pressure gauge can be read to the nearest $\pm 1\,Pa$, then the lowest pressure that will give a 10% accuracy in the pressure reading is 10 Pa. Calculate the air velocity corresponding to a velocity pressure of 10 Pa, assuming an air density of 1.2 kg/m^3.

Solution

$$u = \sqrt{\frac{2p_v}{\rho}}$$

$$= \sqrt{\frac{2 \times 10}{1.2}} = 4.08\,m/s$$

As the great majority of underground openings have air velocities of less than 4 m/s, it is clear that the use of the pitot-static tube for the measurement of air velocity is limited to ventilation ducting and a few high velocity airways; primarily fan drifts and evasees, ventilation shafts, some longwall faces and trunk airways.

One of the difficulties observed when using a pitot-static tube for the spot measurement of pressures or velocities in a turbulent airstream is the oscillation in the readings. A small wad of cotton wool inserted into the flexible pressure tubing between the pitot-static tube and the pressure gauge damps out the short term variations. However, the cotton wool should not be so tightly tamped into the tubing that the gauge reaction becomes unduly slow. Electronic diaphragm gauges are often fitted with an internal damping circuit.

6.2.9   Fixed point traverses

The rotating vane anemometer is an integrating device, accumulating the reading as it is traversed continuously across an airway or duct. Most other instruments for the measurement of air velocity, including the pitot-static tube, do not have this advantage but are confined to giving a single spot reading at any one time. In order to find the mean velocity in an airway from pitot-static tube readings it is, therefore, necessary to take spot measurements at a number of locations over the cross-section.

This procedure is known by the contradictory sounding term 'fixed point traverse'. Differing techniques of conducting such traverses vary in the number of observations, locations of the instrument and treatment of the data. Three of these techniques are described here. In all cases, the fixed point traverse method assumes that the distribution of flow over the cross-section does not vary with time. For permanent monitoring stations, a grid of multiple pitot-static tubes may be left in place.

Method of equal areas

In this method, the cross-section of the duct or airway is divided into subsections each of equal area. Figure 6.4 shows a rectangular opening divided into 25 equal subsections similar in shape to the complete opening. Using a pitot-static tube or anemometer, the velocity at the centre of each subsection is measured. The mean velocity is then simply the average of the subsection velocities.

There are a few precautions that should be taken to ensure satisfactory results. First, if a pitot-static tube is employed then the velocity at each subsection should be calculated. Averaging the velocity pressure before employing equation (6.3) will not give the correct mean velocity. Secondly, it will be recalled that the velocity gradient changes most rapidly near the walls. Hence, accuracy will be improved if the velocities for the subsections adjacent to the walls and, especially, in the corners are determined from a number of readings distributed within each of those subsections. Third, the number of subsections should increase with respect to the size of the airway in order to maintain accuracy. As a guide, an approximation to the recommended

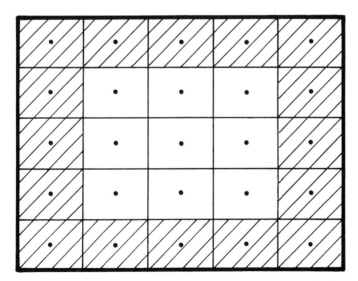

Figure 6.4 Measuring points for a fixed point traverse in a rectangular opening. The velocity in each shaded subsection should be the average from several readings distributed over the subsection.

Table 6.1   Number of measuring points on each diameter
of a circular opening

Diameter of duct (m)	< 1.25	1.25–2.5	> 2.5
Number of points	6	8	12

number of points, n, for a rectangular opening may be calculated from

$$n = 100e^{-8/A} + 23 \qquad (6.4)$$

where e is the base of natural logarithms, 2.7183, and A is the cross-sectional area
(m^2).

The calculated number of points may then be rounded to a value that is convenient
for subdividing the cross-sectional area but should never be less than 24. Correct
positioning of the measuring instrument is facilitated by erecting a grid of fine wires
in the airway to represent the subsections.

In the case of circular openings, the method of equal areas divides the circle into
annuli, each of the same area. Readings should be taken at points across two diameters
and the corresponding velocity profiles plotted. Should those profiles prove to be
skewed then readings should be taken across two additional diameters. The number
of measuring points recommended on each diameter is given in Table 6.1. Figure 6.5
illustrates an eight-point traverse on each of four diameters.

The locations of the points are at the centre of area of the relevant annulus on each
diameter and may be calculated from

$$r = D \sqrt{\frac{2n - 1}{4N}} \quad \text{m} \qquad (6.5)$$

where r = radius of point n from the centre, n = number of the point counted
outwards from the centre, D = diameter of the duct (m) and N = number of points
across the diameter.

Table 6.2 gives locations of points for 6-, 8- and 12-point traverses in terms of
fractions of duct diameter measured from one side.

Where a pitot-static tube traverse is to be conducted across a duct from the
outside then a clamping device should be attached to the outer surface of the duct
to hold the pitot-static tube firmly in place. The positions of measurement should
be marked on the stem of the instrument using Table 6.2 or 6.3. After each relocation
of the measuring head, the pitot-static tube should be yawed slightly from side to
side until the orientation is found that gives the greatest reading of total or velocity
pressure. The head of the instrument is then aligned directly into the airstream.

Log–linear traverse

A more accurate method of positioning points of measurement along the diameters
of a circular duct has been derived from a consideration of the logarithmic law

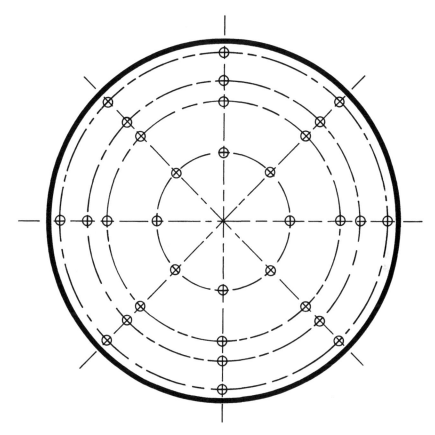

Figure 6.5 Measuring positions for an eight-point transverse on four diameters of a circular duct.

equations that describe the velocity profile for turbulent flow. The effects of observational errors are minimized when the points are located according to this method, known as the log–linear traverse. The corresponding locations are given in Table 6.3.

Velocity contours

One of the difficulties that besets ventilation engineers in measuring large scale airflows is that conditions are often not conducive to good accuracy. 'Textbook' advice is to choose measuring stations well away from obstructions, bends or changes in cross-section. Unfortunately, this is not always possible, especially when measuring airflows at the inlets or outlets of fans. It is not uncommon to find that longitudinal swirl in a fan drift or re-entry in an evasee causes the air to move in the wrong direction within one part of the cross-section. Similarly, in obstructed but high velocity airways underground such as many longwall faces, complex airflow patterns

Table 6.2 Positions of measuring points in a circular duct using the method of equal areas

Number of measuring points on each diameter	*Fractions of one diameter measured from side of duct*											
6		0.044	0.146	0.296		0.704	0.854	0.956				
8	0.032	0.105	0.194	0.323		0.677	0.806	0.895	0.968			
12	0.021	0.067	0.118	0.177	0.250	0.356	0.644	0.750	0.823	0.882	0.933	0.979

Table 6.3 Log–linear traverse positions of measuring points in a circular duct

Number of measuring points on each diameter	*Fractions of one diameter measured from side of duct*											
6		0.032	0.135	0.321		0.679	0.865	0.968				
8	0.021	0.117	0.184	0.345		0.655	0.816	0.883	0.978			
12	0.014	0.075	0.114	0.183	0.241	0.374	0.626	0.759	0.817	0.886	0.925	0.986

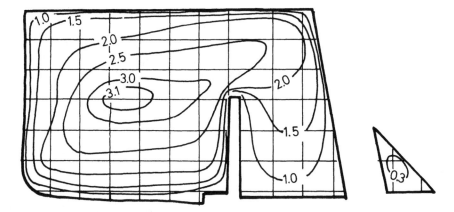

Figure 6.6 Velocity contours at an airflow measuring station (m/s).

may exist. A useful rule of thumb is that the averaging of spot velocities from a pitot-static tube traverse is acceptable if more than 75% of the velocity pressures (p_v) are greater than the maximum p_v divided by 10.

For difficult cases, the construction of velocity contours can provide both a visual depiction of the flow pattern and also a means of quantifying airflow. A scale drawing of the measurement cross-section is made on graph paper. A grid of fine wires is constructed in the airway to define the points of measurement. The number of points should be not less than those defined in the previous subsection. The greater the number of measurement points, the more accurate will be the result. The velocities at the corresponding points of measurement are entered on the graph paper and contour lines of equal velocity (isovels) are constructed. Figure 6.6 shows an example of velocity contours obtained on a longwall face.

The area enclosed by each contour is determined either by planimeter (if the scale drawing is large enough for good planimeter accuracy) or by the rudimentary method of counting squares on the graph paper. Fully automated systems have been devised to obtain the data by scanning the actual cross-section and these produce quantified velocity contour diagrams. However, the expense of such systems is seldom justified other than in research and testing laboratories.

By difference, the area of the band between each contour is evaluated and may be multiplied by the mean of the bounding velocities and the area scale factor to give the airflow for that band. Provided that the outermost contour is close to the walls then the velocity at the walls may be taken as zero. The sum of all band airflows gives the total flow for the airway.

6.2.10 Hot body anemometers

When any heated element is placed in a moving fluid, heat energy will be removed from it at a rate that depends on the rate of mass flow over the element.

In the **hot wire anemometer** a wire element is sited within a small open-ended cylinder to give the instrument a directional bias. The element forms one arm of a Wheatstone bridge circuit. In most hot wire anemometers, the temperature of the element is maintained constant by varying the electrical current passing through it as the air velocity changes. In other designs, the current is kept constant and the temperature (and, hence, electrical resistance) of the element is monitored. Modern hot wire anemometers are compensated for variations in ambient temperature and most also indicate dry bulb temperature. For precise work, readings should be corrected for air density:

$$u = u_i \frac{\rho_c}{\rho_m} \qquad (6.5)$$

where u = true air velocity, u_i = indicated air velocity, ρ_c = air density at calibration (usually $1.2\,\text{kg/m}^3$) and ρ_m = actual air density at time of measurement.

Hot wire anemometers are particularly useful for low velocities and are reliable down to about 0.1 m/s. They are convenient for fixed point traverses in slow moving airstreams. If a hot wire anemometer is to be used in a gassy mine then a check should first be made on the permissibility of that instrument for use in potentially explosive atmospheres.

The **Kata thermometer**, described in section 17.4.3 as a means of measuring the cooling power of an airstream, can also be used as a non-directional device to indicate low air velocities, typically in the range 0.1 to 1 m/s.

The main bulb of the Kata thermometer is heated until the alcohol level is elevated above the higher of the two marks on the stem. When hung in the airstream, the time taken for the alcohol level to fall between the two marks, coupled with the Kata index for the instrument and the air temperature, may be used to determine the non-directional air velocity. The Kata thermometer is seldom used for underground work (except in South Africa) because of its fragility.

6.2.11 Tracer gases

The rate at which injected gases are diluted provides a means of measuring air volume flow without the need for a cross-sectional area. The method is particularly useful for difficult situations such as leakage flow through waste areas, main shafts and other regions of high velocity and excessively turbulent flow, or total flow through composite networks of airways.

Hydrogen, nitrous oxide, carbon dioxide, ozone, radioactive krypton 85 and sulphur hexafluoride have all been used, with the latter particularly suitable for leakage or composite flows. The gas chosen should be chemically inert with respect to the mineralization of the strata.

There are two techniques of using tracer gases for the measurement of airflow. For high velocity airways, the tracer gas may be released at a monitored and steady rate M_g (kg/s). At a point sufficiently far downstream for complete mixing to have occurred, samples of the air are taken to establish the steady state concentration of the tracer gas. Then

$$C = \frac{q_g}{Q} = \frac{M_g}{\rho_g Q} \tag{6.6}$$

or

$$Q = \frac{M_g}{C\rho_g} \quad \frac{m^3}{s} \tag{6.7}$$

where q_g = volume flow of tracer gas (m³/s), Q = airflow (m³/s), C = downstream concentration of tracer gas (fraction by volume) and ρ_g = density of tracer gas at ambient pressure and temperature. It is assumed that the volume flow of tracer gas is negligible compared with the airflow.

In the case of sluggish or composite flows, a known mass of the tracer gas, M (kg), is released as a pulse into the upstream airflow. At the downstream station, the concentration of tracer gas is monitored and a concentration (C) vs. time (t) graph is plotted as shown on Fig. 6.7. Now, the concentration C is given as the ratio of the volume flowrate of gas, q_g, and the airflow, Q:

$$C = \frac{q_g}{Q} \quad \text{(fraction, by volume)}$$

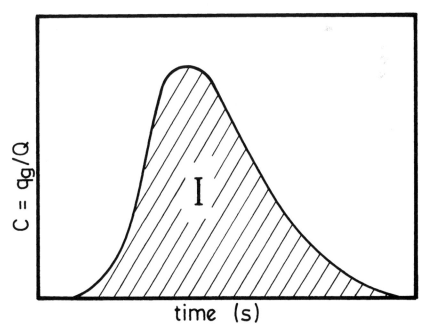

Figure 6.7 Concentration–time curve at a tracer gas monitoring station.

However,

$$q_g = \frac{M_g}{\rho_g} \quad \frac{m^3}{s}$$

where M_g = mass flowrate of gas at the monitoring station (kg/s) and ρ_g = density of gas at the prevailing temperature and pressure, giving

$$C = \frac{M_g}{\rho_g Q}$$

Hence, the complete area under the curve of C against t is

$$I = \int_0^\infty C \, dt$$

$$= \frac{1}{\rho_g Q} \int_0^\infty M_g \, dt$$

However, the total mass of gas released, M, must also be equal to $\int_0^\infty M_g \, dt$, giving

$$I = \frac{M}{\rho_g Q}$$

or

$$Q = \frac{M}{I \rho_g} \quad m^3/s \tag{6.8}$$

6.2.12 Measurement of cross-sectional area

As the vast majority of airflows are determined as the product of a mean velocity and a cross-sectional area, the accuracy of the airflow depends equally on the measured velocity and cross-sectional area. There is little point in insisting upon meticulous procedures for the measurement of mean velocity unless the same care is applied to finding the cross-sectional area.

By far the most common method of measuring airway area is by simple taping. This will give good results where the opening is of regular geometric shape such as a rectangle or circle. Airflow measuring stations should, wherever possible, be chosen where the airway profile is well defined. The frames of removed ventilation control doors provide excellent sites for airflow measurement. Shapes such as arched profiles or trapeziums may be subdivided into simple rectangles, triangles and segments of a circle, and appropriate taped measurements taken to allow the area to be calculated.

Inevitably, there are many situations in which airflows must be determined in less well-defined cross-sections. Several techniques are available for determining the corresponding cross-sectional area. For shapes that approximate to a rectangle, three or more heights and widths may be taped to find mean values of each. Care should be taken in such circumstances to make allowance for rounding at the corners. This often occurs as a result of spalled rock accumulating on the floor at the sides of airways.

A more sophisticated technique is the **offset method** in which strings are erected that define a regular shape within the airway. These are usually two vertical and two horizontal wires encompassing a rectangle. Taping from the wires to the rock walls at frequent intervals around the perimeter allows a plot of the airway profile to be constructed on graph paper.

The **profilometer** is a plane-table device. A vertical drawing board is attached to a tripod in the middle of the airway. Taped measurements made from the centre of the board to points around the rock walls may be scaled down mechanically or manually to reconstruct the airway profile on the drawing board.

The **photographic method** entails painting a white line around the perimeter of the measuring station. A linear scale such as a surveyor's levelling staff is fixed vertically within the defined profile. A camera is located such that it is aligned along a longitudinal centre-line of the airway and with its lens equidistant from all points on the painted line. These precautions reduce perspective errors. The area within the white line may be determined by overlaying the resulting photograph with transparent graph paper.

These time-consuming methods tend to be employed for permanent measuring stations rather than for temporary survey stations.

In all cases, the cross-sectional area of conveyors, ducts or other equipment should be determined and subtracted from the overall area of the airway.

6.3 PRESSURE SURVEYS

The primary purpose of conducting pressure surveys is to determine the frictional pressure drop, p, that corresponds to the airflow, Q, measured in each branch of a complete survey route. There are essentially two methods. The more accurate is the gauge and tube or trailing hose method, in which the two end stations are connected by a length of pressure tubing and the frictional pressure drop is measured directly. The second method, of which there are several variations, involves observing the absolute pressure on a barometer or altimeter at each station.

Although tradition within individual countries tends to favour one or other of the two methods, both have preferred fields of application. In general, where foot travel if relatively easy between measuring stations, the gauge and tube method can be employed. Where access is difficult as in multilevel workings or in shafts then the barometer method may become more practicable.

6.3.1 Gauge and tube surveys

Figure 6.8 illustrates the principles of gauge and tube surveying. A pressure gauge is connected into a length of tubing whose other ends are attached to the total head tappings of pitot-static tubes sited at the end stations. In practice, of course, the tubing

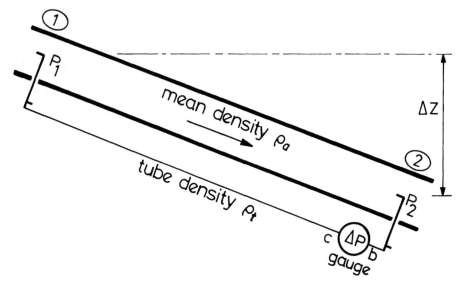

Figure 6.8 Measuring the frictional pressure drop between two stations by the gauge and tube method. c = high pressure tapping, b = low pressure tapping.

and instrumentation are all within the airway. Let us deal first with the essential theory of the method before discussing the practical procedure of gauge and tube surveying.

Theory

From the steady-flow energy equation (3.25) for an airway between stations 1 and 2, and containing no fan,

$$\frac{u_1^2 - u_2^2}{2} + (Z_1 - Z_2)g = \int_1^2 V\,dP + F_{12} \quad \text{J/kg} \tag{6.9}$$

where u = air velocity (m/s), Z = height above mine datum (m), g = gravitational acceleration (m/s²), V = specific volume of air ($= 1/\rho$) (m³/kg) and F_{12} = work done against friction (J/kg). If we assume a linear variation in air density between stations 1 and 2 then we can adopt an arithmetic mean value of density for the airway, $\rho_a = 1/V$. Furthermore, the frictional pressure drop referred to that density is given by equation (2.46) as

$$p_{12} = \rho_a F_{12} \quad \text{Pa} \tag{6.10}$$

Applying these conditions to equation (6.9) gives

$$p_{12} = \rho_a F_{12} = \rho_a \frac{u_1^2 - u_2^2}{2} + \rho_a(Z_1 - Z_2)g - (P_2 - P_1) \quad \text{Pa} \tag{6.11}$$

(See section 3.4.1 for a fuller explanation of this equation.) However, velocity pressure

$$p_v = \frac{\rho_a u^2}{2}$$

and static pressure

$$p_s = \rho_a Zg + P$$

when referred to a common datum.

Hence, equation (6.11) may be written as

$$p_{12} = (p_{v1} + p_{s1}) - (p_{v2} + p_{s2}) \quad \text{Pa}$$
$$= p_{t1} - p_{t2} \quad \text{Pa} \tag{6.12}$$

where p_t = total pressure $(p_v + p_s)$ as sensed by the total head tapping of a pitot-static tube.

This shows that the frictional pressure drop, p_{12}, referred to the mean density in the airway between the pitot-static tubes is given simply as the pressure gauge reading shown as ΔP in Fig. 6.8. If that measured frictional drop is to be referred to a standard value of air density, ρ_{st}, in order to compare or compound it with frictional pressure drops measured in other airways then the correction is given as

$$p_{12}(\text{standardized}) = p_{12}\frac{\rho_{st}}{\rho_a}$$

In the great majority of cases no further calculation is required. This explains why the gauge and tube technique is termed a direct method of measuring the frictional pressure drop in an airway.

To this time, the gauge and tube technique has seldom been used for shafts or highly inclined airways although the often difficult task of measuring shaft resistance would be greatly facilitated by leaving a length of small bore pressure tubing permanently in the shaft. Where the ends of the tube are at significantly different elevations a complication does, however, arise. It is found that, in these circumstances, the reading depends on the location of the gauge and increases as the elevation of the gauge decreases within the airway. This phenomenon occurs because the air within the tubing is stationary and, hence, not affected by friction. The pressure at all points within the tube is, therefore, slightly higher than outside the tube at corresponding elevations. If the temperature and moisture content inside and outside the tube are the same at corresponding points it follows that the mean density in the tube must be a little higher than that in the airway.

If the gauge is to indicate the frictional pressure drop referred to the mean density in the airway, ρ_a then it must be located at the position of mean density. In shafts or other airways of constant slope and resistance, this is very close to the midpoint. However, this is usually difficult to arrange and it is more practicable to locate the pressure gauge either at the top or bottom of the shaft. A correction must then be applied to the reading in order to arrive at the frictional pressure drop referred to the mean air density.

Consider, again, Fig. 6.8 with the gauge at the lower extremity of the tubing. The pressure in the tubing at the high pressure tapping is P_c and that at the low pressure tapping is P_b. The gauge reads $\Delta P = P_c - P_b$. However, the pressure P_c must equal the total pressure at station 1 plus the pressure due to the head of static air within the tube:

$$P_c = P_1 + \frac{\rho_1 u_1^2}{2} + \Delta Z g \rho_t$$

where P = barometric (static) pressure, u = air velocity, ΔZ = difference in elevation between stations 1 and 2 and ρ_t = mean air density within the tube. Also, if the gauge is at the elevation of station 2

$$P_b = P_2 + \frac{\rho_2 u_2^2}{2}$$

Then

$$\Delta P = P_c - P_b = \frac{\rho_1 u_1^2 - \rho_2 u_2^2}{2} + \Delta Z g \rho_t + P_1 - P_2 \qquad (6.13)$$

Now, from equation (6.11), the frictional pressure drop in the airway, referred to the airway mean density, ρ_a, is given as

$$p_{12} = \rho_a \frac{u_1^2 - u_2^2}{2} + \Delta Z g \rho_a + P_1 - P_2 \qquad (6.14)$$

Hence, the 'error' in the gauge reading becomes

$$\varepsilon = \Delta P - p_{12} = \frac{\rho_1 u_1^2 - \rho_2 u_2^2 - (u_1^2 - u_2^2)\rho_a}{2} + \Delta Z g (\rho_t - \rho_a) \qquad (6.15)$$

If we substitute for

$$\rho_a = \frac{\rho_1 + \rho_2}{2} \quad \text{(mean density in airway)}$$

$$\rho_t = \frac{\rho_1 + \rho_c}{2} \quad \text{(mean density in tube)}$$

where ρ_c = density in tube at position c and

$$\rho_c = \rho_2 + \frac{\Delta P}{P_2} \rho_2$$

(for the same temperature at c and b) then, after some algebraic simplification, we obtain

$$\varepsilon = \frac{\Delta Z g \Delta P \rho_2}{2P_2} - \frac{u_1^2 + u_2^2}{4}(\rho_2 - \rho_1) \qquad (6.16)$$

giving

$$p_{12} = \Delta P - \varepsilon = \Delta P \left(1 - \frac{\Delta Z g \rho_2}{2 P_2} \right) + \frac{u_1^2 + u_2^2}{4} (\rho_2 - \rho_1) \qquad (6.17)$$

for a reading taken at the base of the shaft.

This is the full form of the equation that allows the reading on the gauge at the base of the shaft or slope to be corrected to mean density for the airway. The term

$$\frac{\Delta Z g \rho_2}{2 P_2}$$

arises from the difference in mean air density between the airway and the tubing (equation (6.15)), while

$$\frac{u_1^2 + u_2^2}{4} (\rho_2 - \rho_1)$$

is the result of converting the velocity pressures at ρ_1 and ρ_2 to the mean density ρ_a. To be precise, this latter term should be applied even when the airway is level. However, it is normally insignificant and may be ignored for practical purposes, giving

$$p_{12} = \Delta P \left(1 - \frac{\Delta Z g \rho_2}{2 P_2} \right) \qquad (6.18)$$

When the gauge is located at the top of the shaft or slope, similar reasoning leads to

$$p_{12} = \Delta P (\text{top}) \left(1 + \frac{\Delta Z g \rho_1}{2 P_1} \right) + \frac{(u_1^2 + u_2^2)(\rho_2 - \rho_1)}{4} \qquad (6.19)$$

Again, the kinetic energy term involving u values can usually be neglected.

The same equations may be employed for an upcast shaft or ascentional slope. However, for this application, stations 1 and 2 remain the top and bottom locations respectively with air flowing from 2 to 1.

Approximate but simpler relationships for the corrections required due to gauge location in a shaft were derived by Hinsley in 1962:

$$p_{12} = \Delta P \times \frac{P_m}{P_2} \qquad (6.20)$$

for the gauge at the bottom, and

$$p_{12} = \Delta P \times \frac{P_m}{P_1} \qquad (6.21)$$

for the gauge at the top, where P_m = mean barometric pressure in the shaft, $(P_1 + P_2)/2$.

Practical procedure

The procedure for conducting a gauge and tube survey commences by assembling the equipment and calibrating the gauges. For convenience, a list of the required equipment is given here, together with some explanatory comments.

1. Two pitot-static tubes, approximately 1.25 m in length. Shorter instruments may be employed for small airways or for use in ducts.
2. A range of diaphragm pressure gauges varying from a full-scale deflection of 100 Pa to the highest pressure developed by any fan in the system. The gauges should be calibrated in the horizontal position against a primary manometer immediately prior to an important survey. The use of diaphragm gauges rather than inclined manometers has greatly improved the speed of gauge and tube surveying.
3. A continuous length of nylon or good quality plastic tubing between 100 and 200 m in length. The tubing should be mechanically strong so that it can withstand being run over by rubber-tyred vehicles or being dragged under doors without permanent damage. An internal diameter from 2 to 3 mm is convenient. Larger tubing may become difficult to handle while the waiting time of transmission of a pressure wave may become unduly long if the tube is too narrow. The tube should be pressure tested before and after the survey.
4. Short lengths of flexible tubing to connect the pitot-static tubes and gauge to the main tubing. Metal connectors and clamps should also be carried in case it becomes necessary to repair damage to the main tubing.
5. Two or three cans of spray paint for station marking. Chalk or industrial type crayons can also be used.
6. A pocket barometer and a whirling hygrometer.
7. A waterproof field book and pencils.
8. A 100 m flexible measuring tape.
9. Tool kit containing screwdrivers, adjustable spanners (wrenches) and a sharp knife.

The route of the traverse and sites of main junction stations should have been established before commencing the observations (section 6.4). Two persons are required for a gauge and tube survey. It is helpful to have the 100 Pa gauge fixed within a box with a transparent top, and side holes for extended pressure tappings. Straps around the waist and neck of the observer hold the gauge in a horizontal position.

At the starting station, the pressure tubing is unwound and laid out along the airway in the direction of the second main station. At the forward position, the zero setting of the gauge is checked and, if necessary, adjusted by connecting the high and low pressure tappings by a short length of tubing. The gauge is then connected in-line between the main tube and the total head tapping of a leading pitot-static tube as illustrated on Fig. 6.8. At the rear position, the second pitot-static tube is similarly connected to the pressure tubing. The flexible tubing used for connections should be of an internal diameter that fits snugly on to the main tube, the gauge tappings and the pitot-static tube without requiring undue force.

To make the observation, both pitot-static tubes are held facing into the airflow, away from the body of the observer and at a position between one-half and two-thirds the height of the airway. The gauge is observed until the reading becomes constant. This may take two to three minutes depending on the length and diameter of the main tube. Light tapping with the fingers may assist in overcoming any slight frictional resistance of the diaphragm or linkages within the gauge. On completing the gauge reading, the leading observer should indicate that fact to the trailing observer either by caplamp signals or by a tug on the tube. The barometric pressure and wet and dry bulb temperatures are also read and booked by the leading observer together with the distance between observers. In most cases this is the known length of the main tube. For shorter distances, the measuring tape or other means should be used to determine the actual length.

The final duty of the leading observer is to paint or chalk an indicator mark on the rail or airway side. A second tug on the tube or a caplamp signal indicates that it is time to move on. The leading observer walks forward, dragging the tube behind him. When the trailing observer reaches the indicator mark he simply stops, grasping the main tube firmly.

The procedure is repeated for each tube length until the next main (junction) station is reached. The leading observer is kept busy while the trailing observer has little to do other than holding a pitot-static tube at each station or substation and walking forward. However, it is preferable that the observers exchange positions only in alternate shifts rather than during any one day. An experienced team can progress along a traverse route fairly quickly. Indeed, using modern equipment, it is usually the measurement of airflows by the accompanying airflow team rather than frictional pressure drops that dictates the overall speed of the survey (see section 6.4).

Each major junction of airways should be a main station within the gauge and tube traverse. At each of those junctions, the pitot-static tube should be held at the centre of the junction. If high turbulence causes excessive fluctuations on the gauge then the static tapping(s) on the pitot-static tube(s) may be employed. In this case, an anemometer should be held at the position of the pitot-static tube(s) to measure the local velocity. The corresponding velocity pressures should be calculated and applied as a correction to the gauge reading in order to determine the frictional drop in total pressure.

Care should be taken at all times to ensure that the pitot-static tubes do not become clogged by dust or other debris. Similarly, in wet conditions, it is vital to take precautions against water entering any tube. Pitot-static tubes or the open ends of pressure tubing should never be allowed to fall on to the floor during a traverse.

During the course of a pressure traverse, check readings should be taken of the pressure differences across doors between airways. It is convenient to carry a separate 10 m length of flexible tubing for this purpose. It takes only a few seconds to attach a gauge of the required range. If in doubt concerning the range, a high pressure gauge should be used first to establish an approximate pressure difference, then exchanged for a more appropriate instrument. It is usually sufficient to measure the static pressure across a door. Hence, the two ends of the tubing should be protected against the very local air velocities that sometimes occur from leakage close to a door. A practical way of doing this is simply to insert the end of the tube into one's pocket.

6.3.2 Barometer and altimeter surveys

If the absolute static pressures are measured on barometers at the two ends of a subsurface airway then the difference between those two measured pressures will depend on

1. the difference in elevation between the stations,
2. the air velocities and
3. the frictional pressure drop between the two stations at the prevailing airflow.

As the elevations and velocities can be measured independently, it follows that the barometric readings can be used to determine the frictional pressure drop.

The concept of barometers was introduced in section 2.2.4. The instruments used for mine barometric surveys are temperature-compensated microaneroid devices. The facia of the instruments are normally calibrated in kilopascals or other units of pressure. However, it will be recalled that pressure may be quoted in terms of the column of air (or any other fluid) above the point of measurement (see equation (2.8)).

$$P = \rho g h \quad \text{Pa} \tag{6.22}$$

If the air density, ρ, and gravitational acceleration, g, are regarded as constant then the pressure may be quoted in head (metres) or air, h. This relationship is utilized to inscribe the facia of some aneroids in terms of metres (or feet) of air column:

$$h = \frac{P}{\rho g} \quad \text{m} \tag{6.23}$$

The instrument then indicates an approximate elevation or altitude within the earth's atmosphere relative to some datum and, accordingly, is then called an altimeter. For a more accurate elevation, the reading should be corrected for the difference between the actual values of ρ and g and the standard values to which the altimeter has been calibrated. Some sophisticated altimeters have an in built bias which compensates for changes in air density with respect to height. This allows a linear scale for altitude.

In most mining countries, barometric pressure surveys are carried out using direct indicating barometers. In the United States, altimeters are commonly employed. However, if equation (6.22) is used to convert the altimeter readings to pressure units then the two methods become identical. For density compensated altimeters, the conversion takes the form:

$$\text{barometric pressure} = \exp(a - b \times \text{indicated altitude})$$

where the constants a and b depend upon the units of pressure and altitude, and factory settings of the altimeter. It would seem to have been the concept of a head of air and the use of a head of water to indicate frictional pressure drops that led to the early use of altimeters in the United States.

Theory

Again, we commence with the steady flow energy equation for an airway between stations 1 and 2, and containing no fan. In the usual case of polytropic flow, the energy equation gives the work done against friction (see equation (8.1)) as

$$F_{12} = \frac{u_1^2 - u_2^2}{2} + (Z_1 - Z_2)g - R(T_2 - T_1)\frac{\ln(P_2/P_1)}{\ln(T_2/T_1)} \quad \text{J/kg} \quad (6.24)$$

or, for isothermal flow where $T_1 = T_2$

$$F_{12} = \frac{u_1^2 - u_2^2}{2} + (Z_1 - Z_2)g - RT_1\ln(P_2/P_1)$$

where P = barometric pressure (kPa), T = absolute temperature (kelvins), Z = elevation of barometer location (m), u = air velocity at the barometer (m/s) and R = mean gas constant (from equation (14.14)). As all parameters are measurable in these relationships, the work done against friction, F_{12}, can be determined. This, in turn, can be converted into a frictional pressure drop, p_{12}, referred to any given air density, ρ_a (see also, equation (6.10)):

$$p_{12} = \rho_a F_{12} \quad \text{Pa}$$

In the case of an altimeter, the indicated altitude, h, should be multiplied by the calibration value of ρg (equation (6.22)) to determine the barometric pressure, P, before using equation (6.24).

A complication arises if the barometric pressures at the two stations are not read simultaneously. In this case, the surface atmospheric pressure may change during any time interval that occurs between readings at successive stations. If the atmospheric pressure at a fixed control station is observed to increase by ΔP_c during the time elapsed while moving from station 1 to station 2, then the initial value, P_1, should be corrected to

$$P_1 + \Delta P_c \quad (6.25)$$

The correction, ΔP_c, may, of course, be positive or negative. By assuming a series of polytropic processes connecting the control barometer (subscript c) to the traverse barometer (subscript 1), it can be shown that a more accurate value of the correction is given as

$$\Delta P_c \frac{P_1}{P_c} \quad (6.26)$$

Example   The following two lines are an excerpt from a barometer field book:

Station number	Time	Traverse barometer P (kPa)	Temperatures t_d (°C)	t_w (C)	Elevation Z (m)	Velocity u (m/s)	Control barometer P_c(kPa)
1	13:42	103.75	15.6	13.0	2652	2.03	98.782
2	14:05	104.61	17.2	14.2	2573	1.52	98.800

Using the psychrometric equations given in section 14.6, the moisture contents of the air at stations 1 and 2 were calculated to be

$$X_1 = 0.009\,108 \text{ kg/(kg dry air)}$$

$$X_2 = 0.009\,777 \text{ kg/(kg dry air)}$$

Equation (14.14) then indicates the corresponding gas constants as

$$R = \frac{287.04 + 461.5X}{1 + X}$$

giving $R_1 = 288.431$ J/(kg °C) and $R_2 = 288.517$ J/(kg °C) with an arithmetic mean of 288.474 J/(kg °C).

During the time period between taking barometer readings P_1 and P the control barometer registered an increase in atmospheric pressure of

$$\Delta P_c = 98.800 - 98.782 = 0.018 \text{ kPa}$$

Equation (6.26) gives the corrected reading at station 1 as

$$P_1 + \Delta P_c \frac{P_1}{P_c}$$

$$= 103.75 + 0.018 \times \frac{103.75}{98.782}$$

$$= 103.75 + 0.019 = 103.769 \text{ kPa}$$

The steady-flow energy equation (equation (6.24)) then gives

$$F_{12} = \frac{u_1^2 - u_2^2}{2} + (Z_1 - Z_2)g - R(T_2 - T_1)\frac{\ln(P_2/P_1)}{\ln(T_2/T_1)}$$

where

$$T_1 = 273.15 + 15.6 = 288.75\,K$$

$$T_2 = 273.15 + 17.2 = 290.35\,K$$

$$F_{12} = \frac{2.03^2 - 1.52^2}{2} + (2652 - 2573)9.81 - 288.474(17.2 - 15.6)\frac{\ln(104.61/103.769)}{\ln(290.35/288.75)}$$

$$= 0.91 + 744.99 - 674.30\,J/kg$$

$$= 101.6\,J/kg$$

and frictional pressure drop referred to standard density becomes

$$p_{12} = F_{12}\rho = 101.6 \times 1.2$$

$$= 122\,Pa$$

(See also section 8.2.2 and 8.3.3 for further examples.)

Practical procedure

A barometric survey can be conducted with one observer at each station although an additional person at the traverse stations facilitates more rapid progress. The equipment required is as follows:

1. two microaneroid barometers (or altimeters) of equal precision
2. a whirling or aspirated psychrometer
3. two accurate watches
4. an anemometer
5. a 2 m measuring tape
6. two waterproof field books and pencils
7. two or three cans of spray paint to mark station numbers

The microaneroids should be calibrated against a primary barometer prior to an important survey. A calibration cabinet can be constructed with the internal pressure controlled by compressed air feeds and outlet valves. In addition to pressure calibration, the instruments should be checked for temperature compensation and creep characteristics. Modern instruments are stable over the range of temperatures normally encountered in mines and adapt to a change in pressure within a few minutes.

For an underground barometer traverse only the main junctions need be considered as measurement stations. Intermediate substations, as required in the gauge and tube technique, are normally unnecessary. However, for main ventilating shafts, the most accurate results are obtained by taking readings at intervals down the shaft (section 8.2.2).

There are essentially two methods of handling the natural variations in atmospheric pressure that occur during the course of the survey. One technique is to maintain a barometer at a fixed **control station** and to record or log the readings at intervals of about five minutes. The second method is the **leapfrog** procedure in which both barometers are used to take simultaneous readings at successive stations. After each set of readings, the trailing barometer is brought up to the forward station where the two barometers are checked against each other and reset if necessary. The trailing barometer is then moved on to assume the leading position at the next station.

The traverse procedure commences by the observers synchronizing their watches. If the control station method is used, the control should be established in a location that is reasonably stable with respect to temperature and not subject to pressure fluctuations from fans, ventilation controls, hoists or other moving equipment. A location on surface near the top of a downcast shaft and shaded from direct sunlight is usually satisfactory. A recording barometer may be employed at the control station, but only if it provides a precision equivalent to that of the traverse barometer.

At each traverse station the following readings are logged:

1. date, barometer identification and name of observer
2. number and location of station
3. time
4. barometer reading
5. wet and dry bulb temperatures
6. anemometer reading

The location of each station should be correlated with surveyors' plans to determine the corresponding elevation. The traverse barometer should be held at the same height above the floor at each station in order that its elevation can be ascertained to within 0.5 m.

The anemometer should be employed to measure the air velocity at the position of the barometer. There is no need to conduct an anemometer traverse. As shown in the example given in the previous subsection, the effect of air velocity is usually small compared with the other terms in the steady-flow energy equation.

6.4   ORGANIZATION OF PRESSURE–VOLUME SURVEYS

The preceding two sections have discussed the techniques of measuring volume flows, Q, and frictional pressure drops, p, separately. It will be recalled that the results of the two types of surveys will be combined to give the resistence, $R = p/Q^2$, and airpower loss, pQ, of each branch. As airflows and, hence, frictional pressure drops vary with time in an operating subsurface facility, it follows that p and Q should, ideally, be measured simultaneously in any given airway. Typically, there are two observers measuring airflows and another two involved in the pressure survey. The two teams must liaise closely.

6.4.1   **Initial planning**

A pressure–volume survey should be well planned and managed. The practical work for a major survey commences a week or two before the underground observations by assembling, checking and calibrating the equipment. In particular, it is inadvisable to rely on manufacturers' calibrations of vane anemometers or diaphragm pressure gauges. If the equipment required for calibration is unavailable locally then the work may be carried out by a service organization or the instruments returned to the manufacturers for customized calibration. The calibration is normally produced as a table of corrections against indicated readings and taped on to the side of the instrument or carrying case. Interpolation from the table can be carried out at the time of measurement so that the reading, correction and corrected observation can all be logged immediately.

The mine plan should be studied carefully and the routes of the surveys selected. A full mine survey will include each major ventilation connection to surface and the infrastructure of airways that constitute the primary ventilation routes. Subsidiary survey routes may be appended to include individual working districts or to extend a data bank that exists from previous surveys. The routes should be chosen such that they can be formulated into closed traverse paths or loops within the ventilation network of the mine. Branches that connect to the surface close through the pressure sink of the atmosphere. A main loop in a large mine may take several days to survey. However, each main loop should be divided into smaller subsidiary loops each of which can be closed within a single day of surveying.

A reconnaissance of the mine should be carried out, travelling through all airways selected for the primary traverses and establishing the locations of main stations. These are normally at junctions of the ventilation system. Where two or more airways are adjacent and in parallel—and the gauge and tube method is employed for the pressure measurements—then it is necessary to take those measurements in one of the airways only. However, to obtain the total airflow in that composite branch of the network, it will be necessary to take flow measurements at corresponding points in each of the parallel airways. Airflow measuring stations should be selected and marked on the plan and, also, on the walls of the airway.

The subsequent employment of survey data for ventilation network analysis and forward planning (Chapters 7 and 9) should be kept in mind during the management of ventilation surveys. The identification number assigned to each network junction should give an identification of the location of the junction within the mine. In multilevel workings, for example, the first integer of station numbers may be used to indicate the level.

A pre-survey briefing meeting should be held with all observers present. Each observer should be fully trained in survey procedures, use of the instruments and techniques of observation. The traverse routes and system of station identification should be discussed, together with an outline schedule covering the days or weeks required to complete the survey.

6.4.2 Survey management

During production shifts, the airflows and pressure drops in an underground mine are subject to considerable variation due to movement of equipment, changes in resistance in the workings and opening of ventilation doors. Hence, the best time for ventilation surveys is when the mine is relatively quiescent with few people underground. During the period of a survey, the observers should be prepared to work at weekends and on night shifts.

Although the frictional pressure drop and the airflow should, ideally, be measured simultaneously in each leg of the traverse, this is often not practicable. Nevertheless, the teams should stay fairly near to each other so that there is a minimum delay between the two sets of measurements in a given branch. With experienced observers the teams maintain close liaison, assisting one another and always being conscious of the activity of the other team. This avoids the infuriating situation of the pressure team opening doors to take a check reading while the airflow team are in the middle of an anemometer traverse. Friendships have been known to suffer on such occasions.

Immediately following each shift the two teams should check all calculations carried out underground, transcribe the results of that shift's work from the field books to clean log sheets and, also, onto a large-scale copy of a mine map. Positions of measured airflows and pressure drops should be reviewed by both teams to ensure compatibility of those measurement locations and to correlate identification of station numbers. Any difficulties encountered during the shift should be discussed. The final half hour or so of each working day may be spent in reviewing the ground to be covered in the following shift and the allocation of individual duties.

6.4.3 Quality assurance

It is most important that control is maintained over the quality of all aspects of an important ventilation survey, from initial calibration of the instruments through to the production of final results. Field books or booking sheets should be laid out clearly such that persons other than the observers can follow the recording of observations and calculations carried out underground. All calculations should be checked by someone other than the originator. Most of the calculations involved in ventilation surveys are quite simple and may be carried out on a pocket calculator. The exception is for barometric surveys where a verified program for a personal computer is very helpful. Commercially available spreadsheet software can readily be adapted for this purpose.

Adherence to Kirchhoff's laws should be checked both at the time of observations wherever practicable and, also, during the data transposition at the end of each shift. These laws are discussed fully in Chapter 7. Briefly, Kirchhoff I requires that the algebraic sum of airflows entering any junction is zero. Kirchhoff II states that the algebraic sum of standardized pressure drops around any closed loop must also be zero, having taken fans and natural ventilation pressures into account. In a near-level circuit, the closing error of a pressure loop may be expressed as the actual closure divided by the sum of the absolute values of the measured frictional pressure drops around the loop. This should not exceed 5%. The check measurements of pressure differentials across doors are invaluable in tracing or distributing observational errors. In the case of loops involving significant changes in elevation such as shaft circuits, the sum of standardized pressure drops will be a combination of observational errors and natural ventilating effects. The latter may be determined independently from temperature and pressure measurements as discussed in section 8.3.

It is vital that good records are kept of each phase of a survey. The survey team leader should maintain a detailed diary of the activities and achievements of each working day. This should include the clean log sheets of results transcribed from the field books at the end of each shift.

The conclusion of a major survey should see the establishment of a spreadsheet type of data bank or the extension of an existing data bank, holding the frictional pressure drop and corresponding airflow for every branch included in the survey. Other details such as the date of the observation, names of observers, instrument identification and dimensions of the airway may be included. The data bank may then be used to calculate airway resistances, airpower losses and friction factors, and also provides a database from which a computer model of the mine ventilation network can be constructed (Chapter 9). Additionally, the resistances, resistance per unit length and airpower losses may be shown on a colour-coded map in order to highlight sections of airways that are particularly expensive to ventilate.

6.5 AIR QUALITY SURVEYS

While pressure–volume surveys are concerned with the distribution of airflow around a ventilation system, the subsurface environmental engineer must also maintain control of the quality of that air, i.e. the concentrations of gaseous or particulate pollutants, and the temperature and humidity of the air. Such measurements may be made to ensure compliance with mandatory standards and with a regard for the safety and health of the workforce.

Details of the techniques of measuring and quantifying levels of dust, gas concentrations and climatic conditions are given in Chapters 19, 11 and 14 respectively. A set of such measurements made in a systematic manner around a continuous path is known as an air quality survey. This procedure may be employed for two reasons. First, it provides a means of tracking and quantifying the variation in pollutant levels and, secondly, it enables zones of emission of gases, dust, heat and humidity to be identified. Measurements of gas concentration are often made as a normal part of a pressure–volume survey in a gassy mine. Similarly the observations of barometric pressure and wet and dry bulb temperature made during a pressure survey may be used to compute and plot the variations in psychrometric conditions throughout the traverse paths.

FURTHER READING

Air Movement and Control Association (1976) *AMCA Application Manual,* Part 3, *A guide to the Measurement of Fan–System Performance in the Field,* AMCA, 30 West University Drive, Arlington Heights, IL 60004.

Air Movement and Control Association (1985) *American National Standard: Laboratory Methods of Testing Fans for Rating,* AMCA, 30 West University Drive, Arlington Heights, IL 60004.

American Society of Heating, Refrigerating and Air Conditioning Engineers (1985) *Fundamentals Handbook* ASHRAE, 1791 Tullie Circle NE Atlanta, GA 30329.

British Standards Institution (1963) *British Standard 848*, Part 1, *Methods of Testing Fans for General Purposes*, BSI, 2 Park St., London W1.

Hartman, H.L., Mutmansky, J. and Wang, W.J. (1982) *Mine Ventilation and Air Conditioning*, Wiley–Interscience.

Hinsley, F.B. (1962) The assessment of energy and pressure losses due to air-flow in shafts, airways and mine circuits. *Min. Eng.* **121**, (23), 761–77.

Hinsley, F.B. (1964) An enquiry into the principles and mutual interaction of natural and fan ventilation. *Min. Eng.* **124** (49), 63–78.

McPherson, M.J. (1969) A new treatment of mine barometer surveys. *Min. Eng.* **129** (109), 23–34.

McPherson, M.J. (1985) The resistance to airflow on a longwall face. *Proc. 2nd US Mine Ventilation Symp., Reno, NV*, 531–42

McPherson, M.J. and Robinson, G. (1980) Barometric survey of shafts at Boulby Mine, Cleveland Potash Ltd. *Trans. Inst. Min. Metall.* **89** (reproduced in J. *Mine Vent. Soc. S. Afr.* **9** (3).

Mine Ventilation Society of South Africa (1982) *Environmental Engineering in South African Mines*. Mine Ventilation Society of South Africa, Kelvin House, 2 Hollard St., Johannesburg.

National Coal Board (1979) *Ventilation in Coal Mines: A Handbook for Colliery Ventilation Engineers.*

7

Ventilation network analysis

7.1 INTRODUCTION

A vital component in the design of a new underground mine or other subsurface facility is the quantified planning of the distribution of airflows, together with the locations and duties of fans and other ventilation controls required to achieve acceptable environmental conditions throughout the system. Similarly, throughout the life of an underground operation, it is necessary to plan ahead in order that new fans, shafts or other airways are available in a timely manner for the efficient ventilation of extensions to the workings. As any operating mine is a dynamic system with new workings continually being developed and older ones coming to the end of their productive life, ventilation planning should be a continuous and routine process.

The preceding chapters have discussed the behaviour of air or other fluid within an individual airway, duct or pipe. Ventilation network analysis is concerned with the interactive behaviour of airflows within the connected branches of a complete and integrated network. The problems addressed by ventilation network analysis may be formulated quite simply. If we know the resistances of the branches of a ventilation network and the manner in which those branches are interconnected then how can we predict, quantitatively, the distribution of airflow for given locations and duties of fans?

Alternatively, if we know the airflows that we want in specific branches of the network then how can we determine a combination of fans and structure of the network that will provide those required airflows? Ventilation network analysis is a generic term for a family of techniques that enable us to address such questions.

In a given network there are, theoretically, an infinite number of combinations of airway resistances, fans and regulators that will give any desired distribution of flow. Practical considerations limit the number of acceptable alternatives. However, the techniques of network analysis that are useful for modern industrial application must remain easy to use, and sufficiently rapid and flexible to allow multiple alternative solutions to be investigated.

Before the mid-1950s there was no practicable means of conducting detailed and quantitative ventilation network analysis for complete mine systems. Ventilation planning was carried out either using hydraulic gradient diagrams formulated from assumed airflows or, simply, based on the experience and intuition of the ventilation engineer. Attempts to produce physical models of complete mine ventilation systems using air or water as the medium met with very limited success because of difficulties from scale effects. Following earlier research in Holland (Maas, 1950), the first viable electrical analogue computers to simulate ventilation networks were produced in the United Kingdom (Scott *et al.*, 1953) followed rapidly in the United States (McElroy, 1954). These analogues employed electrical current passing through rheostats to simulate airflows. Successive adjustment of the resistances of the rheostats enabled the linear Ohm's law for electrical conductors to emulate the square law of ventilation networks.

Linear resistance analogues become the main automated means of analysing mine ventilation networks in the 1950s and early 1960s. Rapid advances in the electronics industries during that time resulted in the development of direct reading fluid network analogues that replaced the linear resistors with electronic components. These followed a logarithmic relationship between applied voltage and current, producing analogues that were easier to use and much faster in operation (Williams, 1964). However, by the time they became available, ventilation simulation programs for mainframe digital computers had begun to appear (McPherson, 1964). These proved to be much more versatile, rapid and accurate, and their employment soon dominated ventilation planning procedures in major mining countries. Coupled with continued improvements in ventilation survey techniques to provide the data, ventilation network analysis programs resulted in hitherto unprecedented levels of flexibility, precision and economics in the planning, design and implementation of mine ventilation systems.

Throughout the 1970s network programs were developed for large centralized mainframe computers. Their initial use by industry tended to be inhibited by the often pedantic procedures of data preparation together with the costs and delays of batch processing. In the 1980s, the enhanced power and reduced cost of micro-computers led to the evolution of self-contained software packages that allowed very easy interaction between the user and the computer. These incorporated the use of graphics. Ventilation engineers could, for the first time, conduct multiple planning exercises on large networks entirely within the confines of their own offices. The complete processing of data from survey observations through to the production of plotted ventilation plans became automated. Personal computers, printers and plotters proliferated in mine planning offices. Together with the ready availability of software, these led to a revolution in the methodologies, speed and accuracy of subsurface ventilation planning.

In this chapter, we shall introduce the basic laws that govern the distribution of airflow within a network of interconnected branches. The analytical and numerical methods of predicting airflows will be examined before proceeding to a discussion of network simulation packages.

7.2   FUNDAMENTALS OF VENTILATION NETWORK ANALYSIS

Any integrated ventilation system can be represented as a schematic diagram in which each line (branch) denotes either a single airway or a group of openings that are connected such that they behave effectively as a single airway (section 7.3.1). Only those airways that contribute to the flow of air through the system appear on the network schematic. Hence, sealed-off areas of insignificant leakage, stagnant dead-ends and headings that are ventilated locally by ducts and auxiliary fans need not be represented in the network. On the other hand, the tops of shafts or other openings to surface are connected to each other through the pressure sink of the surface atmosphere. The points at which branches connect are known simply as junctions or nodes.

7.2.1   Kirchhoff's laws

Gustav R. Kirchhoff (1824–1887) was a German physicist who first recognized the fundamental relationships that govern the behaviour of electrical current in a network of conductors. The same basic relationships, now known as Kirchhoff's laws, are also applicable to fluid networks including closed ventilation systems at steady state.

Kirchhoff's first law states that the mass flow entering a junction equals the mass flow leaving that junction or, mathematically,

$$\sum_j M = 0 \tag{7.1}$$

where M are the mass flows, positive and negative, entering junction j.

However, it will be recalled that

$$M = Q\rho \quad \text{kg/s} \tag{7.2}$$

where Q = volume flow (m^3/s) and ρ = air density (kg/m^3). Hence

$$\sum_j Q\rho = 0 \tag{7.3}$$

If subsurface ventilation systems, the variation in air density around any single junction is negligible, giving

$$\sum_j Q = 0 \tag{7.4}$$

This provides a means of checking the accuracy of airflow measurements taken around a junction (section 6.4.3).

The simplest statement of **Kirchhoff's second law** applied to ventilation networks is that the algebraic sum of all pressure drops around a closed path, or mesh, in the network must be zero, having taken into account the effects of fans and ventilating pressures. This can be quantified by writing down the steady flow energy equation (3.25), initially for a single airway.

$$\frac{\Delta u^2}{2} + \Delta Zg + W = \int V\,dP + F \quad \frac{J}{kg} \tag{7.5}$$

where u = air velocity (m/s), Z = height above datum (m), W = work input from fan (J/kg), V = specific volume (m³/kg), P = barometric pressure (Pa) and F = work done against friction (J/kg).

If we consider a number of such branches forming a closed loop or mesh within the network then the algebraic sum of all ΔZ must be zero and the sum of the changes in kinetic energy, $\Delta u^2/2$, is negligible. Summing each of the remaining terms around the mesh, m, gives

$$\sum_m \int V\,dP + \sum_m (F - W) = 0 \quad J/kg \tag{7.6}$$

Now the summation of $-\int V\,dP$ terms is the natural ventilating energy, NVE, that originates from thermal additions to the air (section 8.3.1). Hence, we can write

$$\sum (F - W) - NVE = 0 \quad J/kg \tag{7.7}$$

This may now be converted to pressure units by multiplying throughout by a single value of air density ρ:

$$\sum (\rho F - \rho W) - \rho\,NVE = 0 \quad Pa \tag{7.8}$$

However, $\rho F = p$ (frictional pressure drop, equation (2.46)), $\rho W = p_f$ (rise in total pressure across a fan) and $\rho\,NVE = NVP$ (natural ventilating pressure, equation (8.32)), each of these three terms being referred to the same (standard) density. The equation then becomes recognizable as Kirchhoff's second law:

$$\sum (p - p_f) - NVP = 0 \quad Pa \tag{7.9}$$

This is the relationship that is employed as a quality assurance check on a pressure survey (section 6.4.3), or as a means of determining a value for the natural ventilating pressure (equation 8.51)).

7.2.2 Compressible or incompressible flow in ventilation network analysis

Kirchhoff's laws can be applied to fluid networks that conduct either compressible or incompressible fluids. In the former case, the analysis is carried out on the basis of **mass** flow. Equations (7.1) or (7.3) are employed for the application of Kirchhoff's first law and equation (7.7) for Kirchhoff's second law.

There is another set of equations that must be incorporated with Kirchhoff's laws for ventilation network analysis. Those have already been introduced as the square law for each individual branch. When the flow is deemed to be compressible, then the rational form of the square law should be utilized (see equation 2.50)):

$$p = R_t \rho Q^2 \quad \text{Pa}$$

where R_t = rational turbulent resistance (m^{-4}). However, as $p = F\rho$ (equation 2.46)),

$$p = F\rho = R_t \rho Q^2$$

giving

$$F = R_t Q^2 \quad \text{J/kg} \tag{7.10}$$

Kirchhoff's second law for compressible flow (equation (7.7)) becomes

$$\sum_m (R_t Q^2 - W) - \text{NVE} = 0 \quad \text{J/kg} \tag{7.11}$$

As we progress from branch to branch around a closed mesh in a network then it is the algebraic values of the pressure drops, p, or losses of mechanical energy, F, that must be summed. These are both positive in the direction of flow and negative if the flow is moving against the direction of traverse around the mesh. Hence, equation (7.11) may, more appropriately, be written as

$$\sum_m (R_t Q|Q| - W) - \text{NVE} = 0 \quad \text{J/kg} \tag{7.12}$$

where Q = airflow with due account taken of sign ($\pm$ m^3/s) and $|Q|$ = absolute value of airflow and is always positive ($+$ m^3/s). This device ensures that the frictional pressure drop or loss of mechanical energy always have the same sign as airflow.

In the case of incompressible flow the application of Kirchhoff's laws becomes more straightforward. Equations (7.4) and (7.9) give

$$\sum_j Q = 0 \quad \text{(Kirchhoff I)}$$

and

$$\sum_m (RQ|Q| - p_f) - \text{NVP} = 0 \quad \text{(Kirchhoff II)} \tag{7.13}$$

where R = Atkinson resistance (N s^2/m^8).

It will be recalled that the three terms of this latter equation should each be referred to the same (standard) value of air density, normally 1.2 kg/m^3.

Computer programs have been developed for compressible flow networks. These require input data (pressures, temperatures, elevations and air quality parameters) from which variations in air density and natural ventilation effects may be calculated. On the other hand, where compressibility and natural ventilating effects need to be taken into account, there are means by which these can be simulated to an acceptable accuracy by an incompressible flow network program. For these reasons, the great majority of subsurface ventilation planning employ the simpler and faster incompressible flow programs. The more sophisticated and demanding compressible flow programs are required for compressed air (or gas) networks or for specialized applications in subsurface ventilation systems.

The remainder of this chapter will concentrate on incompressible flow network analysis.

7.2.3 Deviations from the square law

The square law was derived in section 2.3.6 from the basic Chezy–Darcy relationship and further developed in section 5.2. Both the rational form of the square law, $p = R_t \rho Q^2$, and the traditional form, $p = RQ^2$, apply for the condition of fully developed turbulence. Furthermore, both the rational resistance, R_t, and the Atkinson resistance, R, are functions of the coefficient of friction, f, for the duct or airway (equations (2.51), (5.4) and (5.1)). Hence, if the flow falls into the transitional or laminar regimes of the Moody chart (Fig. 2.7) then the value of f becomes a function of Reynolds' number. The corresponding values of resistance, R_t, or R, for that airway then vary with the airflow. If the form of the square law is to be retained then the values of resistance must be computed for the relevant values of Reynolds' number.

However, many experimental researchers have found that plotting $\ln(p)$ against $\ln(Q)$ may give a slope that deviates slightly from Chézy–Darcy's theoretical prediction of 2, even for fully developed turbulence. The relationship between frictional pressure drop, p, and volume flowrate, Q, may be better expressed as

$$p = RQ^n \quad \text{Pa} \tag{7.14}$$

where values of the index, n, have been reported in the range of 1.8 to 2.05 for a variety of pipes, ducts and fluids. Similar tests in mine airways have shown that n lies very close to 2 for routes along the main ventilation system but may reduce for leakage flows through stoppings or old workings. This effect is caused primarily by flows that enter the transitional or, even, laminar regimes. In the latter case, n takes the value of 1.0 and the pressure drop–flow relationship becomes (see equation (2.32)

$$p = R_L Q$$

where $R_L = $ laminar resistance.

In the interests of generality, we shall employ equation (7.14) in the derivations that follow.

7.3 METHODS OF SOLVING VENTILATION NETWORKS

There are essentially two means of approach to the analysis of fluid networks. The analytical methods involve formulating the governing laws into sets of equations that can be solved analytically to give exact solutions. The numerical methods that have come to the fore with the availability of electronic digital computers solve the equations through iterative procedures of successive approximation until a solution is found to within a specified accuracy.

In both cases, the primary processes of solution may be based on the distribution of pressures throughout the network or, alternatively, on the distribution of flows.

The former may be preferred for networks that involve many outlet points such as a water distribution network. On the other hand, for networks that form closed systems such as subsurface ventilation layouts it is more convenient to base the analysis on flows.

7.3.1 Analytical methods

Equivalent resistances

This is the most elementary of the methods of analysing ventilation networks. If two or more airways are connected either in series or in parallel then each of those sets of resistances may be combined into a single **equivalent** resistance. Although of fairly limited value in the analysis of complete networks, the method of equivalent resistances allows considerable simplification of the schematic representation of actual subsurface ventilation systems.

In order to determine an expression for a series circuit, consider Fig. 7.1(a). The frictional pressure drops are given by equation (7.14) as

$$p_1 = R_1 Q^n, \qquad p_2 = R_2 Q^n, \qquad p_3 = R_3 Q^n$$

Then for the combined series circuit,

$$p = p_1 + p_2 + p_3 = (R_1 + R_2 + R_3) Q^n$$

or

$$p = R_{ser} Q^n$$

where R_{ser} is the equivalent resistance of the series circuit.

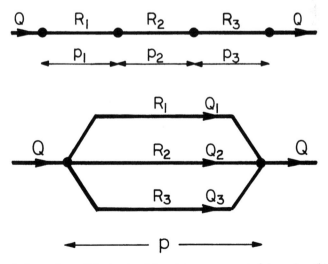

Figure 7.1 Series and parallel circuits: (a) resistances connected in series; (b) resistances connected in parallel.

In general, for a series circuit,

$$R_{ser} = \sum R \tag{7.15}$$

In the case of a parallel circuit, Fig. 7.1(b) shows that each branch suffers from the same frictional pressure drop, p, between the common 'start' and 'end' junctions but passes differing airflows. Then

$$p = R_1 Q_1^n = R_2 Q_2^n = R_3 Q_3^n$$

giving

$$Q_1 = (p/R_1)^{1/n}$$
$$Q_2 = (p/R_2)^{1/n}$$
$$Q_3 = (p/R_3)^{1/n}$$

The three airflows combine to give

$$Q = Q_1 + Q_2 + Q_3 = p^{1/n}\left(\frac{1}{R_1^{1/n}} + \frac{1}{R_2^{1/n}} + \frac{1}{R_3^{1/n}}\right)$$

we may write this as

$$Q = \frac{p^{1/n}}{R_{par}^{1/n}}$$

where R_{par} is the equivalent resistance of the parallel circuit. It follows that

$$\frac{1}{R_{par}^{1/n}} = \frac{1}{R_1^{1/n}} + \frac{1}{R_2^{1/n}} + \frac{1}{R_3^{1/n}}$$

or, in general

$$\frac{1}{R_{par}^{1/n}} = \sum \frac{1}{R^{1/n}} \tag{7.16}$$

In the usual case of $n = 2$ for subsurface ventilation systems,

$$\frac{1}{\sqrt{R_{par}}} = \sum \frac{1}{\sqrt{R}} \tag{7.17}$$

Example   Figure 7.2 illustrates nine airways that form part of a ventilation network. Find the equivalent resistance of the system.

Solution   Airways 1, 2 and 3 are connected in series and have an equivalent resistance of (equation (7.15))

$$R_a = R_1 + R_2 + R_3$$
$$= 0.6 + 0.1 + 0.1$$
$$= 0.8 \quad N s^2/m^8$$

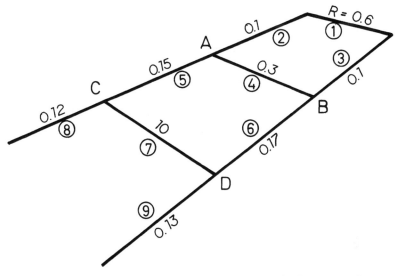

Figure 7.2 Example of a network segment that can be resolved into a single equivalent resistance.

This equivalent resistance is connected across the same two junctions, A and B, as airway 4 and, hence, is in parallel with that airway. The equivalent resistance of the combination, R_b, is given by (equation (7.16))

$$\frac{1}{\sqrt{R_b}} = \frac{1}{\sqrt{0.8}} + \frac{1}{\sqrt{0.3}}$$

from which

$$R_b = 0.1154 \quad N\,s^2/m^8$$

Notice that the equivalent resistance of the parallel circuit is less than that of either component (ref. Buddle, section 1.2). Equivalent resistance R_b is now connected in series with airways 5 and 6 giving a new equivalent resistance of all paths inbye junctions C and D as

$$R_c = 0.1154 \quad + 0.15 + 0.17$$

$$= 0.4354 \quad N\,s^2/m^8$$

Combining this in parallel with airway 7 gives

$$\frac{1}{\sqrt{R_d}} = \frac{1}{\sqrt{0.4354}} + \frac{1}{\sqrt{10}}$$

which yields

$$R_d = 0.298 \quad N\,s^2/m^8$$

The equivalent resistance of all nine airways, R_{equ}, is then given by adding the resistances of airways 8 and 9 in series:

$$R_{equ} = 0.298 + 0.12 + 0.13$$

$$= 0.548 \quad N s^2/m^8$$

If it is simply the total flow into this part of a larger network that is required from a network analysis exercise then the nine individual airways may be replaced by the single equivalent resistance. This will simplify the full network schematic and reduce the amount of computation to be undertaken during the analysis. on the other hand, if it is necessary to determine the airflow in each individual branch then the nine airways should remain as separate resistances for the network analysis.

The concept of equivalent resistances is particularly useful to combine two or more airways that run adjacent to each other. Similarly, although a line of stoppings between two adjacent airways are not truly connected in parallel, they may be grouped into sets of five to ten, each set to be represented as an equivalent parallel resistance, provided that the resistance of the stoppings is large compared with that of the airways. Through such means, the several thousand actual branches that may exist in a mine can be reduced to a few hundred, simplifying the schematic, reducing the amount of data that must be handled, and minimizing the time and cost of running network simulation packages.

Direct application of Kirchhoff's laws

Kirchhoff's laws allow us to write down equation (7.4) for each independent junction in the network and equation (7.9) for each independent mesh. Solving these two sets of equations will give the branch airflows, Q, that satisfy both laws simultaneously.

If there are b branches in the network then there are b airflows to be determined and, hence, we need b independent equations. Now, if the network contains j junctions, then we may write down Kirchhoff I (equation (7.4)) for each of them in turn. However, as each branch is assumed to be continuous with no intervening junctions, a branch airflow denoted as Q_i entering a junction will automatically imply that Q_i leaves a neighbouring junction at the other end of that same branch. It follows that, when we reach the last junction, all airflows will already have been symbolized. The number of independent equations arising from Kirchhoff I is $j - 1$.

This leaves $b - (j - 1)$ or $b - j + 1$ further equations to be established from Kirchhoff II (equation (7.9)). We need to choose $b - j + 1$ independent closed meshes around which to sum the pressure drops.

The direct application of Kirchhoff's laws to full mine circuits may result in several hundred equations to be solved simultaneously. This requires computer assistance. Manual solutions are limited to very small networks or sections of networks. Nevertheless, a manual example is useful both to illustrate the technique and to lead into the numerical procedures described in the following section.

Example   Figure 7.3 shows a simplified ventilation network served by a downcast and an upcast shaft, each passing $100 \, m^3/s$. The resistance of each subsurface branch is shown. A fan boosts the airflow in the central branch to $40 \, m^3/s$. Determine the distribution of airflow and the total pressure, p_b, developed by the booster fan.

Solution   Inspection of the network shows that it cannot be resolved into a series/parallel configuration and, hence, the method of equivalent resistances is not applicable. We shall, therefore, attempt to find a solution by direct application of Kirchhoff's laws.

The given airflows are also shown on the figure with the flow from A to B denoted as Q_1. By applying Kirchhoff I (equation (7.4)) to each junction, the airflows in other branches may be expressed in terms of Q_1 and have been added to the figure.

The shafts can be eliminated from the problem as their airflows are already known. In the subsurface structure we have five branches and four junctions. Accordingly, we shall require

$$b - j + 1 = 5 - 4 + 1 = 2 \quad \text{independent meshes.}$$

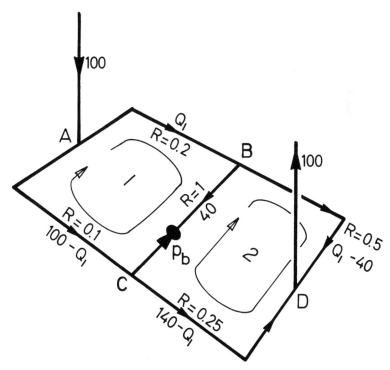

Figure 7.3   Example network showing two meshes and Kirchhoff I applied at each juction.

Figure 7.3 shows that we may choose the meshes along the closed paths ABCA and BDCB.

In order to apply Kirchhoff II (equation (7.9)) to the first mesh, consider the pressure drops around that mesh, branch by branch, remembering that frictional pressure drops are positive in the direction of flow.

Branch	Frictional pressure drop (RQ^2)	Fan
AB	$0.2\, Q_1^2$	—
BC	1.0×40^2	$-p_b$
CA	$-0.1\,(100 - Q_1)^2$	—

Summing up these terms according to Kirchhoff II gives

$$0.2\, Q_1^2 + 1600 - p_b - 0.1\,(100 - Q_1)^2 = 0 \quad \text{Pa} \tag{7.18}$$

Collecting like terms gives

$$p_b = 0.1\, Q_1^2 + 20\, Q_1 + 600 \quad \text{Pa} \tag{7.19}$$

Inspection of Fig. 7.3 commencing from junction B shows that the summation of pressure drops applied to the second mesh may be written as

$$0.5(Q_1 - 40)^2 - 0.25(140 - Q_1)^2 - 1 \times 40^2 + p_b = 0 \quad \text{Pa} \tag{7.20}$$

Notice that in this case, the direction of traverse is such that the pressure falls through the fan, i.e. a positive pressure drop. Substituting for p_b from equation (7.19), expanding the bracketed terms and simplifying leads to

$$0.35 Q_1^2 + 50 Q_1 - 5100 = 0$$

This quadratic equation may be solved to give

$$Q_1 = \frac{-50 \pm \sqrt{50^2 + (4 \times 0.35 \times 5100)}}{2 \times 0.35}$$

$$= 68.83 \text{ or } -211.69 \text{ m}^3/\text{s}$$

The only practicable solution is $Q_1 = 68.83 \text{ m}^3/\text{s}$. The flows in all branches are then given from inspection of Fig. 7.3 as

AB	$68.83 \text{ m}^3/\text{s}$
BD	28.83
BC	40.00
AC	31.17
CD	71.17

The required booster fan pressure, p_b, is given from equation (7.19) as

$$p_b = 0.1(68.83)^2 + 20(68.83) + 600$$
$$= 2450\,Pa$$

7.3.2 Numerical methods

Although the analytical methods of ventilation network analysis were given considerable attention before the development of analogue computers in the 1950s, the multiple simultaneous equations that they produced could not readily be solved by manual means. As shown in the previous example, two mesh problems involve quadratic equations. A little further thought indicates that three mesh problems would involve polynomial equations in Q of order 4. Four meshes would produce powers of 8, and so on (i.e. powers of 2^{m-1}, where m = number of meshes).

Atkinson recognized this problem in his paper of 1854 and suggested a method of successive approximation in which the airflows were initially estimated, then adjusted towards their true value through a series of corrections. Although no longer employed, Atkinson's approach anticipated current numerical methods.

Since the mid 1960s, considerable research has been carried out to find more efficient numerical methods of analysing ventilation networks. Y. J. Wang of West Virginia University in the United States developed a number of algorithms based on matrix algebra and the techniques of operational research.

The method that is most widely used in computer programs for ventilation network analysis was originally devised for water distribution systems by Professor Hardy Cross at the University of Illinois in 1936. This was modified and further developed for mine ventilation systems by D. R. Scott and F. B. Hinsley at the University of Nottingham in 1951. However, it was not until digital computers became more widely available for engineering work in the 1960s that numerical methods became truly practicable.

The Hardy Cross technique

Figure 7.4 shows the system resistance curve for one single representative branch in a ventilation network. If the air flow, Q, is reversed then the frictional pressure drop, p, also becomes negative.

Recalling that a primary purpose of ventilation network analysis is to establish the distribution of flow, the true airflow in our representative branch, Q, will initially be unknown. However, let us assume an airflow, Q_a, that is less than the true value by an amount ΔQ:

$$Q = Q_a + \Delta Q$$

The problem now turns to one of finding the value of ΔQ. We may write the square law as

$$p = R(Q_a + \Delta Q)^2 \tag{7.21}$$

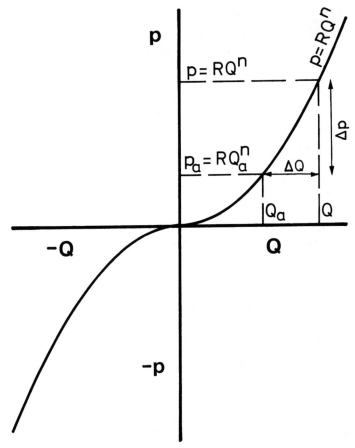

Figure 7.4 In the system resistance curve for an airway, p and Q always have the same sign. The p, Q slope remains non-negative.

or, more generally (section 7.2.3),

$$p = R(Q_a + \Delta Q)^n \tag{7.22}$$

Confining ourselves, for the moment, to the square law, equation (7.21) expands to

$$p = RQ_a^2 + 2RQ_a\Delta Q + R(\Delta Q)^2 \tag{7.23}$$

Furthermore, the frictional pressure drop corresponding to the assumed airflow is

$$p_a = RQ_a^2 \tag{7.24}$$

where the 'error' in p is

$$\Delta p = p - p_a$$

Substituting from equations (7.23) and (7.24) gives

$$\Delta p = 2RQ_a\Delta Q + R(\Delta Q)^2 \tag{7.25}$$

If we can assume that $(\Delta Q)^2$ is small compared with $2Q_a \Delta Q$ then we may write the approximation as

$$\Delta p = 2RQ_a \Delta Q \tag{7.26}$$

(Note that differentiating the square law,

$$p = RQ^2$$

gives the limiting case when $Q_a \to Q$. Then

$$\frac{\Delta p}{\Delta Q} \to \frac{\mathrm{d}p}{\mathrm{d}Q} = 2RQ \tag{7.27}$$

and corresponds to the difference equation (7.26). Then

$$\Delta Q = \frac{\Delta p}{2RQ_a} \tag{7.28}$$

The technique of dividing a function by its first derivative in order to estimate an incremental correction factor is a standard numerical method for finding roots of functions. It is usually known as the Newton or Newton–Raphson method.

If the general logarithmic law is employed, equation (7.22) gives

$$p = RQ_a^n \left(1 + \frac{\Delta Q}{Q_a} \right)^n$$

Expanding by the binomial theorem and ignoring terms of order 2 and higher gives

$$p = RQ_a^n \left(1 + n\frac{\Delta Q}{Q_a} \right)$$

then

$$\Delta p = p - p_a = RQ_a^n n \frac{\Delta Q}{Q_a}$$

giving

$$\Delta Q = \frac{\Delta p}{nRQ_a^{n-1}} \tag{7.29}$$

If we refer back to Fig. 7.4, we are reminded that Δp is the error in frictional pressure drop that was incurred by choosing the assumed airflow, Q_a.

$$\Delta p = RQ^n - RQ_a^n \tag{7.30}$$

Hence, we can write equation (7.29) as

$$\Delta Q = \frac{RQ^n - RQ_a^n}{nRQ_a^{n-1}} \tag{7.31}$$

The denominator in this expression is the **slope** of the p, Q curve in the vicinity of Q_a.

So far, this analysis has concentrated upon one single airway. However, suppose that we now move from branch to branch in a consistent direction around a closed mesh, summing up both the Δp values and the p, Q slopes from equation (7.30):

$$\Sigma \Delta p = \Sigma (RQ^n - RQ_a^n)$$

$$= \Sigma RQ^n - \Sigma RQ_a^n$$

and sum of p, Q slopes $= \Sigma n R Q_a^{n-1}$

We can now write down a **composite** value of ΔQ for the complete mesh in the form of equation (7.31):

$$\Delta Q_m = \frac{\Sigma RQ^n - \Sigma RQ_a^n}{\Sigma n R Q_a^{n-1}} \qquad (7.32)$$

Note that ΔQ_m is no longer the correction to be made to the assumed airflow in any one branch but, rather, is the composite value given by dividing the mean Δp by the mean of the p, Q slopes.

Equation (7.32) suffers from the disadvantage that it contains the unknown true values of airflow, Q. However, the term

$$\Sigma RQ^n$$

is the sum of the corresponding true values of frictional pressure drop around a closed mesh while Kirchhoff's second law insists that this must be zero. Equation (7.32) becomes

$$\Delta Q_m = \frac{-\Sigma RQ_a^n}{\Sigma n R Q_a^{n-1}} \qquad (7.33)$$

Another glance at Fig. 7.4 reminds us that the pressure term in the numerator must always have the same sign as the airflow, while the p, Q slope given in the denominator can never be negative. Hence, equation (7.33) may be expressed as

$$\Delta Q_m = \frac{-\Sigma RQ_a |Q_a^{n-1}|}{\Sigma n R |Q_a^{n-1}|} \qquad (7.34)$$

where $|Q_a^{n-1}|$ means the absolute value of Q_a^{n-1}.

There are a few further observations that will help in applying and understanding the Hardy Cross process. First, the derivation of equation (7.34) has ignored both fans and natural ventilating pressures. Like airways, these are elements that follow a defined p, Q relationship. Equation (7.34) may be expanded to include the corresponding effects:

$$\Delta Q_m = \frac{-\Sigma (RQ_a |Q_a^{n-1}| - p_f - NVP)}{\Sigma (n R |Q_a^{n-1}| + S_f + S_{nv})} \qquad (7.35)$$

where p_f and NVP are the fan pressures and natural ventilating pressures respectively that exist within the mesh, and S_f and S_{nv} are the slopes of the p, Q characteristic curves for the fans and natural ventilating effects. In practice, S_{nv} is usually taken to be zero, i.e. it is assumed that natural ventilating effects are independent of airflow.

For most subsurface ventilation systems, acceptable accuracy is achieved from the simple square law, giving

$$\Delta Q_m = \frac{-\Sigma(RQ_a|Q_a| - p_f - NVP)}{\Sigma(2R|Q_a| + S_f + S_{nv})} \tag{7.36}$$

The Hardy Cross procedure may now be summarized as follows:

1. Draw a network schematic and choose at least $(b - j + 1)$ closed meshes such that all branches are represented (section 7.3.1). Convergence to a balanced solution will be improved by ensuring that each high resistance branch is included in only one mesh.
2. Make an initial estimate of airflow, Q_a, for each branch.
3. Traverse one mesh and calculate the mesh correction factor, ΔQ_m from equation (7.35) or (7.36).
4. Traverse the same mesh, in the same direction, and adjust each of the contained airflows by the amount ΔQ_m.
5. Repeat steps 3 and 4 for every mesh.
6. Repeat steps 3, 4 and 5 until Kirchhoff II is satisfied to an acceptable degree of accuracy, i.e. until the value of $-\Sigma(RQ_i|Q_i^{n-1}| - p_f - NVP)$ is close to zero, where Q_i are the current values of air flow.

A powerful feature of the method is that it is remarkably tolerant to poorly estimated initial airflows, Q_a. This requires some explanation as equation (7.28) and the more general equation (7.29) were both derived on the assumption that ΔQ was small compared with terms involving Q_a. In practice, it is observed that the procedure converges towards a balanced solution even when early values of ΔQ are large. There are two reasons for this. First, the compilation of mesh ΔQ_m tends to dampen out the effect of large ΔQ values in individual branches. Secondly, any tendency towards an unstable divergence is inhibited by the airway p, Q curves always having a consistent (non-negative) slope.

Another advantage of the Hardy Cross technique is its flexibility. Using equation (7.35) for the mesh correction factor allows a different value of the logarithmic index, n, to be used for each airway if required. For subsurface ventilation circuits, this is seldom necessary although special-purpose programs allow either laminar ($n = 1$) or turbulent ($n = 2$) airflow in each branch.

A more general observation is that Kirchhoff's first and second laws are not interdependent. Kirchhoff I ($\Sigma Q = 0$) may be obeyed at every junction in a network that is unbalanced with respect to Kirchhoff II. Indeed, this is usually the case at the stage of selecting assumed airflows, Q_a. Furthermore, if one or more branches are to have their airflows fixed by regulators or booster fans, then those airways may be omitted from the Hardy Cross analysis as they are not to be subjected to ΔQ_m adjustments. Such omissions will result in Kirchhoff I apparently remaining violated at the corresponding junctions throughout the analysis. This does not affect the ability of the system to attain a balanced pressure distribution that obeys Kirchhoff II.

Example   Figure 7.5(a) is the schematic of a simple network and gives the resistance of each branch. A fan produces a constant total pressure of 2000 Pa and the airflow in branch 3 is to be regulated to a fixed airflow of 10 m³/s. Determine the distributions of airflows and frictional pressure drops, and the resistance of the regulator required in branch 3.

Solution   Although the Hardy Cross procedure is widely utilized in computer programs, its tedious arithmetic ensures that it is seldom employed for manual application. In this solution we shall, therefore, follow the procedures that are employed in a typical computer program for ventilation network analysis.

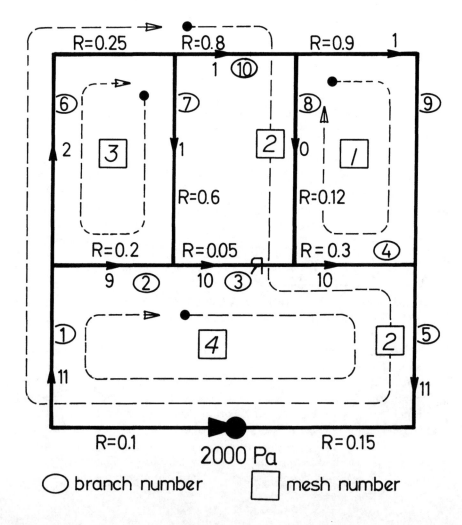

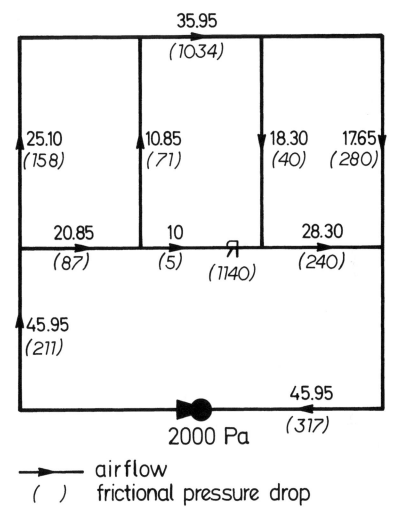

35.95
(1034)

25.10
(158)

10.85
(71)

18.30   17.65
(40)   (280)

20.85   10   28.30
(87)   (5)   (1140)   (240)

45.95
(211)

45.95
(317)

2000 Pa

——→ airflow
( ) frictional pressure drop

Figure 7.5 (a) Example network showing airway resistances, initial assumed airflows and the meshes chosen. (b) Final airflows after nine iterations.

1. *Mesh selection.* There are ten branches labelled in the network. However, branches 1 and 5 are connected in series and may be considered as a single branch for the purposes of mesh selection. Furthermore, branch 3 has a fixed (regulated) airflow and, hence, need not enter into the Hardy Cross analysis. The effective number of branches is, therefore, $b = 8$, while there are $j = 6$ junctions, giving the required minimum number of meshes to be $b - j + 1 = 3$. An arrow is indicated to give each branch a positive direction. This may be chosen as the direction in which the airflow is expected to move although either direction is acceptable. The branch arrow is merely a convenience to assist in traversing closed meshes within the network.

If we write down a list of branch resistances in order of decreasing value, then the three at the top of this list will correspond to branches 9, 10 and 7. We shall choose our first mesh commencing on branch 9 and close the mesh through a convenient route but without using branches 10, 7 or the fixed quantity, branch 3. Figure 7.5(a) shows the loop traversed by mesh 1. The second mesh commences on branch 10 and closes without traversing branches 9, 7, or 3. Similarly, the third mesh commences on branch 7 and involves no other high resistance or fixed quantity branch. The routes of the meshes so chosen are indicated on Fig. 7.5(a).

This technique of mesh selection is sometimes known as the **branch tree** method and leads to efficient convergence towards a balanced network as each high resistance branch appears in one mesh only.

A fourth mesh is selected, commencing on the fixed quantity (branch 3) and closing through any convenient route that does not include other fixed quantities, and irrespective of branch resistance values. This extra mesh will not be employed during the Hardy Cross analysis but will prove useful for the final calculation of the regulator resistance.

2. *Estimate initial airflows.*   This is accomplished in two stages commencing from zero flow throughout the network. First, mesh 4 (the fixed quantity mesh) is traversed in a direction defined by the fixed airflow, adding a ΔQ_m of 10 m^3/s (the required fixed airflow) to each branch around that mesh. As in all subsequent applications of mesh correction factors, ΔQ_m, the actual correction to each branch flow is ΔQ_m multiplied by $+1$ if the branch direction arrow coincides with the path of the traverse, or -1 if the branch direction arrow opposes the path of the traverse. This device gives the required airflow in the fixed quantity branch while ensuring that Kirchhoff I remains true at each junction.

A similar procedure is followed in each of the three 'normal' meshes, but employing a value of $\Delta Q_m = 1$. This merely ensures that the denominator of the mesh correction equation (7.35) or (7.36) will be non-zero during the first iteration.

The airflows at this stage are indicated on Fig. 7.5(a) and show the distribution at the commencement of the Hardy Cross procedure. It should be noted that if the method were to be applied manually then the number of iterations could be reduced by estimating a more realistic initial distribution of airflow.

3. *Calculate mesh correction factor.*   Assuming the flow to be turbulent in each branch, we may utilize the square law and, hence, equation (7.36) to calculate the mesh correction factors. In this example, there are no natural ventilating pressures (NVP $= S_{nv} = 0$) and the fan is assumed to remain at a fixed pressure ($S_f = 0$).

Applying equation (7.36) for mesh 1 gives

$$\Delta Q_{m1} = \frac{-(0.9 \times 1^2 - 0.3 \times 10^2 - 0.12 \times 0^2)}{2(0.9 \times 1 + 0.3 \times 10 + 0.12 \times 0)}$$

Table 7.1 Mesh correction factors and mesh pressure drops converge towards zero as the iterations proceed

Iteration	Sum of pressure drops around Mesh Δp (Pa)			Mesh correction factor ΔQ_m (m^3/s)		
	Mesh 1	Mesh 2	Mesh 3	Mesh 1	Mesh 2	Mesh 3
1	− 29	− 1958	6027	3.73	153.47	− 73.03
2	− 10326	28473	− 4379	73.60	− 64.63	34.17
3	5367	8810	− 1231	− 34.26	− 33.36	16.66
4	1579	2524	− 358	− 16.49	− 14.71	7.43
5	485	613	− 91	− 7.18	− 4.69	2.40
6	128	114	− 15	− 2.25	− 0.97	0.45
7	23	16	− 2	− 0.434	− 0.143	0.056
8	3.2	2.0	0.2	− 0.059	− 0.017	0.007
9	0.4	0.2	− 0.03	− 0.007	− 0.002	0.001

$$= \frac{29.1}{7.8} = 3.73$$

4. *Apply the mesh correction factor.* Mesh 1 is traversed once again, correcting the flows algebraically by adding $\Delta Q_{m1} = 3.73$. The relevant branch flows then become:

$$\text{branch} \quad 9: \quad 1 + (3.73 \times 1) \quad = 4.73 \, \text{m}^3/\text{s}$$
$$\text{branch} \quad 4: \quad 10 + (3.73 \times -1) = 6.27 \, \text{m}^3/\text{s}$$
$$\text{branch} \quad 8: \quad 0 + (3.73 \times -1) = -3.73 \, \text{m}^3/\text{s}$$

5. *Completion of mesh corrections.* Steps 3 and 4 are repeated for meshes 2 and 3 in turn. The corresponding values of ΔQ_{m2} and ΔQ_{m3} are 153.47 and − 73.03 m^3/s respectively.

6. The processes involved in steps 3, 4 and 5 are repeated iteratively until Kirchhoff's second law (numerator of equation (7.36)) balances to within ± 1 Pa for each mesh. The airflow correction factors and sums of pressure drops for each mesh are shown in Table 7.1.

Notice how the initial large oscillations are followed by a controlled convergence towards zero despite the large opening values of ΔQ_m.

The final flow pattern arrived at after these nine iterations is given on Fig. 7.5(b). The corresponding frictional pressure drops are each calculated as $p = RQ^2$ and shown in parentheses against each airflow.

The pressure drop across the regulator, p_{reg}, in branch 3 can be determined from Kirchhoff's second law by summing the known pressure drops around mesh 4:

$$p_{reg} = 2000 - (211 + 87 + 5 + 240 + 317)$$

$$= 1140 \, Pa$$

As the airflow through the regulator is known to be $10 \, m^3/s$, the square law gives the regulator resistance as

$$R_{reg} = p_{reg}/10^2$$

$$= 1.14 \, Ns^2/m^8$$

The actual dimensions of regulator orifice that will give this resistance may be determined using section 5A.5 given at the end of chapter 5.

7.4 VENTILATION NETWORK SIMULATION PACKAGES

7.4.1 Concept of a mathematical model

Suppose that a vehicle leaves its starting station at time t_1 and travels for a distance d at an average velocity of u. The time at which the vehicle arrives at its destination, t_2, is given by

$$t_2 = t_1 + \frac{d}{u}$$

This equation is a trivial example of a mathematical model. It **simulates** the journey in sufficient detail to achieve the objective of determining the arrival time. Furthermore, it is general in that it performs that simulation for any moving object and may be used for any given values of t_1, d and u. Different data may be employed without changing the model.

More sophisticated mathematical models may require many equations to be traversed in a logical sequence so that the result of any one calculation may be used in a later relationship. This logical sequence will often involve feedback loops for iterative processes such as the Hardy Cross procedure.

A **simulation program** is a mathematical model written to conform with one of the computer languages. The basic ambition of a simulation program is to produce numerical results that approximate those that would be given by the real system. There are three considerations that govern the accuracy of a simulation program; first, the accuracy with which each individual process is represented by its corresponding equation (e.g. $p = RQ^2$); secondly, the precision of the data used to characterize the actual system (e.g. airway resistances) and thirdly, the accuracy of the numerical procedure (e.g. the cut-off criterion to terminate cyclic iterations).

Many simulation programs represent not one but a number of interacting features comprising both physical systems and organizational procedures. In this case, several computer programs and data banks may be involved, interacting with each other.

In such cases, the computer software may properly be referred to as a **simulation package** rather than a single program. This is the situation for current methods of ventilation network analysis on desk top personal computers.

7.4.2   Structure of a ventilation simulation package

Figure 7.6 shows the essential structure of a ventilation network analysis package for a microcomputer system. It includes a series of input files with which the user can communicate through a keyboard and screen monitor, either for new data entry or for editing information that is already held in a file. All files are normally maintained on a magnetic disk or other media that may be accessed rapidly by the computer.

The user input file shown in Fig. 7.6 holds data relating to the geometric structure of the network, the location of each branch being defined by a 'from' and 'to' junction number. This same file contains the type of each branch (e.g. fixed resistance or fixed airflow and information relating to branch resistances. The latter may be given (a)

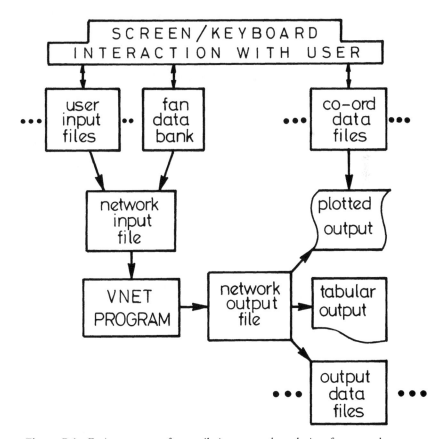

Figure 7.6   Basic structure of a ventilation network analysis software package.

directly by the user, (b) defined as a measured airflow–pressure drop or (c) implied by a friction factor, airway geometry and an indication of shock losses. The user input file allows entry of resistance data in whatever form they are available. Another type of user input file allows the operator to specify the location and the pressure–volume characteristic curves of each fan in the network.

The fan data bank is a convenient alternative means of storing the pressure–volume coordinates for a number of fan characteristic curves. Data from fan manufacturers' catalogues can be held in the fan data bank and recalled for use in the subsequent analysis of any network.

The ventilation network analysis program (denoted as VNET on Fig. 7.6) in most commercial packages is based on the mesh selection process and Hardy Cross technique described and illustrated by example in section 7.3.2. However, these procedures have been modified to maximize the rate of information flow, the speed of mesh selection and to accelerate convergence of the iterative processes. In order to achieve a high efficiency of data transfer, the VNET program may require that the network input data be specified in a closely defined format. However, this is contrary to the need for flexibility in communicating with the user. Hence, Fig. 7.6 shows a network input file acting as a buffer between the user-interactive files and the VNET program. Subsidiary 'management' programs, not shown on Fig. 7.6, carry out the required conversions of data format and control the flow of information between files. Similarly, the output from the VNET program may not be in a form that is immediately acceptable for the needs of the user. Hence, that output is dumped temporarily in a network output file. Under user control it may then be (a) copied into an output data file for longer term storage, (b) displayed in tabular form on the screen or printer and (c) produced as a plotted network on a screen or hard-copy plotter.

If graphical output is required then the user-interactive coordinate data file shown on Fig. 7.6 must first be established. This contains the x, y coordinates of every junction in the network. It is convenient to enter such data directly from a drawn schematic by means of a digitizing pad or plotter. Alternatively, for editing an established coordinate data file it is often preferable to enter numerical values for the amended or additional x, y coordinates.

7.4.3 Operating system for a VNET package

When running a ventilation network analysis package, the user wishes to focus his/her attention on ventilation planning and not to be diverted by matters relating to the transfer of data within the machine. Keyboard entries are necessary for control of the computer operations. However, these should be kept sufficiently simple that they allow a near-automatic response from the experienced user. Messages on the screen should provide a continuous guide on operations in progress or alternative choices for the next operation. For these reasons, a simulation package intended for industrial use must include a **management** or **executive** program that controls all internal data transfers.

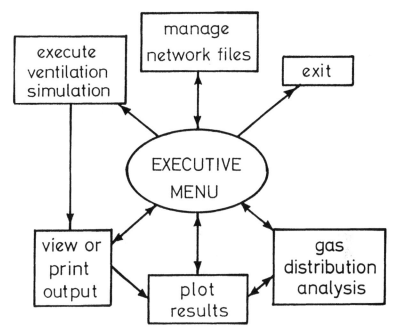

Figure 7.7 Structure of operating system for a ventilation network analysis software package.

The user is conscious of the executive program only through the appearance of a hierarchy of **menus** or **windows** on the screen. Each menu lists a series of next steps from which the user may choose. These include progression to other menus. Hence, with few keyboard entries, and with very little conscious effort, the user may progress flexibly and rapidly through the network exercise.

Figure 7.7 illustrates the executive level (master menu) of one ventilation network package. Choosing the 'execute ventilation simulation' option will initiate operation of the VNET program (Fig. 7.6) using data currently held in the network input file. The 'exit' option simply terminates operation of the network simulator. Each of the other four options leads to other menus and submenus for detailed manipulation of the operating system.

7.4.4 Incorporation of air quality into network simulation package

The primary purpose of a ventilation network analysis package is to predict the distribution of air quantity and pressure differentials throughout the network. However, the quality of the air with respect to gaseous or particulate pollutants, or its psychrometric condition, depends also upon the distribution of airflow. Separate programs designed to predict individual aspects of air quality may, therefore, be incorporated into a more general ventilation package. Branch airflows and/or other data produced as output from the network analysis program may then be utilized as input to air quality simulators.

As an example, the operating system depicted by Fig. 7.7 incorporates a gas distribution analysis. This requests the user to identify the emitting locations and rates of production of any gas, smoke or fumes. It then retrieves the airflow distribution contained in the network output file or any named output data file (Fig. 7.6) in order to compute the distribution and concentrations of that pollutant throughout the network.

A library of magnetic data files coupled with mine airflow and air quality simulation programs provides a powerful tool for the design and control of underground ventilation systems.

7.4.5   Obtaining a ventilation simulation package

The first step in establishing a microcomputer system of ventilation network analysis at any mine or other site is to review the availability of both hardware and software. There are a number of ventilation network packages available for personal computers. The potential user should investigate the following considerations.

Hardware requirements and size of network

The type of package described in sections 7.4.2 and 7.4.3 will require a plotter and printer in addition to the microprocessor, keyboard and screen monitor. The internal memory requirements of the computer will depend upon the scope of the software package and the largest network it is capable of handling.

Cost of software

This can vary very considerably depending on the sophistication of the system and whether it was developed by a commercial company or in the more public domain of a university or national agency.

Speed

The speed of running a network depends on the type of microprocessor, operating electrical frequency, and the size and configuration of the network. Typical run times should be sought from software developers for a specified configuration of hardware.

Scope and ease of use

It is important that, in reviewing simulation packages, the user chooses a system that will provide all of the features likely to be required in the foreseeable future. However, there is little point in expending money on features that are unlikely to be used. Furthermore, it is usually the case that the more sophisticated packages demand a greater degree of user expertise. Demonstration versions may be available to indicate the scope and ease of use of a program package. Other users may be approached for opinions on any given system. Software developers should be requested to give names and addresses of several current users of their packages.

User's manual and back-up assistance

It is vital that any simulation package should be accompanied by a comprehensive user's manual for installation and operation of the system. Additionally, the producer of the software must be willing and anxious to provide back–up assistance to all users on request. This can vary from trouble-shooting on the telephone to replacement of a complete software system.

7.4.6 Example of a computed network

The ventilation networks of actual mines or other subsurface facilities may contain several hundred branches, even after compounding relevant airways into equivalent resistances (section 7.3.1). However, in the interests of brevity and clarity, this example is restricted to the very simple network schematic shown on Fig. 7.8.

In response to screen prompts, the user keys in each item of data shown in Table 7.2. This table is, in fact, a printout of the user input file (section 7.4.2). The pressure–volume coordinates for the fan are either read manually from the appropriate fan

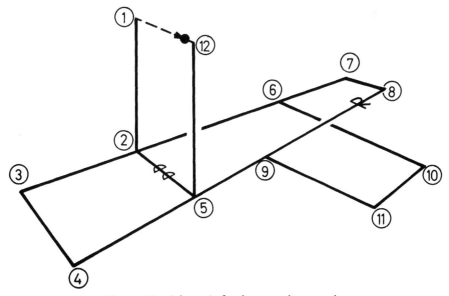

Figure 7.8 Schematic for the example network.

Table 7.2 Input data supplied by user for the case study

★ DESCRIPTIVE DATA ★

File Name: TEXTBOOK
Network Title: EXAMPLE OF VNET PACKAGE RUN
Mine Name: TOTTIE CREEK
Company Name: LAFAYETTE DIGGERS
Comments: ALL DATA IN THIS FILE ENTERED BY USER
SI Units
Surface Reference Junction Number   1

★ FAN DATA ★

Average Efficiency of Fans is 65%, Power Costs are 8.0 cents/kWh

Fan Ref No	From	To	Operating Pressure (kPa)	No. of Char. Pts.
—	12	1	0.500	10

★ FAN CHARACTERISTICS ★

Fan 1	Pressure (kPa)	Airflow (m^3/s)
	0.880	0.00
	0.850	20.00
	0.800	40.00
	0.720	60.00
	0.620	80.00
	0.570	90.00
	0.500	100.00
	0.410	110.00
	0.310	120.00
	0.000	140.00

★ BRANCH DATA ★

```
*************************************************************************
     SI Units: p-Pa  Q-m^3/s  R-Ns^2/m^8  k-kg/m^3  L-meters ★
*************************************************************************
```

No.	From	To	Resist.	Airflow	Press. drop	k	Length	Equiv. length	Area	Per
1	1	2		69.82	13					
2	2	3				0.016	300.0	45.0	12.0	18.0
3	3	4		28.32	94					
4	4	5				0.016	300.0	30.0	12.0	18.0
5	2	5	25.00000							
6	2	6	0.03621							
7	6	7	0.02716							
8	7	8	1.00000							
9	8	9	0.02716	12.00	-Fixed					
10	6	10	0.02714							
11	10	11		17.23	56					
12	11	9	0.01810							
13	9	5	0.02714							
14	5	12				0.004	350.0	100.0	28.3	21.4
15	12	1	0.00000							

Table 7.3 Tabulated output product by the computer

*** OUTPUT DATA ***

Annual costs are based on electricity charges of 8.0 cents per kWh and fan efficiencies of 65.0% Cost given for an NVP represents money saved by natural ventilation

*** Fan Operating Points ***

Fan Ref No.	From	To	Pressure kPa	Quantity m^3/s	Air Power kW	Op. Cost $/years
1 —	12	1	0.511	98.38	50.27	54203

*** Branch Data ***

Branch	From	To	Press. Dp Pascals	Airflow m^3/s	Resist. Ns^2/m^8	AP Loss kW	Op. Cost $/year	
1	1	2	25	98.38	0.00267	2.5	2652	
2	2	3	117	45.19	0.05750	5.3	5700	
3	3	4	239	45.19	0.11720	10.8	11644	
4	4	5	112	45.19	0.05500	5.1	5457	
5	2	5	469	4.33	25.00000	2.0	2190	
6	2	6	86	48.86	0.03621	4.2	4531	
7	6	7	3	12.00	0.02720	0.0	39	
8	7	8	144	12.00	1.00000	1.7	1863	
9★	8	9	169	12.00	1.17966	2.0	2187	-Regulator Required
10	6	10	36	36.86	0.02714	1.3	1431	
11	10	11	256	36.86	0.18863	9.4	10175	
12	11	9	24	36.86	0.01810	0.9	954	
13	9	5	64	48.86	0.02714	3.1	3372	
14	5	12	16	98.38	0.00170	1.6	1697	
15	12	1	0	98.38	0.00000	0.0	0	

Number of Iterations = 9

*** Regular List ***

Branch	From	To	Fixed Quantity (m^3/s)	Regulator Resistance (Ns^2/m^8)
9	8	9	12.0	1.15254

The following table gives the frictional pressure relative to 0 Pascals at junction No. 1. The table may be used to find neutral points and the pressure difference available to produce flow between any two junctions in the network.
** The value 99999 indicates an inaccessible junction **

Junction	Pressure	Junction	Pressure	Junction	Pressure	Junction	Pressure
1	0	2	−25	3	−142	4	−381
5	−493	6	−111	7	−114	8	−258
9	−427	10	−147	11	−403	12	−509

*** NETWORK EXERCISE COMPLETE ***

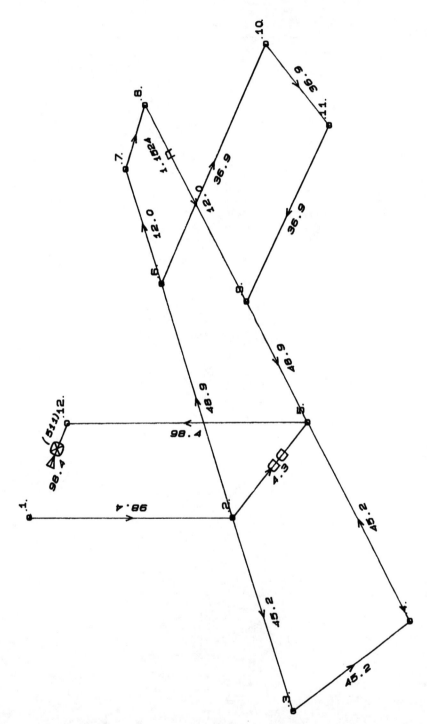

Figure 7.9   Graphical output of airflows (m³/s) produced by the computer. Similar plots may be obtained for other parameters.

characteristic or copied over from a characteristic that has been entered previously into a fan data bank. The branch data section of the user input file illustrates that airway resistances may be entered directly, from survey (p, Q) observations or from airway geometry and friction factor. The software package employed for this example required shock losses to be entered as equivalent lengths (section 5.4.5).

A request by the user to execute the simulation causes the user input file to be translated to the network input file (Fig. 7.6) and the computed results to be deposited in the network output file. These may be inspected on the screen or listed on a printer as shown on Table 7.3.

In order to produce plotted results, the schematic is placed on a digitizing tablet or plotter and a sight-glass positioned over each junction in turn as directed by screen prompts. The resulting coordinate data file is then combined with the computed results, on request by the user, to produce the graphical output shown on Fig. 7.9. Similar plots may be generated for any of the five columns of computed information shown in the branch data section of Table 7.3. Some programs allow the user to select band widths for colour coding of the plots. Similar plots may be generated for air quality parameters if the package includes the relevant programs.

Following the relatively time-consuming process of entering the initial data for a new network, and having inspected the corresponding output, amendments can be made to the user-interactive files and the network re-run rapidly. In this way, errors in data entry can be corrected and further network exercises pursued with only a few keystrokes. It is this efficiency, rapidity and ease of testing many alternatives that has revolutionized the business of ventilation system design. Indeed, it is the skill of the engineer in interrogating computer output and deciding on the next exercise that now governs the speed of ventilation planning rather than the mathematical mechanics of the chosen method of network analysis.

REFERENCES

Mass, W. (1950) An electrical analogue for mine ventilation and its application to ventilation planning. *Geol. Mijnbouw* **12** (April).

McElroy, G. W. (1954) A network analyzer for solving mine ventilation distribution problems. *US Bur. Mines Inf. Circ.* **7704**, 13 pp.

McPherson, M. J. (1964) Mine ventilation network problems (solution by digital computer). *Colliery Guardian* (August 21), 253–254.

Williams, R. W. (1964) A direct analogue equipment for the study of water distribution networks. *Ind. Electron.* **2** 457–459.

FURTHER READING

Atkinson, J. J. (1854) On the theory of the ventilation of mines. *N. Engl. Inst. Min. Eng.* (3), 118.

Cross, H. (1936) Analysis of flow in networks of conduits or conductors. *Bull. Illinois Univ. Eng. Exp. Station*, No. 286.

Hartman, H. L. and Wang, Y. J. (1967) Computer solution of three dimensional mine ventilation networks with multiple fans and natural ventilation. *Int. J. Rock Mech. Sci.* **4**.

McPherson, M. J. (1966) Ventilation network analysis by digital computer. *Min. Eng.* **126** (73), 12–28.

McPherson, M. J. (1984) Mine ventilation planning in the 1980s. *Int. J. Min. Eng.* **2**, 185–227.

Scott, D. R. and Hinsley, F. B. (1951–1952) Ventilation network theory. Parts 1 to 5. *Colliery Eng.* **28**, **29**.

Wang, Y. J. and Saperstein, L. W. (1970) Computer-aided solution of complex ventilation networks. *Soc. Min. Eng. AIME* **247**.

8

Mine ventilation thermodynamics

8.1 INTRODUCTION

Many of the world's practising mine ventilation engineers—perhaps, even, the majority—perform their duties very successfully on the basis of relationships that assume incompressible flow. Some of those engineers may question the need to concern themselves with the more detailed concepts and analyses of thermodynamics. There are, at least, two responses to that query.

First, if the ranges of temperature and pressure caused by variations in elevation and heat transfer produce changes in air density that are in excess of 5%, then analyses that ignore those changes will produce consistent errors that impact adversely on the accuracy of planned ventilation systems. In practical terms, this means that for underground facilities that extend more than 500 m below the highest surface connection, methods of analysis that assume incompressible flow may be incapable of producing results that lie within observational tolerances of accuracy.

However, there is a second and even more fundamental reason why all students of the subject, including practising engineers, should have a knowledge of mine ventilation thermodynamics. Although the incompressible flow relationships are simple to apply, they are necessarily based on an approximation. Air is, indeed, highly compressible. It follows, therefore, that if we are truly to comprehend the characteristics of subsurface ventilation systems, and if we are really to understand the mechanisms that govern the behaviour of large-scale airflow systems, then this can be accomplished only if we have a knowledge of steady-flow thermodynamics.

In this chapter, thermodynamic analyses are carried out on a downcast shaft, underground airways and an upcast shaft. The three are then combined to produce a thermodynamic cycle for a complete mine, first, with natural ventilation only, then for the more usual situation of a combination of fans and natural ventilation. In all cases, the analyses utilize pressure–volume (PV) and temperature–entropy – (Ts) diagrams. This chapter assumes a knowledge of the basic concepts and relationships that are developed in Chapter 3. It is suggested that the reader should study that chapter before proceeding with this one.

One further point—the temperatures and, to a lesser extent, pressures of the air are affected by variations of water vapour in the air. This is given a detailed treatment in Chapter 14. However, for the time being, we shall assume that the airways are dry and are uncomplicated by evaporative or condensation processes.

8.2 COMPONENTS OF THE MINE CYCLE

8.2.1 Elements of the system

A subsurface ventilation system follows a closed cycle of thermodynamic processes and can be illustrated by visual representations of those processes on thermodynamic diagrams that are analogous to the indicator diagrams of heat engines. It was this resemblance that led Baden Hinsley (1900–1988) into the realization that a mine ventilation system is, indeed, a gigantic heat engine. Air enters the system and is compressed and heated by gravitational energy in the downcast shaft. More heat is

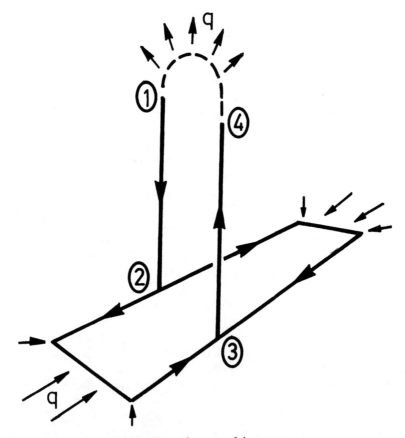

Figure 8.1 Elements of the system.

added to the air from the strata, machines and other sources, and does work as the air expands during its ascent through the upcast shaft. Some of the added heat is converted temporarily to mechanical energy and assists in promoting airflow. In the great majority of mines, this 'natural ventilating energy' is supplemented by input fan work.

When the exhaust air re-enters the pressure sink of the surface atmosphere, it cools to the original entry conditions, closing the cycle. Figure 8.1 illustrates the descending flow through a downcast shaft between stations 1 and 2, level workings 2 to 3, and returning to surface through an upcast shaft 3 to 4. We shall analyse each of the three processes separately before adding the isobaric cooling in the surface atmosphere to complete the cycle.

8.2.2 The downcast shaft

Air enters a downcast shaft at the pressure and temperature of the atmosphere existing at the top of the shaft. As the air falls down the shaft, its pressure increases, just as the pressure on a diver increases as he descends into the ocean. The rise in pressure is caused by the increasing weight of the overlying column of fluid as we plunge deeper into the fluid. However, that simple observation belies the common belief that air always flows from a high pressure region to a connected lower pressure region. In a downcast shaft, exactly the opposite occurs, showing how easily we can be misled by simplistic conceptions.

The process of gravitational compression in the downcast shaft produces an increase in temperature of the air. This is independent of any frictional effects and will be superimposed upon the influence of any heat transfer with the surrounding strata that may occur across the shaft walls. The rate of that heat transfer depends upon the thermal properties of the rock and the difference between the rock temperature and air temperature at any given horizon. Now, while the temperature of the mass of rock surrounding the shaft may change relatively slowly with time, the air temperature at the shaft entrance can change from hour to hour and, especially, between day and night. Because of these surface variations, it is common for the walls and rock surrounding a downcast shaft to absorb heat during the day and to emit heat during the night. The phenomenon continues along the intake airways and tends to dampen out the effects of surface temperature variation as we follow the air into a subsurface facility. This is sometimes called the 'thermal flywheel'. However, for the purposes of this chapter, we shall assume that any heat exchange that takes place in the downcast shaft is distributed equitably so that the process approximates closely to a polytropic law.

If we descend a downcast shaft, stopping every hundred metres or so to take measurements of pressure and temperature, we can plot the PV and Ts diagrams using equation (3.9):

$$V = \frac{RT}{P} \quad \text{m}^3/\text{kg}$$

(Chapter 14 shows how this relationship should be amended for the presence of water vapour.) Equation (3.48) allows us to determine the variation of entropy:

$$s_b - s_a = C_p \ln(T_b/T_a) - R \ln(P_b/P_a) \quad J/(kg\,K)$$

The starting point for the latter equation may be any defined temperature and pressure datum. It is differences in entropy rather than absolute values that are important.

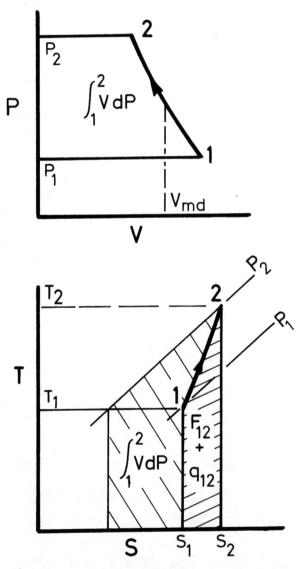

Figure 8.2 *PV* and *Ts* diagrams for a downcast shaft.

The diagrams that emerge have the appearance of those given in Fig. 8.2. In practice, there is often a scatter of points near the end extremities of the shaft due to high turbulence and uneven mixing of thermally stratified airstreams.

The first step in the analysis of the data is to conduct a curve fitting analysis to the points on the PV diagram. The form of the equation to be fitted is the polytropic law

$$P = \frac{C}{V^n}$$

where C is a constant and n the polytropic index. Least-squares regression analyses are widely available for personal computers and hand calculators. The quality of the curve fit should be reviewed. It is sometimes apparent that some points, particularly at shaft ends, should be rejected or down-weighted.

If the value of the polytropic index is higher than the isentropic index, 1.4 for dry air, then the combined effect of added heat and internal friction is positive. If, however, the index is less than 1.4 then sensible heat is being lost from the air to the strata or through conversion into latent heat by evaporation of water. If the measurements have, unwisely, been taken shortly after a significant change in surface temperature, then a variation in n may become apparent by comparing the plotted points with the best-fit polytropic line.

In the majority of cases, satisfactory representation by a polytropic law can be obtained. Where significant deviation occurs then the flow work $\int_1^2 V \, dP$ can be obtained graphically as the area to the left of the curve on the PV diagram.

A thermodynamic analysis of a polytropic compression is given in section 3.5.3. The work done against friction, F_{12}, in the shaft is given by combining the steady flow energy equation (3.25) with the evaluation of the flow work term shown by equation (3.73). This gives

$$F_{12} = \frac{u_1^2 - u_2^2}{2} + (Z_1 - Z_2)g - R(T_2 - T_1)\frac{\ln (P_2/P_1)}{\ln (T_2/T_1)} \quad \text{J/kg} \qquad (8.1)$$

or, alternatively,

$$F_{12} = \frac{u_1^2 - u_2^2}{2} + (Z_1 - Z_2)g - \frac{n}{n-1}R(T_2 - T_1) \quad \text{J/kg}$$

from equation (3.69).

While all terms in these equations are measurable and, in principle, the values at the top (station 1) and the bottom (station 2) of the shaft can be used to determine F_{12}, there are some practical pointers that contribute greatly to the achievement of good results. First, it is often difficult to measure the mean velocity of air in a shaft. It may be preferable to measure the airflow(s) in underground airways and to carry out the necessary summations and corrections for air density to give u_1 and u_2. Fortunately, the magnitude of the kinetic energy term is normally very small compared with the other factors.

The term $Z_1 - Z_2$ is simply the depth of the shaft connecting stations 1 and 2. The local value of gravitational acceleration, g, should be ascertained, taking into account mean elevation and latitude. For the pressures and temperatures P_1, P_2, T_1 and T_2, it is inadvisable to use the raw measurements made at the shaft extremities because of the unstable conditions that may exist at those locations. It is preferable to construct separate plots of pressure and temperature against depth and to use best-fit lines to establish representative pressures and temperatures at depths Z_1 and Z_2.

Equation (8.1) gives the amount of mechanical energy that is converted to heat through frictional processes within the shaft airflow. This can be employed to determine the Chézy–Darcy coefficient of friction, f, for the shaft. From equation (2.46)

$$f = \frac{2d}{4Lu^2} F_{12} \quad \text{(dimensionless)} \tag{8.2}$$

where $d =$ hydraulic mean diameter of the shaft (m), $L =$ length of shaft (m) and $u =$ mean air velocity (m/s).

Although the F term is the primary factor quantifying the work done against friction, it has one major drawback. It cannot be observed directly but must be computed from measurements including those of pressure and temperature. On the other hand, we know that the loss of mechanical energy by friction produces a frictional pressure drop, p, which can be measured directly (Chapter 6). It is, therefore, much easier to conceptualize the notion of a frictional pressure drop, p, than the fundamental frictional work, F, which produces that pressure drop. Furthermore, recalling that ventilation planning for facilities of less than 500 m in depth can be conducted using the simpler relationships of incompressible flow, it is useful to remind ourselves of the relationship between p and F. This was introduced in Chapter 2:

$$p = \rho F \quad \text{Pa} \tag{8.3}$$

Hence, the work done against friction determined from equation (8.1) can be converted to a corresponding frictional pressure drop:

$$p_{12} = \rho F_{12} \quad \text{Pa}$$

Now we have another problem. The density, ρ, varies throughout the shaft, so what value do we select? One option is to employ a mean value, ρ_{md}, where the subscript md denotes 'mean downcast'. For polytropic flow in a dry shaft the variations of both temperature and pressure are near linear with respect to depth. It follows that air density, $\rho = P/RT$, also increases in a near-linear manner. Hence the arithmetic mean of densities measured at equal intervals throughout the shaft may be employed. Again, observers should be cognizant of the unrepresentative values that may be measured at shaft extremities. More generally,

$$\rho_{md} = \frac{\int_1^2 \rho \, dZ}{Z_1 - Z_2} \quad \text{kg/m}^3 \tag{8.4}$$

Using the mean density, and ignoring the small kinetic energy term, the steady flow energy equation can be written as

$$(Z_1 - Z_2)g = V_{md}(P_2 - P_1) + F_{12}$$

where $V_{md} = 1/\rho_{md}$ (mean specific volume)

or

$$p_{12} = F_{12}\rho_{md} = \rho_{md}(Z_1 - Z_2)g - (P_2 - P_1) \quad \text{Pa} \qquad (8.5)$$

This is simply a form of Bernoulli's equation for incompressible flow. We have, however, taken account of compressibility by using a mean value of air density. Equation (8.5) should be highlighted as we shall return to it later as a component in an approximation of natural ventilation pressure.

One further problem remains. If we wish to compare the frictional work done in a series of different airways then the F terms should be employed as these depend only upon the physical characteristics of the airway and the air velocity (equation (8.2)), and are independent of the thermodynamic state of the air. On the other hand, if we compare frictional pressure drops, using equation (8.5), then these depend also on the corresponding values of air density. As the latter may vary considerably around a mine circuit, frictional pressure drops are not a good basis for comparison. (This was the reason for the consistent deviations that were noticed in the results of pressure surveys prior to the development of mine ventilation thermodynamics.) The problem can be met by choosing a 'standard' value of air density, ρ_{st}, to relate p and F in equation (8.3). If the same value is used for all airways, then the comparison between airways becomes a constant multiple of F. The standard value of air density is normally taken as $1.2\,\text{kg/m}^3$ and the corresponding 'frictional pressure drop', p_{st}, should then be identified as being 'standardized', or **referred** to standard density. It should be understood that standardized pressure drops have no direct physical significance other than the fact that they are a constant multiple of work done against friction.

Let us turn now to the Ts diagram on Fig. 8.2. The area under the process line $\int_1^2 T\,ds$ represents the combination of added heat and the internal generation of friction heat $q_{12} + F_{12}$. Fortunately, the three-part steady-flow energy equation (3.25) allows us to separate the two terms. Equation (8.1) gives F_{12}, and

$$q_{12} = H_2 - H_1 - \frac{u_1^2 - u_2^2}{2} - (Z_1 - Z_2)g \quad \frac{\text{J}}{\text{kg}} \qquad (8.6)$$

If the shaft is dry, then $H_2 - H_1 = C_p(T_2 - T_1)$ and, by ignoring the change in kinetic energy, we have

$$q_{12} = C_p(T_2 - T_1) - (Z_1 - Z_2)g \quad \text{J/kg} \qquad (8.7)$$

In many shafts that have reached an effective equilibrium with respect to the temperature of the surrounding strata there is very little heat transfer, i.e. adiabatic conditions with $q_{12} = 0$. Equation (8.7) then gives

$$T_2 - T_1 = \frac{(Z_1 - Z_2)g}{C_p} \quad °\text{C} \qquad (8.8)$$

This well-known equation allows the increase in temperature with respect to depth for a dry shaft to be calculated, in the absence of heat transfer. Inserting the constants 9.81 m/s^2 for g and 1020 J/(kg K) for C_p (value of specific heat of air with a moisture content of 0.0174 kg water vapour per kilogram of dry air) gives

$$T_2 - T_1 = 0.009\,62(Z_1 - Z_2) \quad °C \qquad (8.9)$$

This result is usually rounded off and quoted as a dry bulb temperature **adiabatic lapse rate** of $1 °C$ per 100 m depth. The increase in temperature is a result of potential energy being converted to thermal energy as the air falls through the shaft. The effect is also referred to as **autocompression**.

In practice, the change in temperature often varies considerably from $1 °C/(100 \text{ m})$ because of both heat transfer and reductions in dry bulb temperature caused by evaporation of water.

In section 3.5.3 it was shown that, in addition to areas on the Ts diagram illustrating heat, the same diagram may be used to indicate numerical equivalents of flow work. In particular, on the Ts diagram in Fig. 8.2, the total area under the isobar P_2 between temperatures T_2 and T_1 represents the enthalpy change $H_2 - H_1$.

However, the flow work is given by the steady flow energy equation as

$$\int_1^2 V \, dP = H_2 - H_1 - (q_{12} + F_{12}) \quad \text{J/kg} \qquad (8.10)$$

As the areas representing $H_2 - H_1$ and $q_{12} + F_{12}$ have already been identified on the Ts diagram, their difference gives a representation of the flow work.

Example   This example is taken from an actual survey of a 1100 m deep downcast shaft of diameter 5.5 m. At the time of the survey, the mass flow of air was 353 kg/s. Figures 8.3(a) and 8.3(b) show the barometric pressures and dry bulb temperatures measured at approximately 100 m intervals throughout the shaft.

Variations in pressure and temperature
The increase in pressure with respect to depth is seen to be near linear with a slope of 1.211 kPa per 100 m. The mean air density in the shaft was 1.26 kg/m^3, illustrating the convenient rule of thumb that the rate of pressure increase in kPa (100 m) is approximately equal to the mean density (in SI units). This follows from

$$p = \rho g h = 981 \rho \quad \text{Pa}$$

$$= 0.981 \rho \quad \text{kPa}$$

where $g = 9.81 \text{ m/s}^2$ and $h = 100 \text{ m}$.

The temperature plot should, theoretically, also be linear with respect to depth for adiabatic conditions or where heat transfer is distributed uniformly throughout the length of the shaft. In this case, the actual variation in tempera-

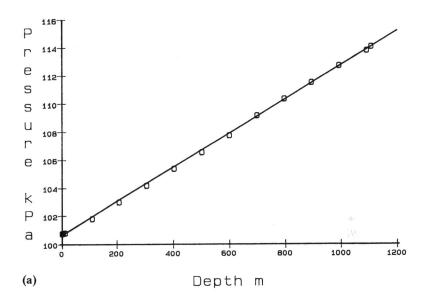

(a)

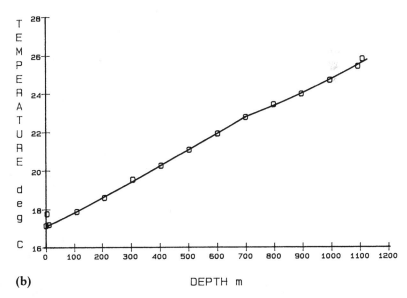

(b)

Figure 8.3 (a) Measurements of pressure in the shaft with a best-fit straight line. (b) Measurements of dry bulb temperature in the shaft with manually drawn line.

ture shows some interesting features. First, the increased scatter of points made at the ends of the shaft is clearly indicated. Secondly, the curve appears to be divided into discrete regions above and below 700 m. Visual observation indicated the existence of free water in the shaft. The lower part passed through an aquifer. In this section tubbing lining was employed and an increased water make produced falling droplets. The effects of enhanced evaporative cooling are shown clearly as a reduction in the slope of the temperature–depth line in the lower part of the shaft. The mean dry bulb temperature lapse rate is 0.827 °C/(100 m) down to 700 m and 0.657 °C/(100 m) at lower elevations, both less than the adiabatic lapse rate of 0.962 °C/(100 m) (equation (8.9)). This shows the important influence of water in the airway.

A third observation on the temperature–depth plot is that, within each of the two sections of the shaft, there is a slight curvature of the line. This is, again, indicative of non-uniform evaporation or heat transfer.

PV diagram, work dissipated against friction and frictional pressure drop

Figure 8.4 shows the *PV* diagram for the shaft. Despite the nonlinearity of the temperature, a curve-fitting exercise produced a good fit with the polytropic equation

$$PV^{1.3007} = 79.433$$

The non-representative points at the shaft ends were omitted for the curve fit. Figure 8.4 shows the curve of this equation superimposed upon the actual observations. The fact that the polytropic index was less than the isentropic 1.4 showed that sensible heat was being removed from the air—again, the combined effect of evaporative cooling and heat transfer.

The measurements may be used to determine the work done against friction within the airstream. Equation (8.1) gives

$$F_{12} = \frac{u_1^2 - u_2^2}{2} + (Z_1 - Z_2)g - R(T_2 - T_1)\frac{\ln (P_2/P_1)}{\ln (T_2/T_1)}$$

where subscripts 1 and 2 refer to the top and bottom of the shaft respectively. Independent measurements and corrections for density gave $u_1 = 12.78$ and $u_2 = 11.63$ m/s. The shaft depth, $Z_1 - Z_2$, was 1100 m and the value of g was 9.807 m/s². If the equation is to be used for the complete shaft then careful consideration must be given to the end values of pressure and temperature. The lack of uniform conditions at the shaft extremities indicates that single measurements of pressure and temperature at those locations are likely to produce erroneous results. The pressure and temperature plots should be examined carefully in order to select values of those parameters that are representative of an extension of trends within the shaft. In this case, the values chosen from the relevant plots were

$$P_1 = 100.79 \text{ kPa} \qquad\qquad P_2 = 113.90 \text{ kPa}$$
$$T_1 = 17.2 + 273.15 = 290.35 \text{ K} \qquad T_2 = 25.5 + 273.15 = 298.65 \text{ K}$$

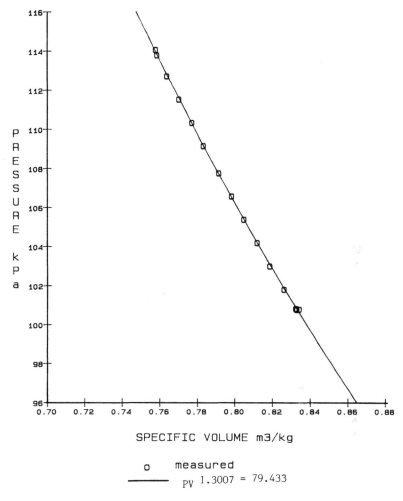

SPECIFIC VOLUME m3/kg

□   measured

―――   pv $^{1.3007}$ = 79.433

Figure 8.4 *PV* diagram for the shaft.

Equation (8.1) then gives

$$F_{12} = \frac{12.78^2 - 11.63^2}{2} + 1100 \times 9.807 - 287.04(25.5 - 17.2) \frac{\ln(113.9/100.79)}{\ln(298.65/290.35)}$$

$$= 14 + 10\,788 - 10\,336 \text{ J/kg}$$

$$= 466 \text{ J/kg}$$

This calculation illustrates that the kinetic energy term is, indeed, small compared with the other terms. More importantly, the potential energy and flow work terms are both large but similar numbers. Hence, small errors in the parameters that contribute to each of those terms can have a significant impact on the calculated value of F_{12}. For this reason, it is preferable to apply equation (8.1) for each increment of the shaft and to sum the individual values of F rather than to use the complete shaft. This approach also accounts for variations in the polytropic index throughout the shaft. When this was done for the current example, the cumulative F_{12} was 493 J/kg.

The flow work may also be calculated from the alternative expression (equation (8.1))

$$\frac{n}{n-1} R(T_2 - T_1) = \frac{1.3007}{0.3007} 287.04(25.5 - 17.2)$$

$$= 10\,305$$

giving

$$F_{12} = 14 + 10\,788 - 10\,305 = 497\,\text{J/kg}$$

We shall continue the analysis using the value of $F_{12} = 493$ J/kg obtained from the incremental approach.

The frictional work can now be converted to a frictional pressure drop referred to any chosen air density. The mean density in the shaft was 1.260 kg/m³, giving a corresponding frictional pressure drop of

$$p_{md} = \rho_{md} F_{12}$$

$$= 1.260 \times 493 = 622\,\text{Pa}$$

For comparison with other airways, the frictional pressure drop referred to standard air density, 1.2 kg/m³, becomes

$$p = \rho_{st} F_{12}$$

$$= 1.2 \times 493 = 592\,\text{Pa}$$

Coefficient of friction and airway resistance

The Chézy–Darcy coefficient of friction is given by equation (8.2):

$$f = \frac{2d}{4Lu^2} F_{12}$$

Using the mean air velocity of 12.20 m/s gives

$$f = \frac{2 \times 5.5}{4 \times 1100 \times (12.20)^2} \times 493$$

$$= 0.008\,28 \quad \text{(dimensionless)}$$

The Atkinson friction factor referred to standard density of 1.2 kg/m³ is

$$k = \frac{\rho_{st} f}{2} = \frac{1.2}{2} \times 0.008\,28$$

$$= 0.0050\,\text{kg/m}^3$$

(The values of f and k are mean values as the type of lining changed through the aquifer.)

The rational resistance of the shaft is given from equation (2.51) as

$$r = \frac{f}{2} \frac{\text{perimeter } L}{A^3} \quad \text{m}^{-4}$$

where perimeter $= \pi 5.5 = 17.28$ m and area $A = (5.5)^2/4 = 23.76$ m^2, giving

$$r = \frac{0.008\,28}{2} \times \frac{17.28}{(23.76)^3} \times 1100 = 0.005\,87 \text{ m}^{-4}$$

The Atkinson resistance referred to standard density becomes

$$R = \rho_{st} r$$

$$= 1.2 \times 0.005\,87 = 0.007\,04 \quad \text{N s}^2/\text{m}^8$$

These resistance values refer to the 1100 m length of open shaft. The shock losses caused by conveyances, inlet and exist losses must be assessed separately and added to give the total shaft resistance.

Heat exchange

The steady-flow energy equation gives the heat transfer to the air as

$$q_{12} = H_2 - H_1 - \frac{u_1^2 - u_2^2}{2} - (Z_1 - Z_2)g$$

The change in enthalpy, taking evaporation into account, may be determined from the methods given in section 14.5.1. If no evaporation or condensation occurred then the change in enthalpy would be given by equation (3.33):

$$H_2 - H_1 = C_p(T_2 - T_1)$$

The specific heat of the air, C_p, would be 1005 J/(kg K) if the air were perfectly dry. Again, Chapter 14 indicates how this can be corrected for the presence of water vapour. The actual value based on mean moisture content for this shaft was 1025 J/(kg K).

If the shaft had been dry then

$$q_{12} = 1025(25.5 - 17.2) - \frac{12.78^2 - 11.63^2}{2} - (1100 \times 9.807)$$

$$= 8507 - 14 - 10\,788$$

$$= -2294 \text{ J/kg}$$

The physical interpretation of this calculation is that, for each kilogram of air, 14 J of heat accrue at the expense of kinetic energy as compression decelerates the air, 10 788 J of thermal energy arise from the loss of potential energy, but the total increase in thermal energy of the air is only 8507 J. Hence, 2294 J of heat must be transferred from the air.

We have conducted this latter calculation on the assumption of a dry shaft. This was not the situation in the actual case study. However, the result we have obtained does have a real meaning. The 2294 J/kg heat loss reflects the reduction in the sensible heat of the air. Most of this was, in fact, utilized to evaporate water and, hence, was returned to the air as latent heat. Chapter 14 elaborates on the meanings of these terms.

In order to convert the heat exchange into kilowatts, we simply multiply by the mass flow of air

$$\frac{-2294}{1000} \times 353 = -810 \qquad \frac{kJ}{kg} \times \frac{kg}{s} = \frac{kJ}{s} \text{ or } kW$$

8.2.3 Level workings

Figure 8.5 shows the PV and Ts diagrams for flow along level workings between stations 2 and 3. The diagrams reflect a decrease in pressure but increases in both specific volume and entropy, indicating heat additions to the air. In practice, a large part of the heat exchange with strata takes place in the working areas where rock surfaces are freshly exposed.

Applying the steady flow energy equation to level workings gives

$$\frac{u_2^2 - u_3^2}{2} + (Z_2 - Z_3)g = \int_2^3 V \, dP + F_{23} = H_3 - H_2 - q_{23} \quad \text{J/kg} \tag{8.11}$$

However, in this case $Z_2 - Z_3 = 0$, giving

$$-\int_2^3 V \, dP = F_{23} - \frac{u_2^2 - u_3^2}{2} \quad \frac{J}{kg} \tag{8.12}$$

Remember that the flow work $\int V \, dP$ is the area to the left of the process curve on the PV diagram and, here, dP is negative. Hence, equation (8.12) shows that work is done by the air and, in level airways, is utilized entirely against friction and accelerating the air. Here again, the change in kinetic energy is usually negligible, leaving the simple relationship

$$-\int_2^3 V \, dP = F_{23} \quad \text{J/kg} \tag{8.13}$$

The heat flow into level workings or airways may also be calculated by applying the condition $Z_2 - Z_3 = 0$ to the steady-flow energy equation:

$$\frac{(u_2 - u_3)^2}{2} + q_{23} = H_3 - H_2 \quad \text{J/kg} \tag{8.14}$$

Again, changes in kinetic energy are normally negligible and, if we apply the

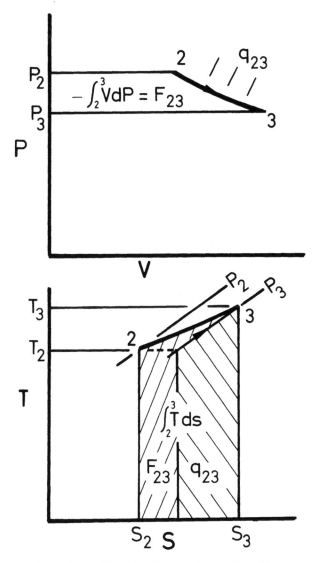

Figure 8.5 *PV* and *Ts* diagrams for level workings.

condition of neither evaporation nor condensation, then

$$q_{23} = C_p(T_3 - T_2) \quad \text{J/kg} \tag{8.15}$$

An interesting situation often arises in well-established return airways where equilibrium has effectively been established between the temperature of the air and the surrounding rock. Heat transfer ceases, giving adiabatic flow, $q_{23} = 0$. However, equation (8.15) shows that if q_{23} is zero then $T_3 = T_2$. The process becomes simultaneously adiabatic and isothermal. This is similar to a throttle in a gas stream, an example favoured by authors of textbooks on thermodynamics. In section 3.4.1, it was shown analytically that the degree of friction has no influence on the temperature variation of the air. The adiabatic–isothermal airway suggests a simple way of comprehending this phenomenon. An ideal adiabatic (isentropic) decompression would produce a fall in temperature. However, frictional heat is generated within the airflow at exactly the correct rate to counteract that fall in temperature.

On the Ts diagram of Fig. 8.5, the $\int_2^3 T \, ds$ heat area under the process line must be equal to the combined effects of added heat and internally generated friction, $q_{23} + F_{23}$. However, the area under the P_3 isobar between temperatures T_3 and T_2 is equal to the change in enthalpy $H_3 - H_2$ or $C_p(T_3 - T_2)$ for a dry airway. We have already shown that $q_{23} = C_p(T_3 - T_2)$. Hence, we can identify separate areas on the Ts diagram that represent F_{23} and q_{23}.

Unlike shafts of depths greater than $500 \, m$, most underground airflow paths involve relatively small changes in density. This is certainly the case for level workings or airways. While such airways play an important role in the mine thermodynamic cycle, the treatment of individual level airways is normally based on incompressible flow. The steady flow energy relationship then reduces to Bernoulli's equation with $Z_2 - Z_3 = 0$:

$$\frac{u_2^2 - u_3^2}{2} = \frac{P_3 - P_2}{\rho_{\mathrm{mw}}} + F_{23} \quad \mathrm{J/kg} \tag{8.16}$$

where the subscript mw denotes 'mean workings'. Neglecting the kinetic energy term, the frictional pressure drop referred to the mean density becomes simply

$$p_{23} = \rho_{\mathrm{mw}} F_{23} = P_2 - P_3 \quad \mathrm{Pa} \tag{8.17}$$

8.2.4 Upcast shaft

Figure 8.6 shows the PV and Ts diagrams for an upcast shaft. As the air ascends the shaft, decompression results in an increase in specific volume despite a decrease in temperature. The latter is shown on the Ts diagram.

As the air returning from most underground facilities remains at a fairly constant temperature (the 'thermal flywheel'), upcast shafts are much less susceptible to variations in heat exchange than downcast shafts. It is common to find that an upcast shaft is operating at near adiabatic conditions. The path line 3 to 4 on the Ts diagram indicates a typical situation, descending from P_3, T_3 to P_4, T_4 and diverted to the right of an isentrope by the effects of friction.

A thermodynamic analysis of the downcast shaft was given in section 8.2.2. The corresponding analysis for the upcast shaft follows the same logic and, indeed, the equations derived for the downcast shaft can also be used, with suitable changes of subscripts, for the upcast shaft. For that reason, details of the analysis will not be repeated. The reader may care to carry out a self-test by attempting to derive the following results and to prove the annotation of areas shown on the Ts diagram.

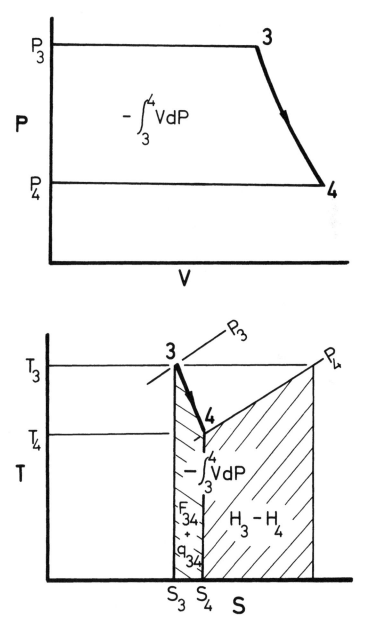

Figure 8.6 *PV* and *Ts* diagrams for an upcast shaft.

For polytropic flow,

$$F_{34} = \frac{u_3^2 - u_4^2}{2} - (Z_4 - Z_3)g + R(T_3 - T_4)\frac{\ln(P_3/P_4)}{\ln(T_3/T_4)} \quad \frac{J}{kg} \tag{8.18}$$

and

$$q_{34} = (Z_4 - Z_3)g - C_p(T_3 - T_4) \quad J/kg \tag{8.19}$$

Assuming constant (mean) density, ρ_{mu},

$$p_{34} = \rho_{mu}(Z_3 - Z_4)g - (P_4 - T_3) \quad Pa \tag{8.20}$$

8.3   THE COMPLETE MINE CYCLE

In the interests of clear explanation, it is convenient to investigate the phenomenon of natural ventilation first, before going on to the more usual situation of a combination of fan and natural ventilation.

8.3.1   Natural ventilation

Suppose that we have a vertical U–tube with fluids of differing densities but the same height in the two limbs. This is an unstable situation. The heavier fluid will displace the lighter and the fluids will move. Motion will be maintained for as long as a difference exists between the mean densities of the fluids filling each limb of the U–tube. It would be difficult to keep this experiment going with two completely different fluids. However, let us choose air as our fluid and apply heat at the base of the U–tube. If any slight perturbation then caused a movement of air so that one limb contained warmer and, therefore, less dense air than the other, then the motion would accelerate until a state of dynamic equilibrium was reached, dependent upon the rate of heat addition. This is the process that causes natural ventilation in mines or any other flow system that involves heat transfer and differences in elevation. It is the same phenomenon that causes smoke to rise up a chimney or convective circulation in a closed cycle, and also explains the effectiveness of the old shaft bottom furnaces described in Chapter 1.

In most cases, heat is added to the air in underground facilities. The air in an upcast shaft is then warmer and less dense than in the downcast shaft. When the shaft bottoms are connected through workings, this constitutes an enormous out-of-balance U-tube. Airflow is promoted and maintained at a rate that is governed by the difference between the mean densities of the air in the two shafts and the depths of those shafts.

If we take the simple case of two shafts of equal depth $Z_1 - Z_2$, one a downcast containing air of mean density ρ_{md}, and the other an upcast of mean density ρ_{mu}, then the pressure at the base of the downcast due to the column of air in the shaft will be $\rho_{md}g(Z_1 - Z_2)$ (equation (2.8)) and the pressure at the base of the upcast due to its column of air will be $\rho_{mu}g(Z_1 - Z_2)$. Hence, the pressure difference across the shaft bottoms available for promoting airflow through the workings will be

$$\text{NVP} = \rho_{md}g(Z_1 - Z_2) - \rho_{mu}g(Z_1 - Z_2) = g(Z_1 - Z_2)(\rho_{md} - \rho_{mu}) \quad \text{Pa}$$
$$(8.21)$$

where NVP = natural ventilating pressure (Pa).

This simple analysis provides the traditional expression for natural ventilating pressure. However, the following thermodynamic analysis given in this section indicates the limitations of equation (8.21) as well as providing a fuller understanding of the mechanisms of natural ventilating effects.

Natural ventilation will occur whenever heat transfer occurs in the subsurface. If the rock is cooler than the air then a natural airflow will occur in the reverse direction. In the case of a geographical region where the diurnal or seasonal ranges of surface air temperature encompass the mean temperature of the strata in a naturally ventilated mine, then reversals of airflow will occur whenever the surface temperature moves through the mean strata temperature. It is for this reason that very few medium size or large mines are now ventilated purely by natural means.

The *PV* and *Ts* diagrams for a naturally ventilated mine are illustrated by combining Figs. 8.2, 8.5 and 8.6 for the downcast shaft, 1–2, level workings, 2–3, and upcast shaft, 3–4, respectively. The result is shown on Fig. 8.7 with the slight curvature of each of the process lines ignored for simplicity.

In the absence of a main surface exhaust fan, the airflow is emitted from the top of the upcast shaft (station 4) at surface atmospheric pressure. We can now close the cycle back to station 1 by considering the process within the pressure sink of the free atmosphere. The exhaust air is emitted from the top of the upcast shaft with some finite velocity. This causes turbulence within the atmosphere. In the majority of cases, the exhaust air is at a temperature higher than that of the surface atmosphere. It cools at constant pressure, transferring heat to the atmosphere, contracts and follows the process lines 4 to 1 on both the *PV* and *Ts* diagrams of Fig. 8.7.

Just as we have combined the separate processes into a closed cycle on the *PV* and *Ts* diagrams, so, also, can we combine the corresponding versions of the steady flow energy equation:

downcast:
$$\frac{u_1^2 - u_2^2}{2} + (Z_1 - Z_2)g = \int_1^2 V\, dP + F_{12} = H_2 - H_1 - q_{12} \quad (8.22)$$

workings:
$$\frac{u_2^2 - u_3^2}{2} + (Z_2 - Z_3)g = \int_2^3 V\, dP + F_{23} = H_3 - H_2 - q_{23} \quad (8.23)$$

upcast:
$$\frac{u_3^2 - u_4^2}{2} + (Z_3 - Z_4)g = \int_3^4 V\, dP + F_{34} = H_4 - H_3 - q_{34} \quad (8.24)$$

atmosphere:
$$\frac{u_4^2 - u_1^2}{2} + (Z_4 - Z_1)g = \int_4^1 V\, dP + F_{41} = H_1 - H_4 - q_{41} \quad (8.25)$$

sum:
$$0 \quad + \quad 0 \quad = \oint V\, dP + \textstyle\sum F = \quad 0 \quad - \textstyle\sum q \quad (8.26)$$

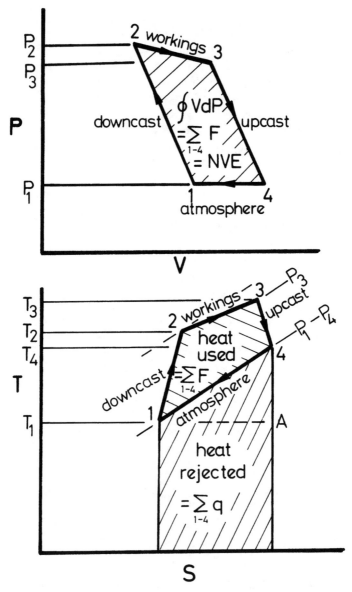

Figure 8.7 *PV* and *Ts* diagrams for a naturally ventilated mine.

Note that when we sum the individual terms, the kinetic energy, potential energy and enthalpy all cancel out.

Let us examine some of the features of the summation equation (8.26). Recalling that each individual $\int V\,dP$ term is represented by the area to the left of the curve on the relevant *PV* diagram, when we sum those areas algebraically, the cyclic integral

$\oint V\,dP$ for the complete loop becomes the enclosed area, 1234, on the PV diagram of Fig. 8.7. Furthermore, the $\int V\,dP$ areas for the workings and upcast shaft are negative (falling pressures). Inspection of the PV diagram shows that the enclosed area is also negative, indicating that net flow work is done by the air. This is the mechanical energy that produces and maintains motion of the air. It has, as its source, an available part of the heat energy that is added to the air and is called the **natural ventilating energy**, NVE.

Further examination of the PV diagram shows that the NVE area depends primarily on two factors: first, the extent to which heat additions in the subsurface expand the specific volume between stations 2 and 3, and secondly, the difference in pressures caused by the depths of the shafts, 1 to 2 and 3 to 4. The greater the heat additions or the depth of the mine then the greater will be the NVE.

Equation (8.26) for the complete cycle gives

$$- V\,dP = \textstyle\sum F \quad \text{J/kg} \tag{8.27}$$

where the symbol $\sum$ denotes summation around a cycle. This shows that the natural flow work done by the air is utilized entirely against friction. The enclosed area, 1234, on the PV diagram also represents the work that is done against friction. Unlike a manufactured heat engine, a mine thermodynamic cycle produces no external work, although it would be perfectly possible to achieve this by introducing a high volume – low pressure turbine into the airstream. The mechanical energy thus produced would be at the expense of flow energy and the rate of airflow would then diminish.

Let us take a closer look at equation (8.25) for the isobaric cooling of the exhaust air. We can make three simplifications to this equation. First, if the tops of both shafts are at the same elevation then $Z_4 = Z_1$. Secondly, we can assume that the surface atmosphere is at rest, $u_1 = 0$ and, thirdly, as the free atmosphere is at constant pressure, $\int_{4}^{1} V\,dP = 0$. Applying these conditions gives

$$\frac{u_4^2}{2} = F_{41} \quad \frac{\text{J}}{\text{kg}} \tag{8.28}$$

This confirms that the kinetic energy of the air issuing from the top of the upcast shaft is used simply against the frictional effects of creating turbulence in the free atmosphere. Equation (8.25) also quantifies the heat that is rejected to the surface atmosphere,

$$- q_{41} = H_4 - H_1 \quad \text{J/kg} \tag{8.29}$$

or, if no evaporation or condensation occurs,

$$- q_{41} = C_p(T_4 - T_1) \quad \text{J/kg} \tag{8.30}$$

However, the summation equation (8.26) gives

$$\textstyle\sum q = 0 \tag{8.31}$$

It follows that the total heat added in the subsurface,

$$\sum_{1-4} q = q_{12} + q_{23} + q_{34}$$

is equal to the heat ultimately rejected to the surface atmosphere, $-q_{41}$, despite the fact that some of that added heat has, temporarily, been converted to mechanical (kinetic) energy in order to create motion of the air.

Turning to the Ts diagram, the heat area below each of the process lines 1–2, 2–3 and 3–4 represents the combination of added heat and frictional heat for the downcast shaft, workings and upcast shaft respectively. It follows that the total area under lines 1–2–3–4 represents $\sum_{1-4} F + \sum_{1-4} q$ for the complete mine. However, the area under the isobar 4–1 represents the heat rejected to the atmosphere, $-q_{41}$, which has already been shown to equal $\sum_{1-4} q$. It follows that the enclosed area 1234 on the Ts diagram represents the work done against friction in the shafts and workings, $\sum_{1-4} F$. As the frictional loss within the atmosphere does not appear on the PV diagram, we have the revealing situation that areas 1234 on both diagrams represent frictional effects in the subsurface. The PV area represents the work done against friction while the Ts area shows that same work, now downgraded by friction, appearing as heat energy.

The area 41A on the Ts diagram lies above the atmospheric temperature line and, therefore, represents rejected heat that, theoretically, is available energy. It could be used, for example, to assist in the heating of surface facilities and can be significant in deep mines. However, that energy is contained in a large volume of air, often polluted by solid, liquid and gaseous contaminants, and with a relatively small temperature differential with respect to the ambient atmosphere to drive heat transfer devices. For these reasons, very few mines attempt to take advantage of the available part of the reject heat.

There are two main methods of quantifying the natural ventilating energy. First, if each of the subsurface processes approximates well to a polytropic law then the work done against friction in each segment may be calculated from the form of equation (8.1)

$$F_{i-1,i} = \frac{u_{i-1}^2 - u_i^2}{2} + (Z_{i-1} - Z_i)g - R(T_i - T_{i-1}) \frac{\ln{(P_i/P_{i-1})}}{\ln{(T_i/T_{i-1})}}$$

where subscript i takes the relevant value 2, 3 or 4. The sum of the resulting F terms then gives the natural ventilation energy, NVE.

Secondly, if polytropic curve fitting shows significant deviations from observed data then each segment can be further subdivided, or the enclosed NVE area on the PV diagram can be determined graphically.

As most modern underground facilities are ventilated by a combination of fans and natural ventilation, it is useful to be able to compare the ventilating potential of each of them. The theoretically correct way of doing this is to express the mechanical work input from the fan in J/kg. This is then directly comparable with the NVE. The fan work can be determined from measurements of pressure, temperature and air velocity across the fan (section 8.3.2.). However, the more usual convention in practice is to quote fan duties in terms of fan pressure and volume flow. A traditional and convenient device is to convert the natural ventilating energy, NVE, into a natural ventilating pressure, NVP:

$$\text{NVP} = \text{NVE} \times \rho \quad \text{Pa} \tag{8.32}$$

Here, again, we have the same difficulty of choosing a value of air density. If frictional pressure drops within individual branches have all been referred to standard density, $1.2 \, \text{kg/m}^3$, then it is appropriate also to refer both natural and fan ventilating pressures to the same standard density. This is, in effect, equivalent to comparing ventilating energies.

Another method, although imprecise, is to sum equations (8.5), (8.17) and (8.20) that gave the frictional pressure drops for the downcast shaft, workings and upcast shaft, respectively, on the basis of the corresponding mean densities:

downcast: p_{12}	$=$	$\rho_{md}(Z_1 - Z_2)g - (P_2 - P_1)$	Pa (see equation (8.5))
workings: p_{23}	$=$	$P_2 - P_3$	Pa (see equation (8.17))
upcast: $\quad p_{34}$	$=$	$-\rho_{md}(Z_4 - Z_3)g - (P_4 - P_3)$	Pa (see equation (8.20))

$$\text{sum:} \quad p_{12} + p_{23} + p_{34} = (Z_1 - Z_2)g(\rho_{md} - \rho_{mu}) \qquad \text{Pa} \tag{8.33}$$

In arriving at this summation, we have assumed that the depths of the two shafts are equal, $Z_1 - Z_2 = Z_4 - Z_3$, and that the atmospheric pressure at the tops of both shafts is the same, $P_1 = P_4$. This is identical to the equation that we arrived at earlier (equation (8.21)) from a consideration of the pressures at the bottoms of the two shafts and which were produced by their corresponding columns of air. Calculation of natural ventilation pressure as

$$\text{NVP} = p_{12} + p_{23} + p_{34} = (Z_1 - Z_2)g(\rho_{md} - \rho_{mu}) \quad \text{Pa} \tag{8.34}$$

is sometimes known as the **mean density method**. While equation (8.34) is simple to use and much quicker than the two methods of determining natural ventilating effects introduced earlier, it should not be regarded as a measure of the thermal energy that assists in the promotion of airflow. The difficulty is that the summation of pressure drops is not referred to a single defined value of air density. Each individual pressure drop is based on a different mean density. Their summation cannot, therefore, be justified as a definitive measure of natural ventilation but remains as a summation of measured frictional pressure drops. This can, in fact, be shown from the steady flow energy equation by ignoring the kinetic energy and fan terms

$$-g \, dZ = V \, dP + dF \quad \text{J/kg} \qquad \text{(see equation (3.25))}$$

or, as $V = 1/\rho$,

$$-g\rho \, dZ = dP + \rho \, dF \quad \text{J/m}^3$$

Integrating around a closed cycle ($\oint dP = 0$) gives

$$-g \oint \rho \, dZ = \oint \rho \, dF \quad \frac{\text{J}}{\text{m}^3} = \text{Pa}$$

The right-hand side of this equation is the sum of the measured frictional pressure drops, $p = \rho F$, while the left-hand side is the algebraic sum of the pressure heads exerted by the columns of air in the system. It is, therefore, an exact form of equations (8.21) and (8.33).

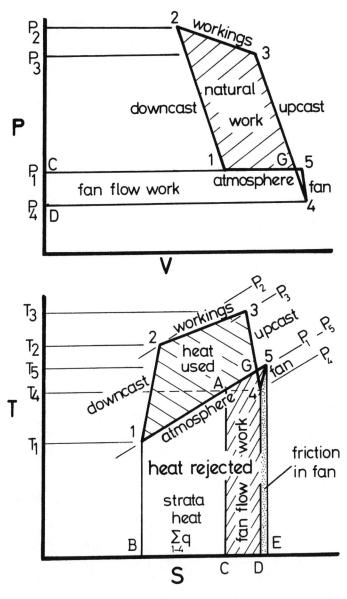

Figure 8.8 *PV* and *Ts* diagrams for a mine ventilated by both natural means and a main exhaust fan.

8.3.2 Combined fan and natural ventilation

Main ventilating fans may be placed at a number of strategic sites in a subsurface system, subject to any restrictions imposed by the governing mining legislation. However, to examine the combination of fan and natural ventilation, we shall use the most common situation—that of a main exhausting fan located at the top of the upcast shaft.

Figure 8.8 gives the corresponding PV and Ts diagrams. The difference between these and the corresponding diagrams for the purely naturally ventilated mine is that station 4, the top of the upcast shaft, is also the inlet to the fan and is at a subatmospheric pressure. On passing through the fan from station 4 to station 5, the air is compressed back to atmospheric pressure.

Concentrating, for the moment, on the segments of the PV diagram that represent the subsurface, the areas to the left of lines 1–2, 2–3 and 3–4 represent work done on or by the air as it progresses through the downcast shaft, workings and upcast shaft respectively. The algebraic summation of those work areas is the negative area 1234DC. This represents the net flow work done by the air between stations 1 and 4. The area to the left of the fan compression curve 4–5 represents the flow work done on the air by the fan. In practice, both the upcast shaft and fan processes are often near adiabatic, i.e. position 5 on the PV diagram is very close to position G. Hence, we can divide the total work done by air into two identifiable zones; area 123G represents the natural ventilating energy and area C154D represents the fan work supplied to the air. The combination of the two provides the total ventilating energy for the facility.

The process lines on the Ts diagram from station 1 to station 4 follow paths similar to those for the naturally ventilated mine except that station 4 is now at a subatmospheric pressure. The fan curve, 4 to 5, is angled slightly to the right of the ideal isentrope. Heat transfer is normally quite small across the fan casing in terms of J/kg because of the high airspeed. Hence, the fan process is essentially a frictional adiabatic. The derivation of the areas quantified on the Ts diagram representing the fan work and associated frictional heart is given in section 3.5.3.

The quantitative analysis of the cycle commences, once again, by stating the component steady flow energy equations and summing them to determine the cycle equation.

$$\text{downcast:} \quad \frac{u_1^2 - u_2^2}{2} + (Z_1 - Z_2)g \quad = \int_1^2 V\,dP + F_{12} = H_2 - H_1 - q_{12} \quad (8.35)$$

$$\text{workings:} \quad \frac{u_2^2 - u_3^2}{2} + (Z_2 - Z_3)g \quad = \int_2^3 V\,dP + F_{23} = H_3 - H_2 - q_{23} \quad (8.36)$$

$$\text{upcast:} \quad \frac{u_3^2 - u_4^2}{2} + (Z_3 - Z_4)g \quad = \int_3^4 V\,dP + F_{34} = H_4 - H_3 - q_{34} \quad (8.37)$$

$$\text{fan:} \quad \frac{u_4^2 - u_5^2}{2} + (Z_4 - Z_5)g + W_{45} = \int_4^5 V\,dP + F_{45} = H_5 - H_4 - q_{45} \quad (8.38)$$

atmosphere: $\dfrac{u_5^2 - u_1^2}{2} + (Z_5 - Z_1)g \qquad = \displaystyle\int_5^1 V\,dP + F_{51} = H_1 - H_5 - q_{51}$ (8.39)

sum: 0 $+0+W_{45}$ $= \displaystyle\oint V\,dp + \sum F = \qquad 0 - \sum q$ (8.40)

Consideration of equations (8.35) to (8.40) reveals a number of significant points.

1. The work areas annotated on the PV diagram are confirmed by a rearrangement of equation (8.40):

$$\sum F = W_{45} \qquad - \oint V\,dP \qquad \text{J/kg} \qquad (8.41)$$

work done   = work supplied + flow work
against friction   by fan   done by air
area 1234DC =   area 45DC + area 123G

2. The changes in kinetic energy and potential energy through the fan are both very small. Ignoring these terms simplifies equation (8.38) to

$$W_{45} = \int_4^5 V\,dP + F_{45} \quad \text{J/kg} \qquad (8.42)$$

The fan input produces flow work and overcomes frictional losses within the fan unit. The fan flow work shows on the PV diagram. However, both flow work and the frictional losses appear on the Ts diagram.

3. The equation for isobaric cooling ($dP = 0$) in the atmosphere (equation (8.39)) gives

$$\frac{u_5^2}{2} = F_{51} \quad \text{J/kg} \qquad (8.43)$$

assuming that the surface atmosphere is at rest ($u_1 = 0$). The kinetic energy of the air exhausting from the fan is dissipated against friction within the atmosphere. This is similar to the result obtained for the naturally ventilated mine (equation (8.28)). However, it now becomes more significant because of the higher velocity of the air issuing from the fan outlet, u_5. The kinetic energy at outlet is a direct loss of available energy from the system. It is, therefore, advantageous to decelerate the airflow in order to reduce that loss. This is the reason that main fans are fitted with expanding evasees at outlet.

4. Summing equations (8.35) to (8.38) for the subsurface and the fan, and taking $u_1 = 0$, gives

$$-\frac{u_5^2}{2} + W_{45} = H_5 - H_1 - \sum_{1-5} q$$

or

$$H_5 - H_1 = W_{45} + \sum_{1-5} q - \frac{u_5^2}{2} \quad \frac{J}{kg} \tag{8.44}$$

showing that the increase in enthalpy between the inlet and outlet airflows is equal to the total energy (work and heat) added, less the kinetic energy of discharge. However, if we inspect equation (8.39) for cooling in the atmosphere, we have

$$H_5 - H_1 = - q_{51} - \frac{u_5^2}{2} \quad \frac{J}{kg} \tag{8.45}$$

where $- q_{51}$ is the heat rejected to the atmosphere.

Comparing equations (8.44) and (8.45) gives

$$- q_{51} = W_{45} + \sum_{1-5} q \quad J/kg \tag{8.46}$$

This latter equation reveals the sad but unavoidable fact that all of the expensive power that we supply to the fans of a ventilation system plus all of the heat energy that is added to the air are ultimately rejected as waste heat to the surface atmosphere. It is possible to recover a little of the exhaust heat in cold climates (section 18.4.2).

8.3.3 Case study

The mine described in this case study was ventilated primarily by a main exhausting fan connected to the top of the upcast shaft and passing an airflow of 127.4 m³/s at the fan inlet. Both the downcast and the upcast shafts were 1219 m deep. Measurements of temperature any pressure allowed Table 8.1 to be established. Throughout the analysis, it was assumed that the air remained dry.

The PV and Ts diagrams were plotted as shown on Figs. 8.9(a) and 8.9(b) respectively. A great deal of information can be extracted from the table and visualized on the diagrams. It is useful to organize the analysis as follows.

General observations on the table of results

The $\int V \, dP$ column shows the large amount of flow work that is done on the air by gravity in the downcast shaft. However, this is more than offset by the work done by the air against gravity as it ascends the upcast shaft.

The addition of heat in the mine was concentrated in the workings, i.e. as the air progressed around the circuit from the downcast to the upcast shaft bottom. This is shown by both the $\int T \, ds$ and entropy columns. In this mine, the entropy increased continuously as the air passed through the shafts and workings. However, this is not always the case. In a mine where heat lost by cooling in any segment exceeds the heat generated by friction then that segment will exhibit a decrease in entropy.

It is interesting to observe, that the difference in barometric pressure across the shaft bottoms,

$$116.49 - 113.99 = 2.50 \quad \text{kPa}$$

is greater than that across the main fan

$$101.59 - 99.22 = 2.37 \quad \text{kPa}$$

This occurs quite frequently in deep mines and has often caused observers some puzzlement. The phenomenon arises from natural ventilating effects in the shaft circuit.

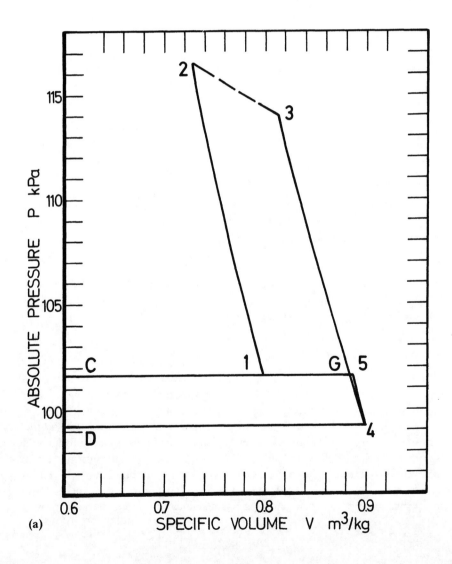

(a)

Polytropic indices

For each segment of the circuit, a to b, the polytropic index, n, may be calculated from

$$\frac{n}{n-1} = \frac{\ln(P_b/P_a)}{\ln(T_b/T_a)} \qquad \text{(see equation (3.72))}$$

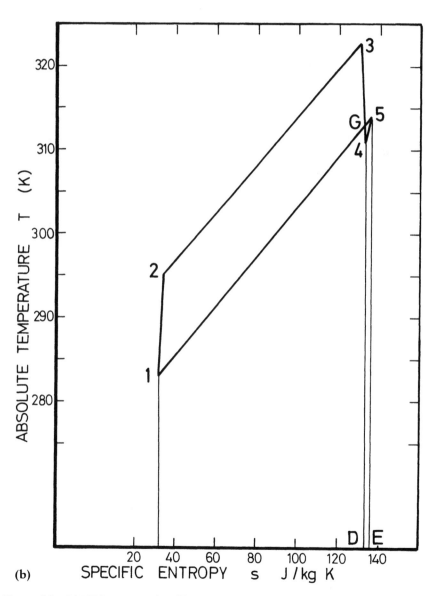

(b)

Figure 8.9 (a) *PV* diagram produced by case study. (b) *Ts* diagram produced by case study.

Table 8.1 Thermodynamic parameters based on representative values of pressure and temperature

Location	Station number	T temperature (K)	P pressure (kPa)	V specific volume (m³/kg) (i)	ρ density (kg/m³)	$\int V\,dP$ flow work (J/kg) (ii)	S entropy (J/(kg K)) (iii)	$\int T\,ds$ $F+q$ (J/kg) (iv)
Top downcast	1	283.15	101.59	0.8000	1.250	11356	31.61	614
Bottom downcast	2	295.06	116.49	0.7270	1.375	1923	33.73	29842
Bottom upcast	3	322.84	113.99	0.8129	1.230	12621	130.39	651
Top upcast	4	310.93	99.22	0.8995	1.112	2117	132.44	1018
Fan outlet	5	314.05	101.59	0.8873	1.127	0	135.70	−31055
Top downcast	1	283.15	101.59	0.8000	1.250		31.61	

In the compilation of the table, the following equations were used. The equation reference numbers indicate the location of their derivations.

(i) $V = \dfrac{RT}{P}$ and $\rho = \dfrac{1}{V}$ where $R = 287.04\ \text{J/kg}$

(ii) $\displaystyle\int_{a}^{b} V\,dP = R(T_b - T_a)\,\dfrac{\ln(P_b/P_a)}{\ln(T_b/T_a)}\ \dfrac{\text{J}}{\text{kg}}$ (equation (3.73))

(iii) $s = s_a - s_o = C_p \ln(T_a/T_o) - R \ln(P_a/P_o)\ \dfrac{\text{J}}{\text{kg}}$ (equation (3.48))

(iv) $\displaystyle\int_{a}^{b} T\,ds = (T_b - T_a)\left[C_p - R\,\dfrac{\ln(P_b/P_a)}{\ln(T_b/T_a)} \right]\ \dfrac{\text{J}}{\text{kg}}$ (equation (3.77))

In these equations $C_p = 1005\ \text{J/(kg K)}$. The subscript 'a' refers to the station number and subscript 'o' refers to the pressure and temperature datums of 100 kPa and 273.15 K respectively.

Segment	Polytropic index, n
Downcast shaft	1.4307
Workings	0.1943
Upcast shaft	1.3715
Fan	1.7330
Atmosphere	0

The effects of friction and added heat cause a PV curve to swing to the right, i.e. to finish with a higher value of V than the corresponding isentropic process. This means that compression curves such as those for the downcast shaft and fan will become steeper and adopt an increased value of n, while decompression curves (workings and upcast shaft) will become shallower and take a lower value of n. The greatly reduced value of n for the workings is simply a reflection of the heat added in that segment. It should be noted, however, that the actual flow through the workings will occur through a network of airways, each with its own series of thermodynamic processes. The value of n calculated here for the workings is, therefore, a composite appraisal.

Fan performance

Table 8.1 shows that the fan produces 2117 J/kg of useful flow work (area 45CD on the PV diagram) and also adds 1018 J/kg of heat (area 45ED on the Ts diagram) to the air. Hence, by ignoring changes in kinetic energy and assuming no heat transfer across the fan casing, the energy input from the fan,

$$W_{45} = \int_4^5 V\,dP + \int_4^5 T\,ds = 2117 + 1018 = 3135 \quad \text{J/kg}$$

The same result is given by the change in enthalpy across the fan

$$W_{45} = C_p(T_5 - T_4) = 1005\,(314.05 - 310.93) = 3136 \quad \text{J/kg}$$

While the input from the fan impeller is mechanical energy, the $\int T\,ds$ part arises from the frictional losses incurred almost immediately by the generation of turbulence. The volume flow of 127.4 m³/s was measured at the fan inlet where the air density was 1.112 kg/m³, giving a mass flow of

$$M = 127.4 \times 1.112 = 141.6 \text{ kg/s}$$

Then

$$\text{useful flow work} = \frac{2117}{1000} \times 141.6 = 300 \frac{\text{kJ}}{\text{kg}} \frac{\text{kg}}{\text{s}} = \frac{\text{kJ}}{\text{s}} \text{ or kW}$$

and

$$\text{fan frictional losses} = \frac{1018}{1000} \times 141.6 = 144 \text{ kW}$$

The polytropic efficiency of the fan may now be expressed as

$$\eta_{poly} = \frac{\int_4^5 V\,dP}{\int_4^5 V\,dP + \int_4^5 T\,ds} \tag{8.47}$$

or

$$\frac{300}{300 + 144} \times 100 = 67.6\%$$

Natural ventilating energy, NVE

The net flow work done by the air against friction in the shafts and workings is the area 1234DC on the *PV* diagram. This is quantified by summing the tabulated $\int V\,dP$ values from station 1 to station 4:

$$-\sum_{1-4} F = \sum_{1-4} \int V\,dP = (11\,356 - 1923 - 12\,621) = -3188 \quad J/kg$$

or

$$\frac{-3188}{1000} \times 141.6 = -451 \quad kW$$

However, the fan produces only 2117 J/kg, or 300 kW, of this flow work. Hence, the balance is provided by natural ventilating energy

$$NVE = -3188 + 2117 = -1071 \quad J/kg$$

(area 123G on both the *PV* and *Ts* diagrams) or

$$-451 + 300 = -151 \quad kW$$

The negative signs show that this is work done by the air.

Natural ventilating pressure, NVP

The natural ventilating energy can be converted to a natural ventilating pressure by multiplying by a specified value of air density. Choosing the standard value of $1.2\,kg/m^3$ gives

$$NVP = NVE \times \rho$$

$$= \frac{1071}{1000} \times 1.2 = 1.285 \quad kPa$$

For comparison with the fan pressure, we must refer that also to standard density, ρ_{st}:

$$p_{st} = \frac{\rho_i}{\rho_{st}} p_i$$

where the subscript 'i' refers to fan inlet conditions giving the standardized fan pressure as

$$p_{st} = \frac{1.112}{1.2} \times 2.37 = 2.196 \quad \text{kPa}$$

The mean density method (section 8.3.1) gives a value of (see equation (8.21))

$$\text{NVP} = (\rho_{md} - \rho_{mu})g(Z_1 - Z_2) \quad \text{Pa}$$

where

$$\rho_{md} = \frac{1.250 + 1.375}{2} = 1.3125 \quad \text{kg/m}^3$$

$$\rho_{mu} = \frac{1.230 + 1.112}{2} = 1.171 \quad \text{kg/m}^3$$

$$g = 9.81 \quad \text{m/s}^2$$

and

$$Z_1 - Z_2 = 1219 \quad \text{m}$$

giving

$$\text{NVP} = (1.3125 - 1.171)9.81 \times 1219$$

$$= 1692 \quad \text{Pa} \quad \text{or} \quad 1.692 \quad \text{kPa}$$

This calculation illustrates that a natural ventilating pressure calculated by the mean density method is not the same as that determined through NVE or a summation of F terms (section 8.3.1).

Heat additions

The total energy addition to the air is given by $W_{45} + \Sigma_{1-5q}$ and is equal to the increase in enthalpy across the system, $H_5 - H_1 = C_p(T_5 - T_1)$. Hence the heat transferred to the air is

$$\sum_{1-5} q = C_p(T_5 - T_1) - W_{45}$$

$$= 1005(314.05 - 283.15) - 3135$$

$$= 31\,055 \qquad\qquad -3135 = 27\,920 \quad \text{J/kg}$$

or

$$\frac{27\,920}{1000} \times 141.6 = 3953 \quad \text{kW}$$

The total heat rejected to the atmosphere is

$$q_{51} = \sum_{1-5} q + W_{45} = 27\,920 + 3135$$

$$= 31\,055 \quad \text{J/kg}$$

(also given directly by $C_p(T_5 - T_1)$), or

$$\frac{31\,055}{1000} \times 141.6 = 4397 \quad \text{kW}$$

This agrees with the value in Table 8.1 for $\int T\,\mathrm{d}s$ within the surface atmosphere.

The sum of the $\int T\,\mathrm{d}s$ values from station 1 to station 4 (in Table 8.1) gives the combined effect of added heat and friction within the shafts and workings. The frictional component can be isolated as

$$\sum_{1-4} F = \int T\,\mathrm{d}s - \sum_{1-4} q$$

$$= 31\,107 - 27\,920 = 3187 \quad \text{J/kg}$$

or

$$\frac{3187}{1000} \times 141.6 = 451 \quad \text{kW}$$

This is in agreement with the value obtained earlier as $\sum_{1-4} \int V\,\mathrm{d}P$. Finally, the NVE may be checked by summing the complete $\int T\,\mathrm{d}s$ column. This gives 1070 J/kg, in agreement with the value determined earlier.

This case study has illustrated the power of thermodynamic analysis. From five pairs of specified temperatures and pressures, the following information has been elicited:

pressure across main fan	2.37 kPa (or 2.196 kPa at standard density)
pressure across shaft bottoms	2.50 kPa
fan flow work	300 kW
fan losses	144 kW
total energy input from fan (impeller work)	444 kW
fan polytropic efficiency	67.6%
work done against friction in the shafts and workings	451 kW
natural ventilating energy	151 kW
natural ventilating pressure at standard density	1.285 kPa
heat added to the air in shafts and workings	3953 kW
heat rejected to surface atmosphere	4397 kW

The case study has also been used to convey the intrinsic coherence of the thermodynamic method; alternative procedures have been employed to cross-check several of the results. Students of the subject, having mastered the basic relationships, should be encouraged to experiment with the analyses and to develop their own procedures. Each exercise will provide new and intriguing insights into thermodynamic logic.

8.3.4 Inclined workings

So far, we have confined ourselves to a level connection between shaft bottoms and drawn a single polytropic curve on the thermodynamic diagrams to represent that

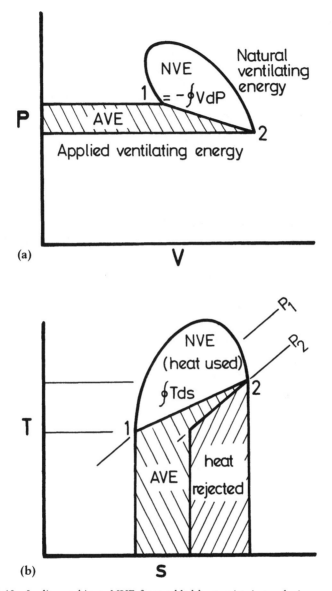

Figure 8.10 In dip workings, NVE from added heat assists in producing ventilation.

connection. However, the PV and Ts diagrams take on a different appearance when the connecting workings between shaft bottoms or any two points on a main intake and main return, are inclined.

A steeply dipping circuit might be ventilated adequately with little or no ventilating pressure applied across it. During a visit underground, one mine manager was quite disconcerted when a ventilation survey team leader pointed out that opening

Mine ventilation thermodynamics

an expensively constructed control door in an outbye cross–cut made no difference to the airflow in an adjoining dip circuit. Just as natural ventilation plays a part in a shaft circuit, it also influences airflow in any other subsurface circuit that involves changes in elevation.

The PV and Ts diagrams for a dip circuit are shown on Fig. 8.10. As the air flows down the descending intake airway(s) the pressure and temperature increase because of gravitational work done on the air. Expansion occurs through the workings and as the air ascends the rising return airways. If a survey team follows a route from station 1, an outbye position in the main intake or downcast shaft bottom, to station 2, a corresponding outbye position in the main return or upcast shaft bottom, then the resulting PV diagram will follow a loop from 1 to 2 whose shape is determined by the three-dimensional geometry of the circuit and the degree of heat transfer. Figure 8.10(a) shows that the $\int V\,\mathrm{d}P$ flow work can be divided into two sections, the applied ventilating energy (AVE) generated by the pressure differential applied across the ends of the circuit, and the natural ventilating energy (NVE) generated by thermal effects within the district:

$$-\int_1^2 V\,\mathrm{d}P = \mathrm{AVE} + \mathrm{NVE} = F_{12} \quad \text{J/kg} \tag{8.48}$$

The elementary $V\,\mathrm{d}P$ strips to the left of the PV curve are dominant for falling pressure $(-\,\mathrm{d}P)$, giving a negative net value $\int_1^2 V\,\mathrm{d}P$.

Suppose we divide the traverse route into a large number of small segments, then the frictional pressure drop along any one segment, ignoring the small changes in kinetic energy, is given as (see equation (8.5))

$$\mathrm{d}p = \rho\,\mathrm{d}F = \rho g\,\mathrm{d}Z - \mathrm{d}P.$$

In order to compare frictional pressure drops in differing segments, we refer the pressure drops to standard density, ρ_{st}:

$$\mathrm{d}p_{\mathrm{st}} = \frac{\rho_{\mathrm{st}}}{\rho}\,\mathrm{d}p = \rho_{\mathrm{st}}\,\mathrm{d}F = \rho_{\mathrm{st}}g\,\mathrm{d}Z - \rho_{\mathrm{st}}\frac{\mathrm{d}P}{\rho}$$

Integrating around the traverse route gives

$$\int_1^2 \mathrm{d}p_{\mathrm{st}} = p_{\mathrm{st}} = \rho_{\mathrm{st}}g\int_1^2 \mathrm{d}Z - \rho_{\mathrm{st}}\int_1^2 \frac{\mathrm{d}P}{\rho}$$

However, $V = 1/\rho$ and, if stations 1 and 2 are at the same level, $\int_1^2 \mathrm{d}Z = 0$, giving

$$p_{\mathrm{st}} = -\rho_{\mathrm{st}}\int_1^2 V\,\mathrm{d}P$$

Equation (8.48) then gives

$$p_{\mathrm{st}} = \rho_{\mathrm{st}}(\mathrm{AVE} + \mathrm{NVE}) \quad \text{Pa} \tag{8.49}$$

or

$$p_{\mathrm{st}} = \mathrm{AVP} + \mathrm{NVP} \quad \text{Pa} \tag{8.50}$$

where AVP = applied ventilating pressure across the circuit (Pa) and NVP = natural ventilating pressure generated within the circuit (Pa), both being referred to standard density.

Equation (8.50) explains how a dip circuit may be ventilated by natural effects even when the applied ventilation pressure is zero.

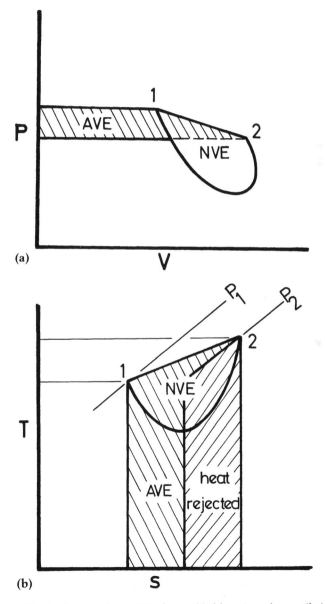

Figure 8.11 In rise workings, NVE from added heat impedes ventilation.

The NVP for the circuit may be determined by measuring the NVE area on the PV diagram (see equation (8.32)),

$$NVP = NVE \times \rho_{st}$$

or from the summation of frictional pressure drops, p_{st}, measured during a survey, each referred to standard density:

$$NVP = p_{st} - AVP \quad Pa \tag{8.51}$$

The total area under the traverse loop 1 to 2 on the Ts diagram of Fig. 8.10(b) represents the sum of frictional heat and added heat, $F_{12} + q_{12}$. However, provided that stations 1 and 2 are at the same level and ignoring changes in kinetic energy, the heat taken out of the district by the airstream is $H_2 - H_1$, or $C_p(T_2 - T_1)$ for dry conditions. This is identified as a heat area on the Ts diagram. The remaining areas represent

$$F_{12} = AVE + NVE \quad J/kg \tag{8.52}$$

The PV and Ts diagrams for a rise circuit are illustrated on Fig. 8.11. In this case the NVE has an opposite sign to the AVE. Hence natural ventilating effects oppose the applied ventilating energy. The analysis follows the same logic as for a dip circuit, leading to the following results:

$$-\int_1^2 V \, dP = AVE - NVE = F_{12} \quad \frac{J}{kg} \tag{8.53}$$

and

$$p_{st} = AVP - NVP \quad Pa \tag{8.54}$$

where, again, the applied and natural ventilating pressure are referred to standard density. These equations show that if the natural ventilating effects become sufficiently high then they will balance the applied ventilating energy or pressure. The flow work that can be done against friction, F_{12}, and the effective ventilating pressure, p_{st}, then both become zero and airflow ceases.

8.3.5 The effect of moisture

In order to highlight underlying principles, this chapter has concentrated on the flow of dry air throughout the system. This is seldom the case in practice. The air entering any facility from the free atmosphere invariably contains water vapour. If this is maintained throughout the system without further evaporation or condensation then the thermodynamic relationships developed in this chapter may continue to be employed. However, the parameters that vary with respect to moisture content require to be redefined in terms of X kg of water vapour associated with each 1 kg of 'dry' air, i.e. in $1 + X$ kg of air–vapour mixture. Subscript m is used to denote moist air.

Gas constant (see equation (14.14)):

$$R_m = \frac{287.04 + 461.5X}{1 + X} \quad \frac{J}{kg\,K}$$

Specific heat at constant pressure (see equation (14.16)):

$$C_{pm} = \frac{1005 + 1884X}{1 + X} \quad \frac{J}{kg\,K}$$

Specific volume (see equation (14.18)):

$$V_m = \frac{287.04 + 461.5X}{1 + X} \frac{T}{P} \quad m^3/(kg\text{ of moist air})$$

Density (see equation (14.21)):

$$\rho_m = \frac{1}{V_m} \quad kg\text{ moist air}/m^3$$

An **apparent** specific volume and density based on 1 kg of dry air are also useful:

$$V_m(\text{apparent}) = 461.5\frac{T}{P}(0.622 + X) \quad m^3/(kg\text{ dry air})$$

$$\rho_m(\text{apparent}) = \frac{1}{V_m(\text{apparent})} \quad kg\text{ dry air}/m^3$$

Enthalpy (see equation (14.40)):

$$H = 1005t_d + X[4187t_w + L + 1884(t_d - t_w)] \quad J/(kg\text{ dry air})$$

where t_d and t_w are wet and dry bulb temperatures (degrees Celsius) respectively and (see equation (14.6))

$$L = \text{latent heat of evaporation at wet bulb temperature}$$

$$= (2502.5 - 2.386t_w)1000 \quad J/kg$$

The reference numbers of these equations indicate the location of their derivation in Chapter 14.

These modifications for moist air make very little difference to the numerical results obtained for a dry mine. However, if free water exists in the subsurface, as it often does, then evaporation and condensation processes can have a major impact on the dry bulb temperature. In particular, the enthalpy change $H_2 - H_1$ is no longer given simply by $C_p(T_2 - T_1)$. Equation (14.40), reproduced above, must be employed.

For a more general analysis, consider the differential form of the steady flow energy equation (3.25) for 1 kg of air (equation (3.25)):

$$-u\,du - g\,dZ + dW = V\,dP + dF = dH - dq \quad J/kg$$

Consider an incremental length of airway within which every kg of air has associated with it X kg of water vapour and w kg of liquid water droplets, each of these increasing by dX and dw respectively through the increment. The vapour moves at the same velocity as the air, u. However, the droplets may move at a different velocity, u_w, and perhaps even in the opposite direction to the air in the case of an upcast shaft. The terms in the steady flow energy equation then become as follows:

1. *Kinetic energy (J/(kg dry air)).*

$$u\,du \quad \text{1 kg of dry air}$$
$$Xu\,du \quad X\text{ kg of water vapour}$$
$$wu_w\,du_w \quad w\text{ kg of droplets}$$

$$\frac{u^2}{2}\,dX \quad \text{energy to accelerate d}X\text{ of water vapour to } u \text{ m/s}$$

$$\frac{u_w^2}{2}\,dw \quad \text{energy to accelerate d}w\text{ of liquid to } u_w \text{ m/s}$$

2. *Potential energy (J/(kg dry air)).*

$$g\,dZ \quad \text{1 kg of dry air}$$
$$Xg\,dZ \quad X \text{ kg of water vapour}$$
$$wg\,dZ \quad w \text{ kg of liquid water}$$

3. *Fan energy.* The energy input terms are now based on 1 kg of 'dry' air. The fan energy input, therefore, remains unchanged at dW but with redefined units of J/(kg dry air).

4. *Flow work.* Provided that the specific volume is the volume of the mixture that contains 1 kg dry air, V_m(apparent), then the flow work, stays at $V\,dP$ J/(kg dry air).

5. *Work done against friction.* Again this remains simply as F but defined in units of J/(kg dry air).

6. *Enthalpy.*

$$dH = C_p\,dT \quad \text{1 kg of dry air}$$

$X\,dH_x \quad X$ kg of water vapour of enthalpy H_x where $H_x = C_w t_w + L + C_{pv}\,(t_d - t_w)$ as given in equation (14.40) above, C_w = specific heat of water, C_{pv} = specific heat of water vapour

$w\,dH_w \quad w$ kg of liquid water of enthalpy H_w where $H_w = C_w t_w$ for droplets at wet bulb temperature

$H_x\,dX \quad$ energy content of dX kg of added vapour

$H_w\,dw \quad$ energy content of dw kg of added droplets

7. *Added heat.* This term remains at q defined in terms of J/(kg dry air).

All of these terms may be collated in the form of equation (3.25), i.e. a three-part steady-flow energy equation. In practice, the kinetic energy terms are normally very small and may be neglected. The steady-flow energy equation for airflow of varying moisture content then becomes

$$
- (g\,dZ + Xg\,dZ + wg\,dZ) + dW
$$
$$
= V\,dP + dF = - (dH + X\,dH_x + w\,dH_w + H_x\,dx + H_w\,dw) - dq
$$
$$
\text{J/(kg dry air)} \tag{8.55}
$$

This equation is cumbersome for manual use. However, it provides a basis for computer programs developed to simulate airflow processes involving phase changes of water. The enthalpy terms encompass the energy content of the air, water vapour, liquid water and any additions of vapour and liquid.

If we apply the mechanical energy terms around a closed cycle ($\oint dZ = 0$) then we obtain

$$
W - \oint V\,dP = \sum F + g\left(\oint X\,dZ + \oint w\,dZ \right) \frac{J}{kg} \tag{8.56}
$$

Comparing this with the corresponding equation for dry conditions (8.41), we see that the fan and natural ventilating energies are now responsible not only for providing the necessary work against friction but also for removing water vapour and liquid from the mine.

FURTHER READING

Professor Baden Hinsley was the primary architect of the development of mine ventilation thermodynamics. He wrote many papers on the subject. Only the first of his major papers is listed here.

Hall, C. J. (1981) *Mine Ventilation Engineering*, Society of Mining Engineers of AIME.

Hemp, R. and Whillier, A. (1982) *Environmental Engineering in South African Mines*, Chapters 2 and 16, Mine Ventilation Society of South Africa.

Hinsley, F. B. (1943) Airflow in mines: a thermodynamics analysis. *Proc. S. Wales Inst. Eng.* **LIX** (2).

McPherson, M. J. (1967) Mine ventilation engineering. The entropy approach. *Univ. Nottingham Min. Mag.* **19**.

9

Ventilation planning

9.1 SYSTEMS ANALYSIS OF THE PLANNING PROCEDURE

In section 1.3.3, we emphasized the importance of integrating ventilation planning with production objectives and overall mine design during the early stages of planning a new mine or other subsurface facility. The team approach should continue for ongoing forward planning throughout the active life of the mine. Compromises or alternatives should be sought which satisfy conflicting demands. For example, the ventilation planner may require a major airway that is too large from the viewpoint of the rock mechanics engineer. An alternative might be to drive two smaller airways but large enough for any equipment that is to move through them. Regular liaison between engineering departments and exchanges during interdisciplinary planning meetings can avoid the need for expensive redesigns and should promote an optimized layout.

Figure 9.1 illustrates an organized system of ventilation planning for an underground mine. The procedure assumes the availability of computer assistance including ventilation simulation software, and eliminates most of the manual techniques and intuitive estimates of older planning methodologies (section 9.7). The initial step is to establish a database in a basic network file. For an existing facility this requires information gained from ventilation surveys. The latter are described in detail in Chapter 6. For a completely new mine, initial layouts should be discussed at the cross-disciplinary planning meetings. It is usual, in this case, that several alternative layouts are required to be investigated by the ventilation design team. The database for a new mine is established from initial estimates of required airway geometry, mining method, types of lining and network layout. Such data will be revised and refined as the design progresses.

In this chapter, we shall discuss the major facets of the planning procedure illustrated on Fig. 9.1.

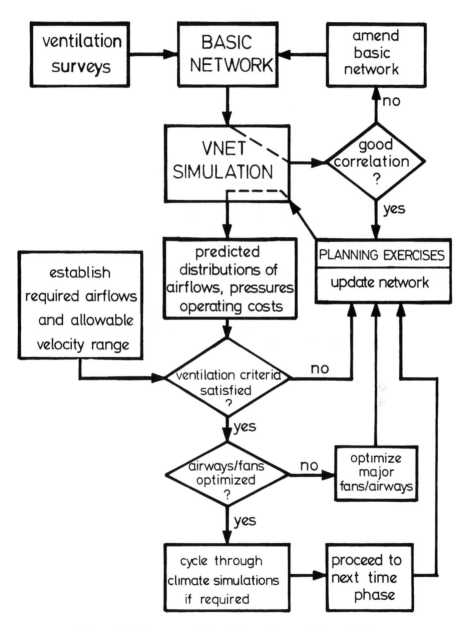

Figure 9.1 Systems analysis of subsurface ventilation planning.

9.2 ESTABLISHMENT OF THE BASIC NETWORK

9.2.1 New mines or other subsurface facilities

It will be recalled from Chapter 7 that a ventilation network consists of a line schematic, each branch representing one or more airways or leakage paths. In general, branches must be of known resistance or designated to pass a specified airflow. For a planned but yet unconstructed facility there are, of course, no airways or ventilation infrastructure that can be surveyed in order to establish actual values of resistance. Recourse must then be made to the methods of estimating airway resistance described in section 5.4, incorporating airway geometry, proposed types of lining (friction factors) and shock losses. Experience and data from other mines that utilize comparable layouts and methods of airway drivage, and operating in similar geological conditions, can be most helpful.

A variety of alternative plans will often require investigation during the initial design of a new mine. Ventilation network analyses should be carried out in order to determine the efficiency and effectiveness of each layout. The results of those analyses may then be considered, together with the other aspects of mine design, at the general planning meetings.

9.2.2 Existing mines

The routes of surveys in the main ventilation infrastructure of an existing mine provide a skeleton network. Additional branches can then be added or incorporated as equivalent resistances to represent airways or leakage paths that were not included in the primary surveys. As illustrated in the computed network example of section 7.4.6, branch resistances may be input to a network simulation program in several ways. This allows considerable flexibility in assembling a representative basic network, commencing from surveyed routes. Essentially, four methods are used to indicate resistance values in the first compilation of a basic network, or to update a basic network file that has previously been established.

1. For surveyed airways, the measured values of frictional pressure drop, p (Pa), and corresponding airflow, Q (m^3/s), may be input to the basic network file. The computer can then calculate the resistance from the square law:

$$R = \frac{p}{Q^2} \quad \text{N s}^2/\text{m}^8$$

2. There are situations in which it is impossible to determine resistance values for existing airways. If, for example, the airflow in an unrestricted branch is low at the time of measurement then so, also, will be the frictional pressure drop. While any pressure difference shown by the gauge should be logged, those lower than 5 Pa may yield resistances of doubtful accuracy. Modern diaphragm gauges will indicate to within ± 0.5 Pa. At 5 Pa, the corresponding error is 0.1 or 10%. Airflow measurements taken by approved methods (Chapter 6) will normally be

accurate to within 5%. Compounding these possible errors gives a maximum corresponding error in airway resistance of

$$\frac{dR}{R} = \pm \frac{dp}{p} \pm 2\frac{dQ}{Q} \tag{9.1}$$

(differentiation of the square law $R = p/Q^2$) giving

$$dR/R = \pm 0.1 \pm (2 \times 0.05)$$

$$= \pm 0.2 \text{ or } 20\%.$$

If, because of low values, the airflow and/or pressure drop cannot be measured to the required accuracies, the airway resistance should be calculated from the Atkinson equation (see equation (5.9))

$$R = kL\frac{\text{per}}{A^3}\frac{\rho}{1.2} \quad \text{N s}^2/\text{m}^8$$

where k = Atkinson friction factor at standard density (kg/m^3), L = length of airway (m), per = perimeter (m), A = cross-sectional area and ρ = air density (kg/m^3).

Representative values of friction factor, k, should, preferentially, be determined from similar airways which have been surveyed in that same mine. In this case, the density correction $\rho/1.2$, is often ignored. Otherwise, values of k may be estimated using one of the methods discussed in section 5.3. In all cases, care should be taken to allow for shock losses and unrepresentative obstructions (appendix in Chapter 5).

In many situations within a given mine, groups of airways are driven to the same dimensions and with a similar friction factor. Typical values of resistance per metre length can be established from measurements taken in, say, intakes, returns and conveyor routes. This facilitates the addition of new or unsurveyed branches to the network.

3. It is often difficult to measure the resistances of doors, stoppings or seals directly, either because of low flows or inaccessibility (Kissell, 1978). In practice, the resistances of doors or stoppings in mines vary from several thousand to, literally, 1 or 2 N s^2/m^8. The reason for this wide range is the highly non-linear relationship between area available for flow and resistance (see equation (5.15)):

$$R \propto 1/d^5$$

where d = hydraulic mean diameter of opening. Hence, the appearance of fractures or apertures in, or around, a previously tight stopping will result in dramatic reductions in its resistance.

The following ranges are suggested for planning purposes and for cross-sectional areas up to 25 m^2:

single doors: 10 to 50 N s^2/m^8 (typical value, 25 N s^2/m^8)

stoppings: 50 to 2000 N s^2/m^8 (typical value 500 N s^2/m^8)

seals: 1000 to 5000 N s^2/m^8

4. Worked-out areas and caved zones are the most difficult to represent on a network schematic. Leakage airflows can often be measured or estimated from survey results. These may be entered as 'fixed quantity' leakage airflows, leaving the computer to evaluate the corresponding resistances. This 'inversion' method remains valid while the number of fixed quantity branches is sufficiently small for unique values of resistance to be calculated. A network that is over-restricted by too many fixed quantity branches should produce a warning message from the VNET simulation package (section 7.4). Care should be taken to remove these 'fixed quantity' airways and to replace them with branches set at the corresponding computed resistances prior to progressing with planning exercises.

9.2.3 Correlation study

Figure 9.1 shows the data pertaining to the initial network schematic stored in the basic network file within computer memory. In order to ensure that the basic network is a true representation of the mine as it stood at the time of the survey, and before that basic network is used to launch planning exercises, it is important that it is subjected to verification through a correlation study. This involves running a ventilation network simulation package (section 7.4) on the data contained in the basic network file and comparing the computed airflows with those that were measured during the surveys. If all the survey measurements had been made instantaneously with perfect observers and perfect instruments, and if no errors had been made in compiling the basic network, then the computed and measured airflows would show perfect agreement. This utopian situation is never attained in practice. Despite the internal consistency checks on airflow and pressure drop measurements made during a survey (section 6.4.3) there will, inevitably, remain some small errors from observational and instrumental sources.

A large potential source of difference between computed and measured airflows is the fact that a VNET airflow distribution is mathematically balanced to a close tolerance throughout the network and represents a steady-state 'snapshot' of the ventilation system. On the other hand, the surveys may have taken several weeks or even months to complete. In a large active mine, updating survey data should be a routine and continuous activity of the mine ventilation department. During the time period of the surveys on which the basic network is established the airflow distribution may have changed as a result of variations in natural ventilating pressures or the resistances offered by the ever-changing work areas.

Lastly, mistakes made in transcribing survey data into the basic network or in key-punching the information into the basic network file will, again, result in disagreement between computed and measured airflows.

The correlation itself is normally carried out by a subsidiary program that lists the computed and observed airflows together with their actual and percentage differences. Leakage airflows of less than $3\,\mathrm{m}^3/\mathrm{s}$ are ignored in this comparison as large percentage errors in such low airflows will usually have little influence on the overall accuracy of the network.

The overall network correlation is quantified as

$$\frac{\sum \text{absolute value of (differences between computed and actual airflows for all surveyed branches)}}{\sum \text{absolute values of (measured airflows for all surveyed branches)}} \times 100 \quad (9.3)$$

A correlation is accepted as being satisfactory provided that (1) no significant airflow branch shows a difference of more than 10% between computed and measured airflows and (2) the overall correlation is also within 10%. However, because of the highly variable conditions that exist along mine traverses, the survey team will often be able to weight certain measurements as being more, or less, reliable than others. Such pragmatic considerations may be taken into account during the correlation study.

If the initial results of a correlation study indicate unacceptable deviations between computed and measured airflows then the basic network must be improved before it can be used as a basis for future planning exercises. The transcription of survey data into the basic network file should be checked. In the case of computed airflows being consistently higher or lower than the corresponding measurements then it is probable that an error has been made in the pressure ascribed to a main fan or may indicate that the resistances of primary airways such as shafts should be verified. Additional or check measurements of surveyed loops may be required.

It is fairly common that disagreements will occur in localized areas of the network. Again, this may indicate that the extent of the surveys had been insufficient to give an adequate representation of the mine network and that further measurements are necessary. More often, however, the problem can be resolved by adjustment of the estimated resistances or fixed airflows allocated to unsurveyed flowpaths. Provided that the main infrastructure of the surveyed network is sound then amendments to those estimated values will direct the simulated network towards an improved correlation. However, if measurements made in the surveyed loops are significantly out of balance then no amount of adjustment of subsidiary airways will produce a well-correlated basic network.

9.3 AIRFLOW REQUIREMENTS AND VELOCITY LIMITS

The estimation of airflow required within the work areas of the mine ventilation network is the most empirical aspect of modern ventilation planning. The majority of such assessments remain based on local experience of gas emissions, dust or heat load and are still often quoted in the somewhat irrational terms of m^3/s per ton of mineral output. Corrections can be applied for variations in the age of the mine, the extent of old workings, distances from shaft bottoms, depth and rates of production. However, as in all empirical techniques, the method remains valid only whilst the proposed mining methods, machinery and geological conditions remain similar to those from which the empirical data evolved. Attempts to extrapolate beyond those circumstances may lead to serious errors in determining required airflows. Fortunately, simulation techniques are available to assist in assessing airflow requirements for both gassy and hot mines.

The characteristics of strata gas emissions, heat flow and dust production are discussed in Chapters 12, 15 and 20 respectively. In this section, we shall confine ourselves to an examination of the methods of determining the magnitudes of airflows required to deal with airborne pollutants.

The overall requirement is that, in all places where personnel are required to work or travel, airflows must be provided in such quantities that will safeguard safety and health, and will also furnish reasonable comfort (section 1.3.1).

The quantity of air required for the purposes of respiration of personnel is governed primarily by the concentration of exhaled carbon dioxide that can be allowed in the mine atmosphere rather than the consumption of oxygen. For vigorous manual work, this demands about $0.01 \, \text{m}^3/\text{s}$ of air for each person and is negligible compared with the quantities of air needed to dilute the other gases, dust, heat and humidity that are emitted into the subsurface atmosphere.

A further requirement is that air volume flows must meet the governing state or national mining laws. Legislation may define minimum airflows that must be provided at specified times and places in addition to threshold limit values for pollutants or the psychrometric condition of the air. The ventilation planner must be familiar with the relevant legislation.

9.3.1 Strata gas

The various methods used to predict the emission rates of methane are described in Chapter 12. Whichever technique is employed, the final calculation of airflow requirement is

$$Q = \frac{100 \, E_g}{C_g} \quad \text{m}^3/\text{s} \tag{9.4}$$

where Q = required airflow (m^3/s), E_g = gas emission rate (m^3/s) and C_g = general body concentration to which gas is to be diluted (percentage by volume). The value of C_g is often taken to be one-half of the concentration at which the law requires action to be taken.

Example It has been predicted that during a 7 h working shift, $2500 \, \text{m}^3$ of methane will be emitted into a working face in a coal mine. If electrical power must be switched off at a methane concentration of 1%, determine a recommended airflow for the face.

Solution The average rate of gas emission during the working shift is

$$E_g = \frac{2500}{7 \times 60 \times 60} = 0.0992 \, \text{m}^3/\text{s}$$

Let us take the allowable concentration for design purposes to be one-half the legal limit. Then

$$C_g = 0.5\%.$$

Equation (9.4) gives the required airflow as

$$Q = \frac{100 \times 0.0992}{0.5} = 19.84$$

or, say, $20\,\text{m}^3/\text{s}$.

9.3.2 Diesel exhaust fumes

There are wide variations in the manner in which different countries calculate the ventilation requirements of mines where diesels are used. The basic stipulation is that there should be sufficient ventilation to dilute exhaust gases and particulates to below each of their respective threshold limit values. One technique, based on engine tests, is to calculate the airflow required to dilute the mass emission of each pollutant to one-half the corresponding TLV (threshold limit value). The maximum of those calculated airflows is then deemed to be the required air quantity. Some countries require analyses of the raw exhaust gases in addition to general body air sampling downstream from diesel equipment. Distinctions may be drawn between short-term exposure and time-weighted averages over an 8 h shift (section 11.2.1).

In addition to exhaust gases, national enforcement agencies may require that other factors be taken into account in the determination of airflow requirements for diesel equipment. These include rated engine power (kW or bhp), number of personnel in the mine or area of the mine, rate of mineral production, engine tests, number of vehicles in the ventilation split and forced dilution of the exhaust gases before emission into the general airstream.

There is no method of determining airflow requirements for diesels that will guarantee compliance with legislation in all countries. Here again, local regulations must always be perused for specific installations.

The criterion that is used most widely for initial estimates of required airflow is based on rated output power of the diesel equipment. For design purposes, many ventilation planners employ 6 to 8 m³/s of airflow over the machine for each 100 kW of rated diesel power, all equipment being cumulative in any one split. However, it should be borne in mind that the actual magnitude and toxicity of exhaust gases depend on the type of engine, conditions of operation and quality of maintenance in addition to rated mechanical power. It is for this reason that mining law refers to gas concentrations rather than power of the diesel equipment. Some manufacturers and government agencies give recommended airflows for specific diesel engines. In hot mines, it may be the heat produced by diesel equipment that sets a limit on its use.

9.3.3 Dust

Pneumoconiosis has been one of the greatest problems of occupational health in the mining industries of the world. Concentrated and long-term research efforts have led to greatly improved understanding of the physiological effects of dusts, methods of sampling and analysis, mandatory standards and dust control measures (Chapters 19 and 20).

There are many techniques of reducing dust concentrations in mines, ranging from water infusion of the solid mineral through to dust suppression by water sprays and air filtration systems. Nevertheless, with current mining methods, it is inevitable that dust particles will be dispersed into the air at all places where rock fragmentation or comminution occurs—at the rock–winning workplace and throughout the mineral transportation route. Dilution of airborne particles by ventilation remains the primary means of controlling dust concentrations in underground mines.

Dispersed dust particles in the respirable range (less than 5 μm diameter) will settle out at a negligible rate. For the determination of dilution by airflow, respirable dust may be treated as a gas. However, in this case it is realistic to estimate dust makes at working faces in terms of grams (or milligrams) per tonne of mineral mined. This will, of course, depend on the method of working and means of rock fragmentation.

The required airflow is given as

$$Q = \frac{E_d}{C_d} \times \frac{P}{3600} \quad \text{m}^3/\text{s} \tag{9.5}$$

where E_d = the emission rate of respirable dust (mg/t), P = rate of mineral production (t/h) and C_d = allowable increase in the concentration of respirable dust (mg/m^3).

Example The intake air entering a working area of a mine carries a mass concentration of 0.5 mg/m^3 respirable dust. Face operations produce 1000 t of mineral over an 8 h shift and add respirable dust particles to the airflow at a rate of 1300 mg per tonne of mineral mined. If the concentration of respirable dust in the return air is not to exceed 2 mg/m^3, determine the required airflow.

Solution Average rate of mineral production:

$$\frac{1000}{8 \times 3600} = 0.0347 \quad \text{t/s}$$

Average rate of emission of respirable dust:

$$0.0347 \times 1300 \quad \frac{\text{t mg}}{\text{s t}}$$

$$= 45.14 \quad \text{mg/s}$$

Dust removal capacity of air:

allowable dust concentration − intake dust concentration

$$= 2 - 0.5 = 1.5 \quad \text{mg/m}^3$$

Required airflow:

$$\frac{45.14}{1.5} \quad \frac{\text{mg}}{\text{s}} \frac{\text{m}^3}{\text{mg}}$$

$$= 30.1 \quad \text{m}^3/\text{s}$$

For larger particles of dust, the treatment is rather different. In this case, it is primarily the velocity rather than the quantity of the airflow that is important. In any airway where dust is produced at a given rate then the distance over which the particles are carried by the airstream depends on the air velocity and the settling rate of the individual dust particles. The latter depends, in turn, on the density, size and shape of the particle as well as the psychrometric condition of the air. The heavier and more spherical particles will settle more rapidly while the particles that are smaller or of greater aspect ratio will tend to remain airborne.

An effect of non-uniform dispersal characteristics is that the mineralogical composition of the airborne and settled dust is likely to vary with distance downstream from the source. The sorting effect of airflow on dust particles explains the abnormally high quartz content of dust that may be found in the return airways of some coal mines (section 20.3.8).

The larger particles will not be diluted by increased airflow and, hence, higher velocities. As the airspeed increases, these particles will remain airborne for longer

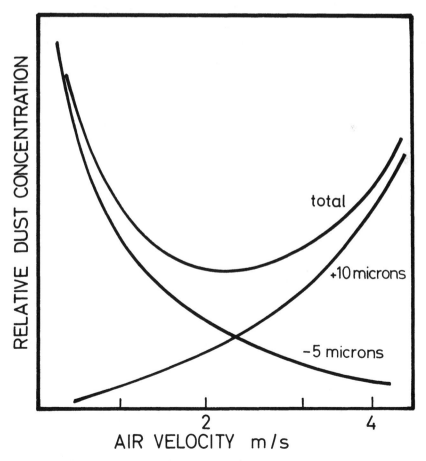

Figure 9.2   Variation of dust concentration with respect to air velocity.

distances before settling out. Furthermore, as the air velocity continues to rise, previously settled particles will be entrained by turbulent eddies into the airstream. The dust concentration will increase.

The effects of these mechanisms are illustrated on Fig. 9.2. A minimum total dust concentration is obtained at a velocity of 2 m/s. Fortunately, the total dust concentration passes through a fairly broad-based curve and air velocities in the range 1 to 4 m/s are acceptable. Above 4 m/s the problem is not so much a health hazard as it is the physical discomfort of large particles striking the skin. In addition to the question of ventilation economics, this limits the volume flow that can be passed through any workplace for cooling or the dilution of other pollutants.

9.3.4 Heat

Most of the pollutants that affect the quality of air in an underground environment enter the ventilating airstreams as gaseous or particulate matter and, hence, may be diluted by an adequate supply of air. The problem of heat is quite different, involving changes in the molecular behaviour of the air itself and its thermodynamic and psychrometric properties. Furthermore, it is normally the case that air entering mine workings from the base of downcast shafts, slopes or intake adits is relatively free of gaseous or particulate pollutants. On the other hand, the condition of the intake air with regard to temperature and humidity depends on the surface climate, and the depth of the workings. Whilst the removal of heat by ventilation remains the dominant method of maintaining acceptable temperatures in the majority of mines, quantifying the effects of varying airflow is by no means as straightforward as for gases or dust.

The complication is that air velocity and, by implication, airflow is only one of several variables that affect the ability of a ventilating airstream to produce a physiologically acceptable mine climate. Calculations of the climatic effect of variations in airflow are necessarily more involved than for gases or particulates.

The heat energy content of air, defined in terms of kilojoules of heat associated with each kilogram of dry air, is known as **sigma heat**, S. This concept is discussed fully in section 14.5.3. Sigma heat depends only on the wet bulb temperature of the air for any given barometric pressure.

The stages of determining the airflow required to remove heat from a mine or section of a mine are as follows.

1. Evaluate the sigma heat of the air at inlet, S_1, using equations (14.44) to (14.47).
2. Evaluate the highest value of sigma heat, S_2, that can be accepted in the air leaving the mine or section of the mine. This threshold limit value may be specified in terms of one of the indices of heat stress (Chapter 17) or simply as a maximum acceptable (cut-off) value of wet bulb temperature.
3. Estimate the total heat flux, q_{12} (kW), into the air from all sources between inlet and outlet (Chapter 15). This may involve simulation studies for additions of strata heat.

4. The required airflow, Q, is then given as

$$Q = \frac{q_{12}}{\rho(S_2 - S_1)} \quad \frac{m^3}{s} \tag{9.6}$$

where ρ = mean density of the air (kg/m³). The factor $Q\rho(S_2 - S_1)$ is sometimes known as the heat removal capacity (HRC) of a given airflow, Q.

It should be noted that as S_1 approaches S_2, or for high values of added heat, q_{12}, the required airflow may become excessive. In this case, either q_{12} must be reduced or air cooling plant installed (Chapter 18).

Example   Air enters a section of a mine at a wet bulb temperature, t_w, of 20 °C and a density, ρ, of 1.276 kg/m³. The mean barometric pressure is 110 kPa. It has been determined that 2 MW of heat are added to the air in the section. If the wet bulb temperature of the air leaving the section is not to exceed 28 °C and no air coolers are to be used, determine the required airflow.

Solution   From equations (14.44) to (14.47), sigma heat is given as

$$S = \frac{0.622 e_{sw}}{P - e_{sw}}(2502.5 - 2.386 t_w) + 1.005 t_w \quad \frac{kJ}{kg} \tag{9.7}$$

where saturation vapour pressure at wet bulb temperature is calculated as

$$e_{sw} = 0.6106 \exp\left(\frac{17.27 t_w}{237.3 + t_w}\right) \quad kPa$$

At intake, $t_w = 20\,°C$, giving

$$e_{sw1} = 0.6106 \exp\left(\frac{17.27 \times 20}{237.3 + 20}\right)$$

$$= 2.3375 \quad kPa$$

and

$$S_1 = \frac{0.622 \times 2.3375}{110 - 2.3375}(2502.5 - 2.386 \times 20) + (1.005 \times 20)$$

$$= 53.25 \quad kJ/kg$$

At return, using the same equations with t_w set at the cut-off value of 28 °C gives

$$S_2 = 82.03 \quad kJ/kg$$

Required airflow at the intake density by equation (9.6) is

$$Q_1 = \frac{q_{12}}{\rho(S_2 - S_1)}$$

$$= \frac{2000}{1.276\,(82.03 - 53.25)} \quad \frac{kJ\,m^3\,kg}{s\,\,kg\,\,kJ}$$

$$= 54.5 \quad m^3/s$$

In this section, we have assumed that the heat added to the air can be determined. Although this may be straightforward for machine heat and other controlled sources (section 15.3), it is more difficult to quantify heat emitted from the strata. For deep mines, the design of combined ventilation and air-conditioning systems has been facilitated by the development of climatic simulation programs. These may be incorporated into the planning procedure as shown at the bottom of Fig. 9.1. A more detailed discussion of the interaction between ventilation network and climatic simulation programs is given in section 16.3.5.

9.3.5   Workshops and other ancillary areas

In addition to areas of rock fragmentation, there are many additional locations in mines or other subsurface facilities that require the environment to be controlled. These include workshops, stationary equipment such as pumps or electrical gear, battery charging and fuel stations, or storage areas. For long-term storage of some materials in repository rooms, it may not be necessary or even desirable to maintain a respirable environment. However, temperatures and humidities will often require control.

In mines, it is usual for workshops and similar areas to be supplied with fresh intake air and to regulate the exit flow into a return airway either directly or through a duct.

The airflow requirement should, initially, be determined on the basis of pollution from gases, dust and heat as described in sections 9.3.1 to 9.3.4. However, in the case of large excavations, this may give rise to exessively low velocities with zones of internal and uncontrolled recirculation. In this situation, it is preferable to employ the technique of specifying a number of air changes per hour.

Example   An underground workshop is 40 m long, 15 m wide and 15 m high. Estimate the airflow required (a) to dilute exhaust fumes from diesel engines of 200 kW mechanical output and (b) on the basis of ten air changes per hour.

Solution   (a) Employing an estimated airflow requirement of 8 m³/s per 100 kW of rated diesel power (section 9.3.2) gives

$$\frac{200}{100} \times 8 = 16 \quad m^3/s$$

(b) Volume of room = 40 × 15 × 15 = 9000 m³. At ten air changes per hour,

$$airflow = \frac{9000 \times 10}{60 \times 60} = 25 \quad m^3/s$$

Hence, the larger value of 25 m³/s should be employed.

Table 9.1 Recommended *maximum* air velocities

Area	Velocity (m/s)
Working faces	4
Conveyor drifts	5
Main haulage routes	6
Smooth lined main airways	8
Hoisting shafts	10
Ventilation shafts	20

9.3.6 Air velocity limits

The primary consideration in the dilution of most pollutants is the volume flow of air. However, as indicated on Fig. 9.1, the velocity of the air (flowrate/cross-sectional area) should also be determined. Excessive velocities not only exacerbate problems of dust but may also cause additional discomfort to personnel and result in unacceptable ventilation operating costs.

Legislation may mandate both maximum and minimum limits on air velocities in prescribed airways. A common lower threshold limit value for airways where personnel work or travel is 0.3 m/s. At this velocity, movement of the air is barely perceptible. A more typical value for mineral-winning faces is 1 to 3 m/s. Discomfort will be experienced by face personnel at velocities in excess of 4 m/s (section 9.3.3), because of impact by large dust particles and, particularly, in cool conditions.

Table 9.1 gives a guide to upper threshold limit values recommended for air velocity.

In wet upcast shafts where condensation or water emissions result in airborne droplets, the air velocity should not lie within the range 7 to 12 m/s. Water blanketing may occur in this range of velocities. The resulting variations in shaft resistance cause an oscillating load on main fans and can produce large intermittent cascades of water falling to the shaft bottom.

9.4 PLANNING EXERCISES AND TIME PHASES

The dynamic nature of a mine or any other evolving network of subsurface airways requires that the infrastructure of the ventilation system be designed such that it can accommodate major changes during the life of the undertaking. For a completely new mine, the early ventilation arrangements will provide for the sinking of shafts, drifts or adits, together with the initial development of the primary underground access routes. When the procedures illustrated on Fig. 9.1 are first applied to an existing mine, the establishment and correlation of a basic network file will often reveal weaknesses and inefficiences in the prevailing ventilation system. The initial planning exercises should then be directed towards correcting those deficiencies while, at the same time, considering the future development of the mine.

Following the establishment of a new or revised ventilation design, further planning investigations should be carried out for selected stages of future development. Although this is commonly termed 'time phasing', it might more accurately be regarded as representing phases of physical development rather than definitive periods of time. Schedules have a habit of slipping or are subject to considerable revision during the course of mining.

9.4.1 Network planning exercises

Referring once again to Fig. 9.1, the first step in a ventilation network planning exercise is to establish the airflows required in all places of work, travel or plant location (section 9.3). Air velocity limits should also be set for the main ventilation flowpaths (section 9.3.6). Additional constraints on the direction and magnitudes of pressure differentials across doors, bulkheads or stoppings, or between adjoining areas of the facility, may also be imposed. This is of particular importance in nuclear waste repositories.

The ventilation design teams should review the existing network (correlated basic network, or previous time phase) and agree upon a series of alternative designs that will satisfy revised airflow objectives and meet constraints on velocity and pressure differentials. The network schematic should then be modified to represent each of those alternative designs in turn. The network modifications may include

1. sealing worked-out areas and opening up new districts,
2. adding main airways and lengthening existing airways,
3. adding and/or removing air crossings, doors and stoppings,
4. adding, removing or amending fixed quantity branches,
5. adding or sealing shafts and other surface connections,
6. adding, removing or relocating main or booster fans, or changing to different fan characteristic curves in order to represent adjustments to fan speed or vane settings, and
7. amending natural ventilation pressures.

It is worth reminding ourselves, at this stage, that current network simulation programs do not perform any creative design work. They simply predict the airflow and pressure distributions for layouts that are specified by the ventilation planner. While computed results will usually suggest further or alternative amendments, it is left to the engineer to interpret the results, to ensure that design objectives and constraints are met, to compare the efficiency and cost effectiveness of each alternative design and to weigh the practicality and legality of each proposed scheme.

Each alternative layout chosen by the design engineers should be subjected to a series of VNET simulations and network amendments until either all the prescribed criteria are met or it becomes clear that the system envisaged is impractical. Tables, histograms and spreadsheets should be prepared in order to compare fan duties (pressures and air quantities), operating costs, capital costs and the practical advantages/ disadvantages of the alternative layouts.

9.4.2 Time phases

During the operating life of an underground mine, repository or other subsurface facility, there will occur periods when substantial changes to the ventilation system must be undertaken. This will occur, for example, when

1. a major new area of the mine is to be opened up or an old one sealed,
2. workings become sufficiently remote from surface connections that the existing fans or network infrastructure are incapable of providing the required face airflows at an acceptable operating cost, or
3. two main areas of the mine (or two adjoining mines) are to be interconnected.

'Time phase' studies should be carried out on alternative schematics to represent stages before and after each of these changes. Additional scenarios may be chosen at time intervals between major changes or to represent transitional stages. The latter often impose particularly heavy duties on the ventilation system.

Time phase exercises should be conducted to cover the life of the mine, or as far into the future as can reasonably be predicted, assuming a continued market for the mined product.

It is, perhaps, intuitive to conduct time phase exercises in an order that emulates the actual planned chronological sequence. This may not always be the most sensible order for the network exercises. For example, if it appears inevitable that a new shaft will be required at some time in the future, it may be prudent first to investigate the time phase when that shaft has become necessary and then to examine earlier time phases. This assists the planners in deciding whether it would be advantageous to sink the new shaft at a prior time rather than to wait until it becomes absolutely necessary.

The time phase exercises will produce a series of quantified ventilation networks for each time period investigated. These should be reviewed from the viewpoint of continuity between time phases and to ensure that major additions to the ventilation infrastructure have an acceptable and cost-effective life. Indeed, the need for such continuity should remain in the minds of the planners throughout the complete investigation, although not to the extent of stifling viable alternative designs at later stages of the time phase exercises.

The result of such a review should be a series of selected layouts that represent the continuous development of the ventilation system throughout the projected life of the mine. Again, tables and histograms should be drawn up to show the variations of fan duties and costs, this time with respect to chronological order.

9.4.3 Selection of main fans

Care should be taken that any main fan purchased should be capable of producing the range of pressure–volume duties demanded of it throughout its projected life. The variation in required fan duties resulting from time phase exercises should be considered, particularly where a fan is to be relocated at some time during its life.

For a new purchase, fan manufacturers' catalogues should be perused and, if necessary, discussions held with those manufacturers, in order to ensure that the fan selected is capable of providing the required range of duties. Fortunately, the availability of variable pitch axial fans and inlet vane control on centrifugal fans allows a single fan to provide a wide range of duties with a fixed speed motor. A more detailed discussion of fan specifications is given in section 10.6.2.

9.4.4 Optimization of airflow systems

The network exercises carried out for any one time phase should incorporate a degree of optimization concerning the layout of airways, and the locations and duties of new shafts and fans. For a single major flowpath such as a proposed new shaft, a detailed optimization study can be carried out (section 9.5.5). However, the larger questions that can arise involve a balance between capital and operating costs and might include a choice between additional airways and fans of greater power, or the minimization of total power costs incurred by alternative combinations of main and booster fans. Special-purpose optimization programs are available to assist in the resolution of such questions (Calizaya *et al.*, 1988).

9.4.5 Short-term planning and updating the basic network

This chapter is concerned primarily with the major planning of a subsurface ventilation system. The time phases are chosen to encompass periods of significant change in the ventilation layout. However, much of the daily work of a mine ventilation department involves planning on a much shorter time scale.

As working districts advance or retreat, the lengths of the airways that serve them may change. Also, strata stresses can result in reductions of the cross-sectional areas of those airways. Additional roof support may be required, resulting in increased friction factors. Stoppings, regulators, doors and air-crossings may deteriorate and allow greater leakage. All of these matters cause changes in the resistances of individual airways and, hence, of the complete network. Additionally, wear of fan impellers can result in variations in the corresponding pressure–volume characteristic curves. For these reasons, a basic network file that correlated well with survey data at the start of any given time phase may become less representative of the changing real system.

In addition to regular maintenance of all ventilation plant and controls, further surveys should be carried out on a routine basis and, in particular, whenever measured airflows begin to deviate significantly from those predicted for that time phase. Visual inspections and frequent liaison with shift supervisors are also invaluable for the early identification of causes of deviation from planned airflows. Relationships with production personnel will remain more congenial if the latter are kept well advised of the practical consequences of blocking main airflow routes with supplies or waste material.

As new survey data become available, the basic network file should be updated accordingly. If the basic network file is maintained as a good representation of the current mine then it will prove invaluable in the case of any emergency that involves the ventilation system.

In the event that such updates indicate permanent and significant impact on the longer term plans then the network investigations for the future time phases should be re-run and the plans amended accordingly. In the majority of cases this may become necessary because of changes in mine production plans rather than from any unpredicted difficulties with ventilation.

9.5 VENTILATION ECONOMICS AND AIRWAY SIZING

The subsurface ventilation engineer must be capable of dealing with two types of costs:

1. Capital costs that require substantial funds at a moment in time, or distributed over a short time period. A main fan installation or sinking a new shaft come into this category.
2. Working, or operating costs, these representing the expenditure of funds on an ongoing basis in order to keep a system operating. Although consumable items, maintenance and even small items of equipment may be regarded as working costs, ventilation system designers often confine the term to the cost of providing electrical power to the fans.

The problem that typically arises concerns the combination of capital and operating costs that will minimize the real total cost to the company or mining organization; for example, whether to purchase an expensive but efficient fan to give low operating costs or, alternatively, an inexpensive fan that will necessitate higher operating costs. Another question may be whether the money saved initially by sinking a small diameter ventilation shaft is worth the higher ongoing costs of passing a required airflow through that shaft. These are examples of the types of questions that we shall address in this section.

In the following examples and illustrations, we shall employ the dollar ($) sign to indicate a unit of money. However, in all cases, this may be replaced by any other national unit of currency.

9.5.1 Interest payments

Money can be regarded as a commodity that may be circulated to purchase goods or services. If we borrow money to buy a certain item then we are, in effect, renting the use of that money. We must expect, therefore, that in addition to returning the borrowed sum, we must also pay a rental fee. The latter is termed the **interest payment.** Hence, if we borrow $100 for a year at an annual interest rate of 9%, then we must repay $109 one year from now. Even if we owned the money and did not have to borrow, spending the $100 now would mean that we are prevented from earning interest on that sum by lending it to a bank or other financial institution. Using money always involves an interest penalty.

If we borrow an amount of money, P (principal) at a fractional interest rate i (e.g. at 9%, $i = 0.09$) then we will owe the sum S where

now	$S = P$
after 1 year	$S = P(1 + i)$
after 2 years	$S = P(1 + i)(1 + i) = P(1 + i)^2$
after 3 years	$S = P(1 + i)^3$
after n years	$S = P(1 + i)^n$

$$(9.8)$$

Figure 9.3 gives a visual indication of the effects of compounding interest each year.

Example 1 $250,000 is borrowed for eight years at an annual interest rate of 10%. Determine the total sum that must be repaid at the end of the period, assuming that no intervening payments have been made.

Solution

$$S = 250\,000(1 + 0.1)^8$$
$$= \$535\,897$$

Example 2   How long does it take for an investment to double in dollar value at an annual interest rate of 12 per cent?

Solution   In this example, the final sum is twice the principal, $S = 2P$. Hence, equation (9.8) becomes

$$2P = P(1 + 0.12)^n$$

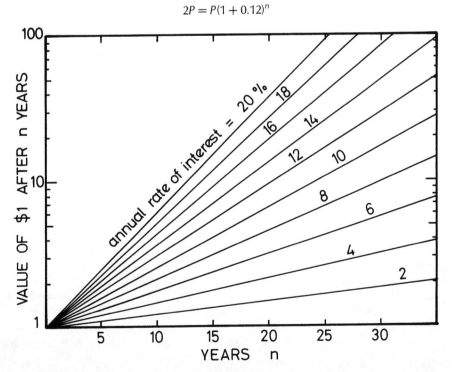

Figure 9.3   Effect of compounding interest.

giving

$$n = \frac{\ln(2)}{\ln(1.12)} = 6.12 \text{ years}$$

9.5.2 Time value of money, present value

It is clear from a consideration of interest that the value of a set sum of money will vary with time. At 9% interest, $100 now is worth $109 in one year and $118.81 two years from now. The effect of national inflation will, or course, partially erode growth in the value of capital. To determine real purchasing power, rates of interest should be corrected for inflation. In this section we are assuming that this has been done and quoted interest rates have taken inflation into account.

Present value of a lump sum

The time variation of the value of money makes it inequitable to compare two sums that are borrowed or spent at different times. We need a common basis to which both sums can be related. One method of doing this is to determine what principal or capital, P, we need to invest now in order that it will grow to a desired sum, S, in a specified number of years. This is given by a simple transposition of equation (9.8)

$$P = \frac{S}{(1+i)^n} \tag{9.9}$$

When determined in this way, the value of P is known as the **present value** of the future sum S.

If all future investments or expenditures are reduced to present values then they can accurately be compared.

Example Three fans are to be installed; one immediately at a price of $260,000, one in five years at an estimated cost of $310,000 and the third eight years from now at $480,000. Determine the total expenditure as a present value if the annual interest rate is 10%.

Solution From equation (9.9), $P = S/(1.1)^n$.

Fan	Time n	Purchase price S	Present value P
1	Now	$260,000	$260 000
2	5 years	$310,000	$192 486
3	8 years	$480,000	$223 924
			$676 410

In this example, the first fan is actually the most expensive.

Present value of regular payments

In the case of operating costs, payments must be made each year. Such future payments may also be expressed as present values in order to compare and compound
them with other expenditures.

We consider operating costs, S_o, to be paid at the end of each year. Then at
the end of the first year ($n = 1$), equation (9.9) gives

$$P_{0,1} = \frac{S_o}{1 + i}$$

where $P_{0,1}$ = present value of the first year's annual operating cost S_o, at an interest
rate of i. Similarly, the present value of the same operating cost, S_o, in the second
year ($n = 2$) is

$$P_{0,2} = \frac{S_o}{(1 + i)^2}$$

It follows that the total present value, P_o, of operating costs, S_o, paid at each
year-end for n years becomes

$$P_o = S_o \left[\frac{1}{1 + i} + \frac{1}{(1 + i)^2} + \frac{1}{(1 + i)^3} + \cdots + \frac{1}{(1 + i)^n} \right] \qquad (9.10)$$

The term in square brackets is a geometric progression which can be summed
through binomial expansions and collection of similar terms to give

$$P_o = \frac{S_o}{i} \left[1 - \frac{1}{(1 + i)^n} \right] \qquad (9.11)$$

Figure 9.4 gives a graphical representation of this equation for $S_o = \$1$.

Example 1   Electrical power costs at a mine are estimated to be $850\,000$ in
each of the next 12 years. Determine the present value of this expenditure at
11% interest.

Solution   With $S_o = \$850\,000$, $i = 0.11$ and $n = 12$ years, equation (9.11) gives

$$P_o = \frac{850\,000}{0.11} \left[1 - \frac{1}{(1 + 0.11)^{12}} \right]$$

$$= \$5\,518\,503$$

Example 2   Tenders received from manufacturers indicate that a fan priced
at $170\,000$ will cost $220\,000$ per year to run, while a fan with a purchase
price of $265\,000$ will require $190\,000$ per year in operating costs. If the
costing period is five years at an annual interest of 7%, determine which fan is
most economical.

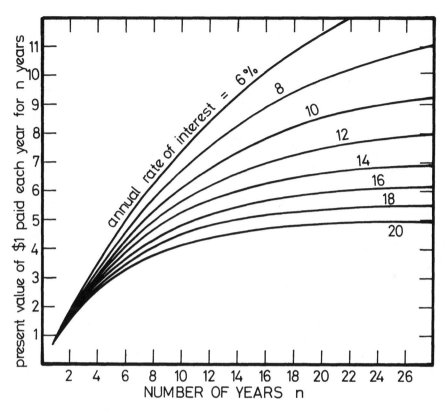

Figure 9.4 Present value of annual operating costs.

Solution   Using equation (9.11) with $n = 5$ years, $i = 0.07$ and S_o set at $220\,000$ and $190\,000$ respectively allows the present value of the operating costs to be determined.

Fan	Current purchase price ($)	Annual operating cost ($/year)	Present values		
			Purchase price ($)	Operating cost over 5 years ($)	Total ($)
1	170 000	220 000	170 000	902 043	1 072 043
2	265 000	190 000	265 000	779 038	1 044 038

In this example, fan 2 is shown to be the more economic choice.

During the analysis of time phase exercises (section 9.4.2) it may be necessary to establish the present value of operating costs for periods that do not commence at the present time. By applying equation (9.11) to periods of *m* years and *n* years respectively, and taking the difference, the present value of operating costs for the future period n to m years becomes

$$P_{o,nm} = \frac{S_o}{i}\left[\frac{1}{(1+i)^n} - \frac{1}{(1+i)^m}\right] \tag{9.12}$$

Example   This example shows an excerpt from a spreadsheet used to calculate the present values of capital and operating expenditures. Present values are calculated from equation (9.9) for capital (lump sum) expenditures and equation (9.10) for operating costs. The annual rate of interest is 10%.

Period	Capital expenditure ($)	Operating cost ($/year)	Present value ($)
1. Now	500 000		500 000
2. Now to end of 4 years ($n = 0, m = 4$)		620 000	1 965 317
3. At end of 4 years	235 000		160 508
4. End of 4 years to end of 10 years $n = 4, m = 10$)		680 000	2 022 797
Total			4 648 622

Spreadsheets of this type can be set up on a personal computer to give a very rapid means of costing the results of alternative network exercises.

9.5.3. Equivalent annual cost

When capital is borrowed, it is often more convenient to repay it, including interest, in equal installments each year, rather than as a lump sum paid at the end of the complete time period. This is a common method for domestic purchases such as a house (mortgage payments) or an automobile. In industry, this **equivalent annual cost** facilitates budgetary planning and enables major items of capital expenditure, such as a new shaft, to be spread evenly over a number of years. The method of equivalent annual cost allows capital expenditures to be compared with operating costs on a year by year basis and, for producing mines, also enables those capital expenditures to be expressed in terms of cost per tonne of mineral.

The equivalent annual cost is given by a re-interpretation of equation (9.11). P is the capital that must be invested now (present value) in order to be able to make a regular payment, S, each year. The statement may also be made in reverse, i.e. S is the regular payment, or annual equivalent cost, to be met each year in order to pay off the capital and interest on a borrowed amount, P. Transposing equation (9.11) gives the equivalent annual cost, EAC, as

$$\text{EAC}(=S) = \frac{Pi}{1 - 1/(1 + i)^n} \tag{9.13}$$

Example A mine shaft is to be sunk at a cost of $5.6 million. The life of the shaft is estimated to be 15 years during which time the average planned rate of mineral production is 1.6 million tonnes per year. If the annual interest rate is 8.75%, determine (a) the equivalent annual cost of the shaft in $ per year and (b) corresponding production cost in $ per tonne mined.

Solution (a) Using equation (9.13), with $P = \$5.6$ million, $i = 0.0875$ and $n = 15$ years, gives the equivalent annual cost as

$$\text{EAC} = \frac{5\,600\,000 \times 0.0875}{1 - 1/(1.0875)^{15}}$$

$$= \$684{,}509$$

(b) The corresponding production cost is

$$\frac{684\,509}{1\,600\,000} = \$0.428 \text{ per tonne}$$

9.5.4 Ventilation operating costs

A fan unit, comprising an electric motor, transmission and impeller, converts electrical energy into air power. The latter is reflected as kinetic energy of the air and a rise in total pressure across the fan. Air power delivered by a fan was quantified in Chapter 5 as

$$p_{ft} \times Q \quad \text{W}$$

(from equation (5.6)) where p_{ft} = rise in total pressure across the fan (Pa) and Q = airflow (m^3/s). (See also Chapter 10 for the effects of compressibility.)

However, the electrical power taken by the fan motor will be greater than this as losses occur inevitably in the motor, transmission and impeller. If the overall fractional efficiency of the unit is η, then the electrical input power to the motor will be

$$\frac{p_{ft} \times Q}{\eta} \quad \text{W}$$

Electrical power charges are normally quoted in cost per kilowatt hour. Hence, the cost of operating a fan for 24 h per day over the 365 days in a year is

$$S_o = \frac{p_{ft} Q}{1000\eta} e \times 24 \times 365 \text{ \$/year} \tag{9.14}$$

where e is the cost of power ($ per kW h). In practice, the fan pressure, p_{ft}, is often quoted in kPa, obviating the need for the 1000.

Equation (9.14) also applies for the annual cost of ventilating an individual airway. In this case, p_{ft} simply becomes the frictional pressure drop across the airway, p, at the corresponding airflow, Q, and η the overall efficiency of the fan primarily responsible for ventilating that airway. In cases of multiple fans, a weighted mean average of fan efficiency may be employed.

Example An underground airway is driven at a capital cost of $1.15 million dollars. During its life of eight years, it is planned to pass an airflow, Q, of 120 m³/s at a frictional pressure drop, p, of 720 Pa. The main fans operate at an overall efficiency, η, of 72%. If the annual rate of interest is 9.5% and the average cost of electrical power is $0.06 per kWh, determine the annual total cost of owning and ventilating the airway.

Solution The equivalent annual cost of owning the airway is given by equation (9.13) with $P_c = 1.15 million, $i = 0.095$ and $n = 8$ years:

$$EAC = \frac{1\,150\,000 \times 0.095}{1 - 1/(1.095)^8}$$

$$= \$211\,652 \text{ per year}$$

The annual operating cost is given by equation (9.14) with $p_{ft} = 720$ Pa, $Q = 120$ m³/s, $\eta = 0.72$ and $e = $0.06 per kWh:

$$S_o = \frac{720 \times 120}{1000 \times 0.72} \times 0.06 \times 24 \times 365$$

$$= \$63\,072 \text{ per year}$$

The total yearly cost is given by adding the cost of owning the airway and the annual operating cost of ventilation:

$$\text{total annual cost } C = EAC + S_o = 211\,652 + 63\,072$$

$$= \$274,724 \text{ per year}$$

9.5.5 Optimum size of airway

There are several factors that influence the size of a subsurface airway, including

1. the airflow to be passed through it,
2. the cost of excavation,
3. limitations on air velocity (section 9.3.5),
4. the span that can be supported adequately, and
5. the size of equipment required to travel through the airway.

This section considers the first two of these matters. However, the size of a planned new shaft or airway must satisfy the other constraints.

As the size of an airway is increased then its resistance and, hence, ventilation operating costs will decrease for any given airflow (section 5.4.1). However, the capital cost of excavating the airway increases with size. The combination of capital and operating costs will be the total cost of owning and ventilating the airway. The most economic or optimum size of the airway occurs when that total cost is a minimum. The costs may be expressed either as present values or in terms of annual (equivalent) costs.

In order to quantify the optimum size, it is necessary to establish cost functions for both capital and operating costs that relate those expenditures to airway size.

Capital cost function

The business of arriving at a cost of excavating a mine shaft or major airway often involves protracted negotiations between a mining company and a contractor. Even when the task is to be undertaken in-house, the costing exercise may still be extensive.

There are, essentially, two components—fixed costs and variable costs. The fixed costs are independent of airway size and may include setting up and removal of equipment such as temporary headgear, hoisting facilities or a shaft drilling rig. Other items such as pipe ranges, air ducts and tracklines are usually independent of airway size but are a function of airway length.

The variable costs are those that can be expressed as functions of the cross-sectional area or diameter of the shaft or airway. These may include the actual cost of excavation and supports.

The capital cost, P_c, may then be expressed as the cost function

$$P_c = C_f + \phi(AL) \tag{9.15}$$

where C_f = fixed costs and $\phi(AL)$ = function of cross-sectional area, A, and length, L.

In a simple case, the capital cost function may take the form

$$P_c = C_f + aV + bL \tag{9.16}$$

where V = volume excavated, $A \times L\,(m^3)$, and a and b are constants.

For mechanized excavations, there may be discontinuties in the capital cost function due to step increases in the size and sophistication of the equipment.

Operating cost function

Equation (9.14) gave the annual ventilation cost of an airway to be

$$S_o = pQ\frac{e}{1000\eta} \times 24 \times 365 \quad \$/year \tag{9.17}$$

However, from equation (2.50),

$$p = R_t\rho \quad Q^2$$

where R_t = rational resistance of the airway (m^{-4}) and ρ = air density (kg/m^3),

giving

$$S_o = R_t \rho Q^3 \frac{e}{\eta} \times \frac{24 \times 365}{1000}$$

This demonstrates that the operating cost varies with airway resistance, R_t, density, ρ, and the cube of the airflow, Q.

Substituting for R_t from equation (2.51) gives

$$S_o = \frac{fL \, \text{per}}{2A^3} \rho Q^3 \times \frac{e}{\eta} \times \frac{24 \times 365}{1000} \quad \text{\$/year} \tag{9.18}$$

where f = coefficient of friction (dimensionless), L = length of airway (m), per = perimeter (m) and A = cross-sectional area (m²).

In order to achieve the corresponding result with the Atkinson friction factor, $k_{1.2}$, substituting for p from equation (5.8) gives the operating cost function as

$$S_o = k_{1.2} L \frac{\text{per}}{A^3} \frac{\rho}{1.2} Q^3 \times \frac{e}{\eta} \times \frac{24 \times 365}{1000} \quad \text{\$/year} \tag{9.19}$$

Case study

During the design of a proposed circular shaft, the following data were generated.

Shaft sinking:
    equipment set up and decommissioning costs $650 000
    excavation cost $290/m³
    fittings and lining $460/m

Physical data:
    depth of shaft $L = 700$ m
    effective coefficient of friction $f = 0.01$
    (friction factor $k_{1.2} = 0.6f = 0.006 \, \text{kg/m}^3$)
    mean air density $\rho = 1.12 \, \text{kg/m}^3$
    airflow $Q = 285 \, \text{m}^3/\text{s}$
    fan efficiency $\eta = 0.65$
    life of shaft $n = 15$ years

Additional financial data:
    annual rate of interest   10%
    average cost of electrical power $e = \$0.075$ per kW h

Determine the optimum diameter of the shaft.

Solution

Task 1: Establish the capital cost function
From the data given, the capital cost function can be expressed in the form of

equation (9.16)

$$P_c = 650\,000 + 290V + 460L$$

where $V = AL = \dfrac{\pi D^2}{4} \times 700$ (m^3) and D = shaft diameter (m) giving

$$P_c = 650\,000 + \left(290 \times 700 \times \frac{\pi D^2}{4}\right) + (460 \times 700)$$

The capital cost function then simplifies to

$$P_c = 972\,000 + 159\,436D^2 \quad \$ \qquad (9.20)$$

This is the present value of the capital cost of sinking the shaft. The complete analysis may be carried out either in terms of present values or in annual (equivalent) costs. The latter method is employed for this case study. Hence, the capital cost of shaft sinking must be spread over the 15 year life using equation (9.13) with $i = 0.1$ and $n = 15$.

Equivalent annual cost of shaft sinking is

$$EAC = \frac{(972\,000 + 159\,436D^2) \times 0.1}{1 - 1/(1.1)^{15}}$$

$$= 127\,793 + 20\,962D2 \quad \$/\text{year} \qquad (9.21)$$

Task 2: Establish the operating cost function
This is accomplished by substituting the given data into equation (9.18) or (9.19). Annual operating cost is

$$S_o = \frac{0.01 \times 700 \times \pi D}{2(\pi D^2/4)^3} \times 1.12 \times 285^3 \times \frac{0.075}{0.65} \times \frac{24 \times 365}{1000}$$

$$= 594.775 \times 10^6/D^2 \quad \$/\text{year} \qquad (9.22)$$

Task 3: Establish the total cost function
Addition of the capital and operating cost functions from equations (9.21) and (9.22) respectively gives the total annual cost function as

$$C = 127\,793 + 20\,962D^2 + \frac{594.775 \times 10^6}{D^2} \quad \$/\text{year} \qquad (9.23)$$

Task 4: Determine the optimum diameter
This can be accomplished in either of two ways.

1. *By graphical means.* The three cost functions have been plotted on Fig. 9.5. The minimum point on the total annual cost curve occurs at an optimum diameter of approximately 4.9 m.

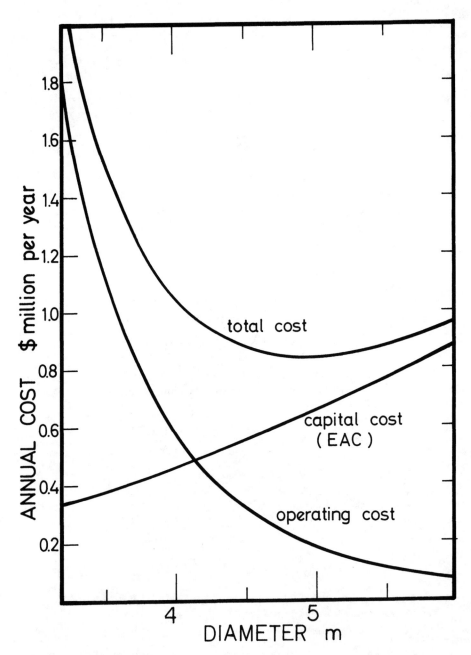

Figure 9.5 Annual cost functions for the circular shaft used in the case study.

An advantage of constructing the cost curves is that it gives a visual indication of the behaviour of the cost functions. In particular, the operating and total costs escalate rapidly if the airway diameter is reduced significantly below the optimum point. However, the total cost curve remains fairly shallow above the optimum point. This is a typical result and illustrates that the shaft size may be escalated to a standard diameter above the optimum to facilitate shaft sinking. Indeed, experience has shown the wisdom of sizing a shaft above the optimum. This allows subsequent flexibility for modifications to the airflow or mine production.

2. *By analytical means.* If the total cost function, C, has been expressed as a convenient function of cross-sectional area, A, or diameter, D, then it can be differentiated to find the optimum location when the slope dC/dA or dC/dD becomes zero.

In this case study, differentiation of equation (9.23) gives

$$\frac{dC}{dD} = 2 \times 20\,962D - \frac{5 \times 594.775 \times 10^6}{D^6}$$

At $dC/dD = 0$

$$41\,924D^7 = 2973.875 \times 10^6$$

giving

$$D = 4.93\,m.$$

This is an accurate value of the optimum diameter read from Fig. 9.5 as the approximation 4.9 m. In practice, the recommended diameter would be rounded up to at least 5 m.

9.5.6 Incorporation of shaft design into network planning exercises

Figure 9.1 shows the optimization of major airways and fans to be a part of the ventilation planning process. However, it may be impractical to carry out such optimization for every investigatory cycle of a ventilation network exercise. The sizing of a shaft is normally conducted only after network analysis has established a satisfactory distribution of airflows underground and can be combined with the shaft design procedures described in section 5.4.6.

The demand made upon a mine shaft may include access for personnel and equipment, the hoisting of planned tonnages of rock and the passing of specified airflows. We can now define a set of guidelines to assist in the management of a combined network analysis and shaft design exercise.

1. Assess the duties required for rock hoisting (tonnes per hour), number of personnel to be transported, time allowed at shift changes and the size, weight and frequency of hoisting materials and equipment.
2. Determine alternative combinations of conveyance sizes and hoisting speeds.
3. Conduct ventilation network investigations, initially on the basis of an estimated shaft resistance until a satisfactory distribution of airflow is achieved. This will establish the required airflow in the shaft.

4. Assess the dimensions of proposed shaft fittings including pipes, cables, guides and buntons.
5. Conduct an optimization exercise to find the size of shaft that will pass the required airflow at the minimum combination of operating costs and the capital expense of shaft construction (section 9.5.5).
6. Check the free area air velocity. If this exceeds 10 m/s in a hoisting shaft or 20 m/s in a shaft used only for ventilation then the cost of enlarging the shaft should be reviewed. It will be recalled that the total cost vs. diameter curve is usually fairly flat above its minimum point (section 9.5.5).
7. Determine the coefficient of fill, C_f, for the largest of the proposed conveyances (section 5.4.6). If this exceeds 30% for two or more conveyances, or 50% for a single-conveyance shaft, then the dimensions of the skip or cage should be reviewed or the size of the shaft, again, be re-examined.
8. Calculate the maximum relative velocity between the airflow and the largest conveyance,

$$(u_a + u_c)/(1 - C_f)$$

(see section 5.4.6 for nomenclature).
If this exceeds 30 m/s then additional precautions should be taken to ensure aerodynamic stability of the moving conveyances. In any event, the relative velocity should not exceed 50 m/s.
9. Assess the air velocities at all loading/unloading stations and, if necessary, redesign the excavations to include air bypasses or enlarged shaft stations.
10. Determine the total resistance of the shaft using the methods described in section 5.4.6. Examine all feasible means of reducing the resistance including the streamlining of buntons and the aerodynamic design of intersections. For shafts of major importance, construct and test physical models of lengths of the shaft and main intersections.
11. Re-run ventilation network analyses with the established value(s) of shaft resistance in order to determine the final fan pressures required.

It is clear that there is a considerable amount of work involved in the design, sizing and costing of a proposed mine shaft or major airway. Fortunately, microcomputer program packages are available that reduce the effort to little more than assembling the essential data. These are particularly valuable for carrying out sensitivity studies in order to assess the effects of changes in design or financial constraints.

9.6 BOOSTER FANS

9.6.1 The application of booster fans

In deep mines or where workings have become distant from surface connections the pressures required to be developed by main fans become very high in order to maintain acceptable face airflows. This leads to practical difficulties at airlocks and during the transportation of personnel, mined mineral and materials. More serious, however, is the fact that higher pressures at main fans inevitably cause greater leakage throughout the entire system. Any required fractional increase in face airflows will necessitate the same fractional increase in main fan volume flows for any given system resistance. Hence, as both fan operating power and costs are proportional to the product of fan pressure and airflow, those costs can rapidly become excessive as a large mine continues to expand. In such circumstances, the employment of booster fans provides an attractive alternative to the capital penalties of driving new airways, enlarging existing ones, or providing additional surface connections. Legislation should be checked for any national or state restrictions on the use of booster fans.

Unlike the main fans which, in combination, handle all of the mine air, a booster fan installation deals with the airflow for a localized area of the mine only. The primary objectives of a booster fan are

1. to enhance or maintain adequate airflow in areas of the mine that are difficult or uneconomic to ventilate by main fans, and
2. to redistribute the pressure pattern such that air leakage is minimized.

A modern booster fan installation, properly located, monitored and maintained, creates considerable improvements in environmental conditions at the workplace, and can allow the extraction of minerals from areas that would otherwise be uneconomic to mine. It has frequently been the case that the installation of underground booster fans has resulted in improved ventilation of a mine while, at the same time, producing significant reductions in total fan operating costs. However, these benefits depend on skilled system design and planning. An inappropriate use of booster fans can actually raise operating costs if, for example, fans act in partial opposition to each other. Furthermore, if booster fans are improperly located or sized then they may result in undesired recirculation.

This section is directed towards the planning, monitoring and control of a booster fan installation.

9.6.2 Initial planning and location

An initial step in planning the incorporation of a booster fan into an existing subsurface system is to obtain or update data by conducting ventilation surveys throughout the network (Chapter 6) and to establish a correlated basic network (section 9.2). Two sets of network exercises should then be carried out.

First, a case must be established for adding a booster fan to the system. Network exercises should investigate thoroughly all viable alternatives. These may include

1. adding or upgrading main fans,
2. enlarging existing airways and/or driving new ones including shafts or other surface connections,
3. redesigning the underground layout to reduce leakage and system resistance—for example, changing from a U-tube to a through-flow system (section 4.3), and
4. reducing face resistance by a redesign of the face ventilation system or the replacement of line brattices with auxiliary ducts and fans.

A second series of network exercises should be carried out to study the location and corresponding duty of the proposed booster fan. Conventional VNET programs can be used for this purpose. Each of the differing feasible locations is simulated in a series of computer runs in order to compare flow patterns and operating costs. Alternatively, optimization programs have been developed (Calizaya *et al.*, 1988) that select the optimum combination of main and booster fan duties, based on the minimization of operating costs for each booster fan location selected by the user. In either case, each predicted flow pattern should be checked against the constraints of required face airflows and velocity limits, as for any other network exercise (Fig. 9.1). Additionally, there are several other checks that should be carried out for a booster fan investigation.

Practical constraints on booster fan location

Although there may be a considerable number of branches in the network within which a booster fan may, theoretically, be sited in order to achieve the required flow enhancement, the majority of those may be eliminated by practical considerations.

Booster fans should, wherever possible, avoid locations where airlocks would interfere with the free movement of minerals, materials or personnel. Furthermore, the availability of dedicated electrical power and monitoring circuitry, or the cost of providing such facilities to each potential site, should be considered. If a booster fan is located remotely from frequently travelled airways, then particular vigilance must be maintained to ensure that it receives regular visual inspections in addition to continuous electronic surveillance (section 9.6.3).

Leakage and recirculation

For any given pressure developed by a booster fan in a specified circuit, there will exist a fan position where the summation of leakages inbye and outbye the fan is a minimum without causing undesired recirculation. In a U–tube circuit (Fig. 4.4), this location occurs where the operation of the fan achieves a zero pressure differential between intake and return at the inlet or outlet of the fan. Conventional wisdom is to locate the booster fan at this 'neutral point'. In practice, there are some difficulties. The neutral point is liable to move about quite considerably because of

1. variations in resistance due to face operations and advance/retreat of workings,
2. changes in the pressures developed by the booster fan or any other fan(s) in the system, or
3. variations in the system resistance due to movement of vehicles or longer-term changes in airway cross-section.

It is preferable to examine the complete leakage pattern predicted for each proposed booster fan location.

The question of recirculation must be examined most carefully. If the purpose of the booster fan is to induce a system of controlled partial recirculation (section 4.5), then the predicted airflow pattern must conform to the design value of percentage recirculation. However, undesired or uncontrolled recirculation must be avoided. At all times, legislative constraints must be observed.

Installing a booster fan in one area of a mine will normally cause a reduction in the airflows within other districts of the mine. In extreme cases, reversals of airflow and unexpected recirculations can occur. While examining network predictions involving booster fans, care should be taken to check airflows in all parts of the network and not simply in the section affected most directly by the booster fan.

In addition to their ability to affect airflows, booster fans may be regarded as a very effective means of managing the pressure distribution within the network. It is this feature that enables booster fans to influence the leakage characteristics of a subsurface ventilation system. A properly located and sized booster fan can be far more effective in controlling leakage across worked-out areas than sealants used on airway sides or stoppings. Conversely, a badly positioned booster fan can exacerbate leakage problems. These matters are of particular consequence in mines liable to spontaneous combustion. Hence, the pressure differences predicted across old workings or relaxed strata should be checked carefully.

Steady-state effects of stopping the booster fan

For each proposed location for a booster fan, network simulations should be run to investigate the effect of stopping that fan. In general, the result will be that the airflow will fall in that area of the mine and increase in others. (A cross-cut recirculation booster fan can cause the opposite effect as discussed in section 4.5.3.) In all cases, the ventilation must remain sufficient, without the booster fan, to ensure that all persons can evacuate the mine safely and without undue haste. The implication is that main fans alone must always be able to provide sufficient airflows for safe travel within the mine, and that booster fans merely provide additional airflow to working faces for dilution of the pollutants caused by breaking and transporting rock. It is good practice to employ two or more fans in parallel for a booster fan installation. This allows the majority of the airflow to be maintained if one fan should fail (section 10.5.2). Further, considerations regarding the transient effects of stopping or starting fans are discussed in section 9.6.3.

Economic considerations

The cost–benefits of a proposed booster fan with respect to time should be analysed. A major booster fan installation should be assured of a reasonable life, otherwise it may be preferable to tolerate temporary higher operating costs. On the other hand, a cluster of standard axial fans connected in a series/parallel configuration (section 10.5) can provide an inexpensive booster fan solution to a short-term problem. The financial implications of such questions can be quantified by combining total fan operating costs and the capital costs of purchasing and installing the booster fans, using the methods described in section 9.5.

9.6.3 Monitoring and other safety features

When a booster fan is installed in a subsurface ventilation system, it becomes an important component in governing the behaviour of that system. However, unlike other components such as stoppings, regulators or air crossings, the booster fan is actively powered and has a high speed rotating impeller. Furthermore, it is usually much less accessible than a surface fan in case of an emergency condition. For these reasons a subsurface booster fan should be subjected to continuous surveillance. The old way of doing this was to post a person at the booster fan location at all times when it was running. This must have been one of life's more tedious jobs. Some installations reverted to the use of television cameras. A bank of display monitors covering a number of sites could be observed by a single observer at a central location. The problems associated with such methods were that they depended on the vigilance of human beings and were limited to visual observations. The task may now be undertaken much more reliably and efficiently by electronic surveillance, employing transducers to sense a variety of parameters. These transducers transmit information in a near-continuous manner to a central control station, usually located on surface, where the signals are analysed by computer for display, recording and audio-visual alarms when set limits are exceeded. Alarms and, if required, displays may be located both at the control station and also at the fan site.

The monitors used at a booster fan may form part of a mine-wide environmental monitoring system which can, itself, be integrated into a general communication network for monitoring and controlling the condition and activities of conveyors, roof supports, excavation machinery and other equipment. However, sufficient redundancy should be built into the booster fan surveillance such that it remains operational in the event of failures in other parts of the mine monitoring system.

The parameters to be monitored at, or near, a booster fan installation include

1. gas concentration (e.g. methane) in the air approaching the fan,
2. carbon monoxide and/or smoke both before and after the fan,
3. pressure difference across the fan measured, preferably, at the adjoining airlock,
4. an indication of airflow in the fan inlet or outlet,
5. bearing temperatures and vibration on both the motor and the fan impeller, and
6. positions of the airlock and anti-reversal doors.

All monitors should be fail-safe, i.e. the failure of any transducer should be detected by interrogative signals initiated by the computer, resulting in automatic transfer to a companion transducer and display of a warning message and accompanying printed report. It follows that all transducers must be provided with at least one level of back-up devices.

The monitoring function is particularly important where controlled partial recirculation is practiced or at times of rapid transient change in the ventilation system. The latter may occur by an imprudent operation of doors in other areas of the mine or at a time when either the booster fan itself or other fans in the system are switched on or off. Any sudden reduction in air pressure caused by such activities can cause gases held in old workings or relaxed strata to expand into the ventilating airstream (section 4.2.2).

A booster fan should be provided with electrical power that is independent of the main supply system to the mine. However, electrical interlocks should be provided according to a predetermined control policy (section 9.6.4). This might include

1. automatic stoppage of the booster fan if the main fan(s) cease to operate,

2. stoppage of the booster fan if all of the local airlock doors are opened simultaneously, and
3. isolation of all electrical power inbye if the booster fan ceases to operate.

The second item on this list may also be required in reverse. Hence, for example, the booster fan could not be started if the local airlock doors were all open. However, those doors should open automatically when the booster fan stops, in order to re-establish conventional ventilation by the main fans. The question of interlocks may be addressed by legislation and may also be influenced by existing electrical systems. All electrical equipment should be subject to the legislation pertaining to that mine.

If the fan is to operate in a return airway, then the design will be improved if the motor is located out of the main airstream and ventilated by a split of fresh air taken from an intake airway. In this regard, it is also sensible to ensure that there are no flammable components within the transmission train.

Anti-reversal doors, flaps or louvres should be fitted to a booster fan in order to prevent reversal of the airflow through the fan when it stops.

All materials used in the fan, fan site and within some 50 m of the fan should be non-flammable. Furthermore, no flammable materials should be used in the construction of the local airlock, stoppings or air crossings. Finally, an automatic fire suppression system should be provided at the fan station.

9.6.4 Booster fan control policy

The development of monitoring systems provides a tremendous amount of data. During the normal routine conditions that occur for the vast majority of the time, all that is required is that the monitoring system responds to any manual requests to display monitored parameters, and to record information on electronic or magnetic media. However, the ability to generate and transmit control signals raises the question of how best to respond to unusual deviations in the values of monitored parameters.

The vision of complete automatic control of the mine environment produced a flurry of research activity in the 1960s and 1970s (Mahdi, Rustan, Aldridge). This resulted in significant advances in the development of transducers and data transmission systems suitable for underground use. However, a complete closed loop form of control has not been implemented for practical use. The difficulty is that there are a number of variables which, although of vital importance in an emergency situation, may not be amenable to direct measurement. These may include the locations of personnel and any local actions they may have taken in an attempt to ameliorate the situation.

During the design of a monitoring and control system for a booster fan, careful consideration should be given to the action to be initiated by the system in response to deviations from normal conditions. The control policy that is established should then be incorporated into the computer software (Calizaya *et at.*, 1989).

The simplest control policy is to activate audio-visual alarms and to allow control signals to be generated manually. Manual override of the system must be possible at all times. In order to examine additional automatic responses, unusual deviations in each of the monitored parameters listed in the previous subsection are considered here.

Methane (or other gas of concern)

Four concentration levels may be chosen or mandated by law:

1. concentration at which personnel are to be withdrawn;
2. concentration at which power inbye the fan is to be cut off;
3. concentration at which power to the booster fan should be cut;
4. concentration below which no action is to be initiated.

Levels 1 and 2 are usually fixed by law. Electrical power inbye the fan should be isolated automatically when level 2 is reached. If the fan is powered by an electric motor which is within the airstream then the power to be fan should also be cut when level 2 is reached. However, this may cause even more dangerous conditions for any personnel that are located inbye. Having the motor located in a fresh air split allows greater flexibility. At the present time, it is suggested that when level 3 is exceeded manual control should be established.

Level 4 is relevant only for those systems where control of the volume flow of air passing through the fan is possible. This may be achieved by automatic adjustment of impeller blade angle, guide vanes or motorized regulators. Level 4 may be set at one-half the concentration at which power inbye the fan must be cut. Variations in the monitored gas concentration that occur below level 4 are to be considered as normal and no action is taken. Variations between levels 4 and 3 may result in a PID (proportional/integral/differential) response. This means that control signals will be transmitted to increase the airflow to an extent defined by the level of the monitored concentration, how long that level has existed and the rate at which it is increasing. The objective is to prevent the gas concentration reaching level 3. The maximum increase in airflow must have been predetermined by network analysis such that the ventilation of the rest of the mine is maintained at an acceptable level.

Smoke or carbon monoxide

The system should be capable of distinguishing hazardous levels of smoke or carbon monoxide from short-lived peaks caused by blasting or nearby diesel equipment. A monitoring system that initiates alarms frequently and without good cause will rapidly lose its effectiveness on human operators.

There is no simple answer to the response to be initiated if these alarms indicate a real fire. Similar alarms arising from detectors in other parts of the mine may allow the probable location of the fire to be determined. Permitting the booster fan to continue running may accelerate the spread of the fire and products of combustion. On the other hand, stopping the fan might allow toxic gases to spread into areas where personnel may have gathered, particularly if damage to the integrity of air crossings or stoppings is suspected. The only realistic action in these circumstances is to revert to manual control.

Pressure differential and airflow

An increase in the pressure developed by the fan will usually indicate an increase in resistance inbye. This might be caused by normal face operations, partial blockage of an airway by a vehicle or other means, or obstruction of the fan inlet grill. No action is required other than a warning display and activation of an alarm if the fan pressure approaches stall conditions. A rapid fall in fan pressure accompanied by an increase in airflow is indicative of a short circuit and local recirculation. A warning display should be initiated together with checks on the positions of airlock doors followed, if necessary, by a manual inspection of the area. A simultaneous rapid fall in pressure and airflow suggests a problem with the fan itself. The power supply should be checked and, again, the fan should be inspected manually. Such circumstances will usually be accompanied by indications of the source of the difficulty from other transducers.

Bearing temperatures, vibration

Indications of excessive (or rising) values of bearing temperatures or vibration at either the motor or the fan impeller should result in immediate isolation of power to the fan and all electrical equipment inbye the fan.

Positions of airlock doors

An airlock door which is held open for more then a predetermined time interval should initiate a warning display and, if uncorrected, should warrant a manual inspection.

9.7 TRADITIONAL METHOD OF VENTILATION PLANNING

The methods of subsurface ventilation planning described in the previous sections of this chapter rely upon the availability of computers and appropriate software. A question that arises is how such planning can be carried out without high speed computational aid. For this, we may revert back to the traditional methodology that was well developed prior to the computer revolution. Although the older techniques cannot begin to match the speed, versatility and detail of computer assisted planning, they do retain a role in estimating generic values for airflows, fan pressures and air quality at the early 'conceptual design' stage of a proposed new mine or extension to an existing facility.

The traditional approach proceeds along the following sequence:

1. Determine air volume flows required in working areas. This can employ the methods discussed in section 9.3. However, at a preliminary stage of planning, empirical values of airflow based on rate of tonnage may be used. In this case, care must be taken to ensure that the basis of the empirical guidelines is compatible with the intended mining method and geology.

2. Assess the airflow requirements for development areas, mechanical or electrical plant and workshops, and estimate the volume flows that pass through abandoned workings, stoppings and other leakage paths. The estimation of leakage flows relies strongly upon the experience and intuition of the ventilation engineer. Unfortunately, in many mines the volumetric efficiency (section 4.2.3) is fairly low. Inaccuracies in the estimate of airflow through individual leakage paths, while of little consequence in themselves, will accumulate into major errors in the main ventilation routes. The square law, $p = RQ^2$, then produces twice the corresponding percentage error in the frictional pressure drop. The treatment of leakage flows is probably the single greatest cause of imprecision in the traditional method of ventilation planning.

3. Indicate the estimated airflows on a mine plan and compound them to show air flowrates, Q, through every major airway.

4. Using the given airflows and proposed sizes of airways, determine the corresponding air velocities. If these exceed limiting values (section 9.3.6), the need for larger or additional airways is indicated.

5. Assess the resistance, R, of each branch along the main ventilation routes, either from estimated friction factors and airway geometries or on the basis of local empirical data.

6. Using the square law, $p = RQ^2$, or charts, determine the frictional pressure drop, p, for each main branch and indicate these on the mine plan.

7. Commencing from the top of a downcasting surface connection, trace a path along intake airways to the most distant workings, through those workings, and back to the surface via return airways. Sum the frictional pressure drops around the complete traverse. This exercise is repeated for a number of such traverses to incorporate various working areas. The loop showing the greatest summation of frictional pressure drops then gives an approximation of the main fan pressure required to ventilate the mine. Subsidiary circuits may be controlled by regulators or upgraded by booster fans. Pressure gradient diagrams can be employed to give a visual indication of the cumulative pressure drops.

The traditional approach is similar, in principle, to that used for the design of duct systems in buildings. It is simple in concept and requires little computational aid. Unfortunately, it suffers from some severe drawbacks:

1. It relies strongly upon the experience of the engineer and his empirical knowledge of the distribution patterns. A mine is very different from a duct system in a building, not only in scale but also because of the dynamic nature of mining operations, and the tremendous variability in the geometry of airflow paths with respect to time as well as location.

2. The highly interactive and non-linear relationships that exist between ventilation parameters is largely ignored. Leakage airflows through caved strata, old workings or across stoppings, doors and air crossings are dependent upon the geometry of the flow paths, the pressure differential and the degree of turbulence. Ascribing fixed values, or even fixed proportions of available airflow, to leakage can do no

more than achieve rough approximations. In many mines, the majority of total airflow passes through leakage paths. Errors in estimated leakages will accumulate and be reflected in the corresponding main airflow routes and, because of the non-linearity of the laws of airflow can result in large errors in the cumulative pressure drops.

3. There is a basic lack of reality in a procedure that estimates an airflow pattern and then backtracks to find a fan pressure that will produce an airflow capable of being manipulated into the required distribution. In practice, when a fan is switched on, a complex configuration of interdependent pressure drops and airflows is set up throughout the network, the ventilating pressures producing airflows and the airflows, in turn, producing frictional pressure drops. In effect, the airflows shown by the traditional method are simply the initial estimates based on desired airflows in the workplaces and assumed leakages. No attempt is made to simulate the actual airflows that will occur when the fans are switched on.

4. There is very little opportunity to study alternative options in order to optimize the effectiveness and operating economics of the ventilation system.

5. Assuming the airflow distribution allows little flexibility in investigating the effects of fan duty/position, or the adjustment or resiting of regulators, doors and booster fans.

REFERENCES

Calizaya, F. *et al.* (1988) A computer program for selecting the optimum combination of fans and regulators in underground mines. *Trans. 4th Int. Mine Ventilation Congr. Brisbane,* 141–50.

Calizaya, F. *et al.* (1989) Guidelines for implementation of a booster fan monitoring and control system in coal mines. *Proc. 4th US Mine Ventilation Symp. Berkeley, CA,* 382–91.

Kissell, F. N. (1978) Some new developments in mine ventilation from the U.S. Bureau of Mines. *Mine Ventilation Soc. S. Afr.* **31**, 85–9.

FURTHER READING

Aldridge, M. D. and Nutter, R. S. (1980) An experimental ventilation control system. *Proc. 2nd Int. Mine Ventilation Congr., Reno, NV,* 230–38.

Dunmore, R. (1980) Towards a method of prediction of firedamp emission for British coal mines. *Proc. 2nd Int. Mine Ventilation Congr., Reno, NV,* 351–64.

Lambrechts, J. de V. and Howe, M. J. (1982) Mine ventilation economics. *Environmental Engineering in South African Mines,* Chapter 33, Mine Ventilation Society of South Africa.

Mahdi, A. and McPherson, M. J. (1971) An introduction to automatic control of mine ventilation systems. *Min. Technol.* (May).

McPherson, M. J. (1984) Mine ventilation planning in the 1980s. *Int. J. Min. Eng.* **2**, 185–227.

McPherson, M. J. (1988) An analysis of the resistance and airflow characteristics of mine shafts. *Trans. 4th Int. Mine Ventilation Congr. Brisbane.*

Rustan, A. and Stöckel, I. (1980) Review of developments in monitoring and control of mine ventilation systems. *Proc. 2nd. Int. Mine Ventilation Congr., Reno, NV,* 223–29.

10

Fans

10.1 INTRODUCTION

A fan is a device that utilizes the mechanical energy of a rotating impeller to produce both movement of the air and an increase in its total pressure. The great majority of fans used in mining are driven by electric motors, although internal combustion engines may be employed, particularly as a standby on surface fans. Compressed air or water turbines may be used to drive small fans in abnormally gassy or hot conditions, or where an electrical power supply is unavailable.

In Chapter 4, mine fans were classified in terms of their location, main fans handling all of the air passing through the system, booster fans assisting the through-flow of air in discrete areas of the mine and auxiliary fans to overcome the resistance of ducts in blind headings. Fans may also be classified into two major types with reference to their mechanical design.

A **centrifugal fan** resembles a paddle wheel. Air enters near the centre of the wheel, turns through a right angle and moves radially outward by centrifugal action between the blades of the rotating impeller. Those blades may be straight or curved either backwards or forwards with respect to the direction of rotation. Each of these designs produces a distinctive performance characteristic. Inlet and/or outlet guide vanes may be fitted to vary the performance of a centrifugal fan.

An **axial fan** relies on the same principle as an aircraft propeller, although usually with many more blades for mine applications. Air passes through the fan along flowpaths that are essentially aligned with the axis of rotation of the impeller and without changing their macro-direction. However, later in the chapter we shall see that significant vortex action may be imparted to the air. The particular characteristics of an axial fan depend largely on the aerodynamic design and number of the impeller blades together with the angle they present to the approaching airstream. Some designs of axial impellers allow the angle of the blades to be adjusted either while stationary or in motion. This enables a single-speed axial fan to be capable of a wide range of duties. Axial fan impellers rotate at a higher blade tip speed than a centrifugal fan of similar performance and, hence, tend to be noisier. They also suffer from a pronounced stall characteristic at high resistance. However, they are more compact, can easily be combined into series configurations and may have their direction of rotation reversed, although at greatly reduced performance. Both types of fan are used as main fans for mine ventilation systems while the axial type is favoured for underground locations.

In this chapter, we shall define fan pressures and examine some of the basic theory of fan design, the results of combining fans in series and parallel configurations, and the theory of fan testing.

10.2 FAN PRESSURES

A matter that has often led to confusion is the way in which fan pressures are defined. In section 2.3.2 we discussed the concepts of total, static and velocity pressures as applied to a moving fluid. That section should be revised, if necessary, before reading on. While we use those concepts in the definitions of fan pressures, the relationships between the two are not immediately obvious. The following definitions should be studied with reference to Fig. 10.1(a) until they are clearly understood.

1. **Fan total pressure**, FTP, is the increase in total pressure, p_t, (measured by facing pitot tubes) across the fan:

$$FTP = p_{t2} - p_{t1} \qquad (10.1)$$

2. **Fan velocity pressure**, FVP, is the average velocity pressure at the fan outlet only:

$$p_{v2} = p_{t2} - p_{s2}$$

3. **Fan static pressure**, FSP, is the difference between the fan total pressure and fan velocity pressure,
   or

$$FSP = FTP - FVP \qquad (10.2)$$

$$FSP = p_{t2} - p_{t1} - (p_{t2} - p_{s2}) = p_{s2} - p_{t1} \qquad (10.3)$$

The reason for defining fan velocity pressure in this way is that the kinetic energy imparted by the fan and represented by the velocity pressure at outlet has, traditionally, been assumed to be a loss of useful energy. For a fan discharging directly to atmosphere this is, indeed, the case. As the fan total pressure, FTP, reflects the full increase in mechanical energy imparted by the fan, the difference between the two, i.e. fan static pressure, has been regarded as representative of the useful mechanical energy applied to the system.

Notwithstanding that historical explanation, it is important to remember that, in subsurface ventilation planning, it is total pressures that must be used to quantify frictional pressure drops in airways. Accordingly, it is fan total pressures that should be employed in ventilation network exercises.

The interpretations of fan pressures that are most convenient for network planning

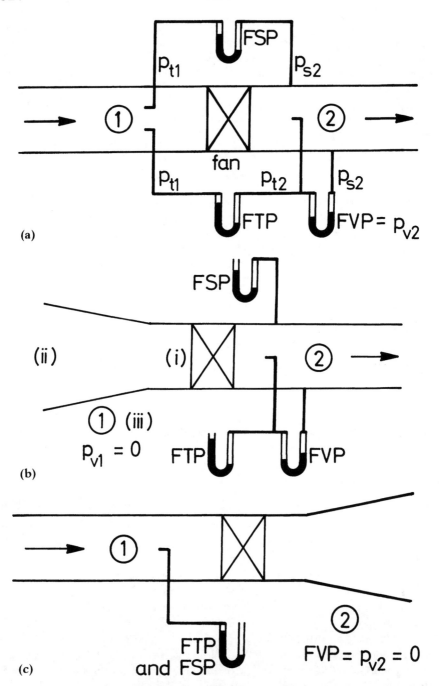

Figure 10.1 Illustrations of fan pressure: (a) fan with inlet and outlet ducts; (b) fan and inlet unit in a forcing system; (c) fan and outlet unit in an exhausting system.

are further illustrated on Fig. 10.1. In the case of a fan located within an airway or ducted at both inlet and outlet (Fig. 10.1(a)), the fan static pressure, FSP, can be measured directly between a total (facing) tube at inlet and a static (side) tapping at outlet. A study of the diagram and equation (10.3) reveals that this is, indeed, the difference between FTP and FVP.

Figure 10.1(b)) shows the situation of a forcing fan drawing air from atmosphere into the system. A question that arises is where to locate station 1, i.e. the fan inlet. It may be considered to be

  i. immediately in front of the fan,
 ii. at the entrance to the inlet cone, or
iii. in the still external atmosphere.

These three positions are labelled on Figure 10.1(b). If location i is chosen then the frictional and shock losses incurred as the air enters and passes through the cone must be assessed separately. At location ii the fan and inlet cone are considered as a unit and only the shock loss at entry requires additional treatment. However, if location iii is selected then the fan, inlet cone and inlet shock losses are all taken into account. It is for this reason that location iii is preferred for the purposes of ventilation planning. Figure 10.1(b) shows the connection of gauges to indicate the fan pressures in this configuration.

The same arguments apply for a fan that exhausts to atmosphere (Fig. 10.1(c)). If the outlet station is taken to be in the still external atmosphere then the fan velocity pressure is zero and the fan total and fan static pressure become equal. In this configuration the fan total (or static) pressure takes into account the net effects of the fan, frictional losses in the outlet cone and the kinetic energy loss at exit.

During practical measurements, it is often found that turbulence causes excessive fluctuations on the pressure gauge when total pressures are measured directly using a facing pitot tube. In such cases, it is preferable to measure static pressure from side tappings and to add, algebraically, the velocity pressure in order to obtain the total pressure. The mean velocity can be obtained as flowrate divided by the appropriate cross-sectional area. Particular care should be taken with regard to sign. In the case of an exhausting fan (Fig. 10.1(c)), the static and velocity pressures at the fan inlet have opposite signs.

Another practical problem arises when fan manufacturers publish characteristic curves in terms of fan static pressure rather than the fan total pressure required for ventilation planning. This is understandable as such manufacturers may have no control over the types of inlet and outlet duct fittings or the conditions at entry or exit to inlet/outlet cones. Where fan velocity pressures are quoted then they are normally referred to a specific outlet location, usually either at the fan hub or at the mouth of an evasee.

It is clear that care must be taken in using fan manufacturer's characteristic data for ventilation network planning. A simple guide for main surface fans is that it is the total pressure of the airflow on the inbye (system) side that should be employed for network exercises. This takes into account cone and entry/exit losses.

10.3 IMPELLER THEORY AND FAN CHARACTERISTIC CURVES

An important aspect of subsurface ventilation planning is the specification of pressure–volume duties required of proposed main or booster fans. The actual choice of particular fans is usually made through a process of perusing manufacturers' catalogues of fan characteristic curves, negotiation of prices and costing exercises (section 9.5). The theory of impeller design that underlies the characteristic behaviour of differing fan types is seldom of direct practical consequence to the underground ventilation planner. However, a knowledge of the basics of that theory is particularly helpful in discussions with fan manufacturers and in comprehending why fans behave in the way that they do.

This section of the book requires an elementary understanding of vector diagrams. Initially, we shall assume incompressible flow but will take compressibility of the air into account in the later section on fan performance (section 10.6.1).

10.3.1 The centrifugal impeller

Figure 10.2 illustrates a rotating backward-bladed centrifugal impeller. The fluid enters at the centre of the wheel, turns through a right angle and, as it moves outwards radially, is subjected to centrifugal force resulting in an increase in its static pressure. The dotted lines represent average flowpaths of the fluid relative to the moving blades. Rotational and radial components of velocity are imparted to the fluid. The corresponding outlet velocity pressure may then be partially converted into static pressure within the surrounding volute or fan casing.

At any point on a flowpath, the velocity may be represented by vector components with respect to either the moving impeller or to the fan casing. The vector diagram on Fig. 10.2 is for a particle of fluid leaving the outlet tip of an impeller blade. The velocity of the fluid relative to the blade is W and has a vector direction that is tangential to the blade at its tip. The fluid velocity also has a vector component in the direction of rotation. This is equal to the tip (peripheral) velocity and is shown as u. The vector addition of the two, C, is the actual or **absolute** velocity. The **radial** (or meridional) component of velocity, C_m, is also shown on the vector diagram.

Theoretical pressure developed by a centrifugal impeller

A more detailed depiction of the inlet and outlet vector diagrams for a centrifugal impeller is given on Fig. 10.3. It is suggested that the reader spend a few moments examining the key on Fig. 10.3 and identifying corresponding elements on the diagram.

In order to develop an expression for the theoretical pressure developed by the impeller, we apply the principle of angular momentum to the mass of fluid moving through it.

If a mass, m, rotates about an axis at a radius, r, and at a tangential velocity, v, then it has an **angular momentum** of mrv. Furthermore, if the mass is a fluid that is

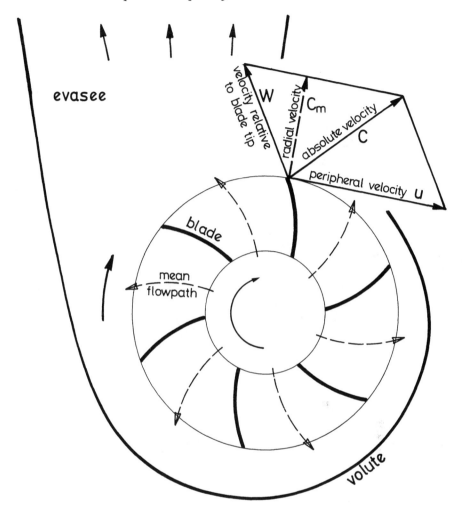

Figure 10.2 Idealized flow through a backward-bladed centrifugal impeller.

continuously being replaced then it becomes a mass flow, dm/dt, and a torque, T, must be maintained that is equal to the corresponding continuous rate of change of momentum.

$$T = \frac{dm}{dt}(rv) \quad \text{N m or J} \tag{10.4}$$

In the case of the centrifugal impeller depicted in Fig. 10.3, the peripheral component of fluid velocity is C_u. Hence the torque becomes

$$T = \frac{dm}{dt}(rC_u) \quad \text{J} \tag{10.5}$$

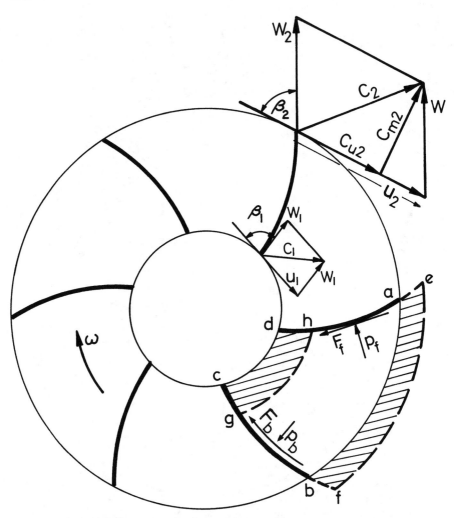

Figure 10.3 Velocities and forces on a centrifugal impeller: subscript 1, inlet; subscript 2, outlet; ω = angular velocity (rad/s), C = absolute fluid velocity (m/s), u = peripheral speed of blade tip (m/s), W = fluid velocity relative to vane (m/s), C_m = radial component of fluid velocity (m/s), C_u = peripheral component of fluid velocity (m/s), β = vane angle, p_f = pressure on front side of vane (Pa), p_b = pressure on back side of vane (Pa), F_f = shear resistance on front side of vane (N/m^2), F_b = shear resistance on back side of vane (N/m^2).

Consider the mass of fluid filling the space between two vanes and represented as abcd on Fig. 10.3. At a moment, dt, later it has moved to position efgh. The element abfe leaving the impeller has mass dm and is equal to the mass of the element cdhg entering the impeller during the same time. The volume represented by abgh has, effectively, remained in the same position and has not, therefore, changed its angular momentum. The increase in angular momentum is that due to the elements abfe and cdhg. Then, from equation (10.5) applied across the inlet and outlet locations,

$$T = \frac{dm}{dt}(r_2 C_{u2} - r_1 C_{u1}) \quad J \tag{10.6}$$

Extending the flow to the whole impeller instead of merely between two vanes gives dm/dt as the total mass flow, or

$$\frac{dm}{dt} = Q\rho \quad \frac{kg}{s}$$

where $Q =$ volume flow (m^3/s) and $\rho =$ fluid density (kg/m^3), giving

$$T = Q\rho\,(r_2 C_{u2} - r_1 C_{u1}) \quad J \tag{10.7}$$

Now the power consumed by the impeller, P_{ow}, is equal to the rate of doing mechanical work,

$$P_{ow} = T\omega \quad W \tag{10.8}$$

where $\omega =$ speed of rotation (rad/s):

$$P_{ow} = Q\rho\omega\,(r_2 C_{u2} - r_1 C_{u1}) \quad W \tag{10.9}$$

However, $\omega r_2 = u_2 =$ tangential velocity at outlet and $\omega r_1 = u_1 =$ tangential velocity at inlet. Hence

$$P_{ow} = Q\rho\,(u_2 C_{u2} - u_1 C_{u1}) \quad W \tag{10.10}$$

The power imparted by a fan impeller to the air was given by equation (5.56) as

$$p_{ft} Q \quad W$$

where $p_{ft} =$ rise in total pressure across the fan. In the absence of frictional or shock losses, this must equal the power consumed by the impeller, P_{ow}. Hence

$$p_{ft} = \rho\,(u_2 C_{u2} - u_1 C_{u1}) \quad Pa \tag{10.11}$$

This relationship gives the theoretical fan total pressure and is known as **Euler's equation**.

The inlet flow is often assumed to be radial for an ideal centrifugal impeller, i.e. $C_{u1} = 0$, giving

$$p_{ft} = \rho\,u_2 C_{u2} \quad Pa \tag{10.12}$$

Euler's equation can be re-expressed in a manner that is more capable of physical interpretation. From the outlet vector diagram

$$W_2^2 = C_{m2}^2 + (u_2 - C_{u2})^2$$
$$= C_{m2}^2 + u_2^2 - 2u_2 C_{u2} + C_{u2}^2$$

or

$$2u_2 C_{u2} = u_2^2 - W_2^2 + C_{m2}^2 + C_{u2}^2$$
$$= u_2^2 - W_2^2 + C_2^2 \quad \text{(Pythagoras)}$$

Similarly for the inlet,

$$2u_1 C_{u1} = u_1^2 - W_1^2 + C_1^2$$

Euler's equation (10.11) then becomes

$$p_{ft} = \rho \left(\frac{u_2^2 - u_1^2}{2} - \frac{W_2^2 - W_1^2}{2} + \frac{C_2^2 - C_1^2}{2} \right) \quad \text{Pa} \qquad (10.13)$$

centrifugal   effect of   change in
effect   relative velocity   kinetic energy

gain in static pressure + gain in velocity pressure

Theoretical characteristic curves for a centrifugal impeller

Euler's equation may be employed to develop pressure–volume relationships for a centrifugal impeller. Again, we must first eliminate the C_u term. From the outlet vector diagram on Fig. 10.3,

$$\tan \beta_2 = \frac{C_{m2}}{u_2 - C_{u2}}$$

giving

$$C_{u2} = u_2 - \frac{C_{m2}}{\tan \beta_2}$$

For radial inlet conditions, Euler's equation (10.12) then gives

$$p_{ft} = \rho u_2 C_{u2} = \rho u_2 \left(u_2 - \frac{C_{m2}}{\tan \beta_2} \right) \quad \text{Pa} \qquad (10.14)$$

However,

$$C_{m2} = \frac{Q}{a_2} = \frac{\text{volume flowrate}}{\text{flow area at impeller outlet}}$$

Equation (10.14) becomes

$$p_{ft} = \rho u_2^2 - \frac{\rho u_2}{\tan \beta_2 \, a_2} Q \quad \text{Pa} \qquad (10.15)$$

For a given impeller rotating at a fixed speed and passing a fluid of known density, ρ, u, a and β are all constant, giving

$$p_{ft} = A - BQ \quad \text{Pa} \qquad (10.16)$$

where constants

$$A = \rho u_2^2$$

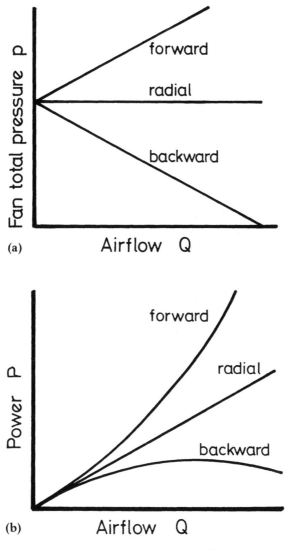

Figure 10.4 Theoretical characteristic curves for a centrifugal fan: (a) pressure–volume characteristics; (b) power–volume characteristics.

and

$$B = \frac{\rho u_2}{a_2 \tan \beta_2}$$

The flowrate, Q, and, hence, the pressure developed vary with the resistance against which the fan acts. Equation (10.16) shows that, if frictional and shock losses are ignored, then fan pressure varies linearly with respect to the airflow.

We may apply this relationship to the three types of centrifugal impeller.

1. *Radial bladed.* $\beta_2 = 90°$ and $\tan \beta_2 = $ infinity, giving $B = 0$. Then

$$p_{ft} = \text{constant } A = \rho u_2^2$$

i.e. theoretically, the pressure remains constant at all flows (Fig. 10.4(a)).

2. *Backward bladed.* $\beta_2 < 90°$, $\tan \beta_2 > 0$, and

$$p_{ft} = A - BQ$$

i.e. theoretically, the pressure falls with increasing flow.

3. *Forward bladed.* $\beta_2 > 90°$, $\tan \beta_2 < 0$, and

$$p_{ft} = A + BQ$$

i.e. theoretically, pressure rises linearly with increasing flow.

This latter and rather surprising result occurs because the absolute velocity, C_2, is greater than the impeller peripheral velocity, u_2, in a forward-bladed impeller. This gives an **impulse** to the fluid which increases with greater flowrates. (In an actual impeller, friction and shock losses more than counteract the effect.)

The theoretical pressure–volume characteristic curves are shown on Figure 10.4(a).

The theoretical relationship between impeller power and airflow may also be gained from equation (10.16):

$$P_{ow} = p_{ft} Q = AQ - BQ^2 \quad \text{W} \qquad (10.17)$$

The three power–volume relationships then become as follows.

1. *Radial bladed.* $B = 0$, and

$$P_{ow} = AQ \quad \text{(linear)}$$

2. *Backward bladed.* $B > 0$, and

$$P_{ow} = AQ - BQ^2 \quad \text{(falling parabola)}$$

3. *Forward bladed.* $B < 0$, and

$$P_{ow} = AQ + BQ^2 \quad \text{(rising parabola)}$$

Forward bladed fans are capable of delivering high flowrates at fairly low running speeds. However, their high power demand leads to reduced efficiencies. Conversely, the relatively low power requirement and high efficiencies of backward-bladed impellers make these the preferred type for large centrifugal fans. The theoretical power-volume curves are shown on Fig. 10.4(b).

Actual characteristic curves for a centrifugal impeller

The theoretical treatment of the preceding subsections led to linear pressure–volume relationships for radial, backward and forward bladed centrifugal impellers. In an actual fan, there are, inevitably, losses which result in the real pressure–volume curves lying below their theoretical counterparts. In all cases, friction and shock losses produce pressure–volume curves that tend toward zero pressure when the fan runs on open circuit, that is, with no external resistance.

Figure 10.5 shows a typical pressure–volume characteristic curve for a backward-bladed centrifugal fan. **Frictional losses** occur because of the viscous drag of the fluid on the faces of the vanes. These are denoted as F_f and F_b on Fig. 10.3. A **diffuser effect** occurs in the diverging area available for flow as the fluid moves through the impeller. This results in a further loss of available energy. In order to transmit mechanical work, the pressure on the front face of a vane, p_f, is necessarily greater than that on the back, p_b. A result of this is that the fluid velocity close to the trailing face is higher than that near the front face. These effects result in an asymmetric distribution of fluid velocity between two successive vanes at any given radius and produce an **eddy loss**. It may also be noted that, at the outlet tip, the two pressures p_f and p_b must become equal. Hence, although the tip is most important in its influence on the outlet vector diagram, it does not actually contribute to the transfer of mechanical energy. The transmission of power is not uniform along the length of the blade.

The shock (or separation) losses occur particularly at inlet and reflect the sudden turn of near 90° as the fluid enters the eye of the impeller. In practice, wall effects impart a vortex to the fluid as it approaches the inlet. By a suitable choice of inlet

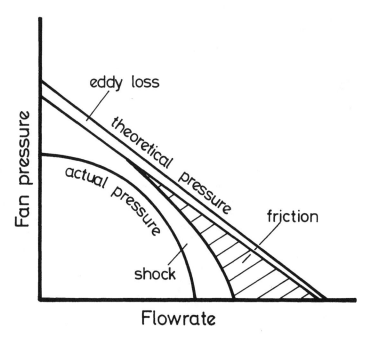

Figure 10.5   Effect of losses on the pressure–volume characteristic of a backward-bladed centrifugal fan.

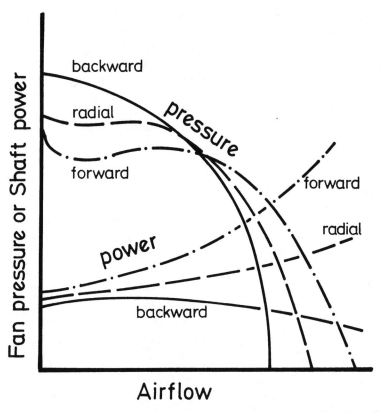

Figure 10.6   Actual pressure and shaft power characteristics for centrifugal impellers.

blade angle, β_1 (Fig. 10.3), the shock losses may be small at the optimum design flow. An inlet cone at the eye of the impeller or fixed inlet and outlet guide vanes can be fitted to reduce shock losses.

In the development of the theoretical pressure and power characteristics, we assumed radial inlet conditions. When the fluid has some degree of pre-rotation, the flow is no longer radial at the inlet to the impeller. The second term in Euler's equation (10.11) takes a finite value and, again, results in a reduced fan pressure at any given speed of rotation.

The combined effect of these losses on the three types of centrifugal impeller is to produce the characteristic curves shown on Fig. 10.6. The non-overloading power characteristic together with the steepness of the pressure curve at the higher flows are major factors in preferring the backward impeller for large installations.

10.3.2   The axial impeller

Axial fans of acceptable performance did not appear until the 1930s. This was because of a lack of understanding of the behaviour of airflow over axial fan blades. The early axial fans had simple flat-angled blades and produced very poor efficiencies. The growth of the aircraft industry led to the development of aerofoils for the wings of aeroplanes. In this section, we shall discuss briefly the characteristics of aerofoils. This facilitates our comprehension of the behaviour of axial fans. The theoretical treatment of axial impellers may be undertaken from the viewpoint of a series of aerofoils or by employing either vortex or momentum theory. We shall use the latter in order to remain consistent with our earlier analysis of centrifugal impeller although aerofoil theory is normally also applied in the detailed design of axial impellers.

Aerofoils

When a flat stationary plate is immersed in a moving fluid such that it lies parallel with the direction of flow then it will be subjected to a small drag force in the direction of flow. If it is then inclined at a small angle, α, to the direction of flow then that drag force will increase. However, the deflection of streamlines will cause an increase in pressure on the underside and a decrease in pressure on the top surface. This pressure differential results in a lifting force. On an aircraft wing, and in an axial fan impeller, it is required to achieve a high lift, L, without unduly increasing the drag, D. The dimensionless **coefficients** of lift, C_L, and drag, C_D, for any given section are defined in terms of velocity heads

$$L = \rho \, \frac{C^2}{2} \, A \, C_L \quad \text{N} \tag{10.18}$$

$$D = \rho \, \frac{C^2}{2} \, A \, C_D \quad \text{N} \tag{10.19}$$

where $C =$ fluid velocity (m/s) and $A =$ a characteristic area usually taken as the underside of the plate or aerofoil (m^2).

The ratio of C_L/C_D is considerably enhanced if the flat plate is replaced by an aerofoil. Figure 10.7 illustrates an aerofoil section. Selective evolution in the world of nature suggests that it is no coincidence that the aerofoil shape is decidedly fish like. The line joining the facing and trailing edges is known as the **chord** while the **angle of attack**, α, is defined as that angle between the chord and the direction of the approaching fluid.

A typical behaviour of the coefficients of lift and drag for an aerofoil with respect to angle of attack is illustrated on Fig. 10.8. Note that the aerofoil produces lift and a high C_L/C_D ratio at zero angle of attack. The coefficient of lift increases in a near-linear manner. However, at an angle of attack usually between 12° and 18°, breakaway of the boundary layer occurs on the upper surface. This causes sudden loss of lift and an increase in drag. In this **stall** condition, the formation and propagation of turbulent vortices causes the fan to vibrate excessively and to produce additional low frequency noise. A fan should never be allowed to run continuously in this condition as it can cause failure of the blades and excessive wear in the drive shaft and other transmission components.

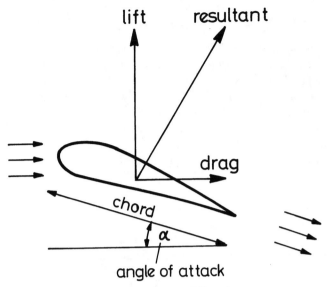

Figure 10.7 An aerofoil section.

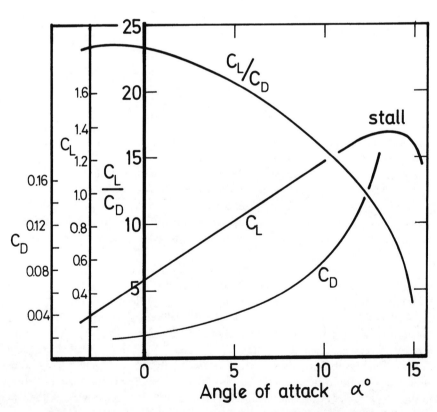

Figure 10.8 Typical behaviour of lift and drag coefficients for an aerofoil.

Theoretical pressure developed by an axial fan

Consider an imaginary cylinder coaxial with the drive shaft and cutting through the impeller blades at a constant radius. If we produce an impression of the cut blades on two-dimensional paper then we shall produce a drawing similar to that of Fig. 10.9(a).

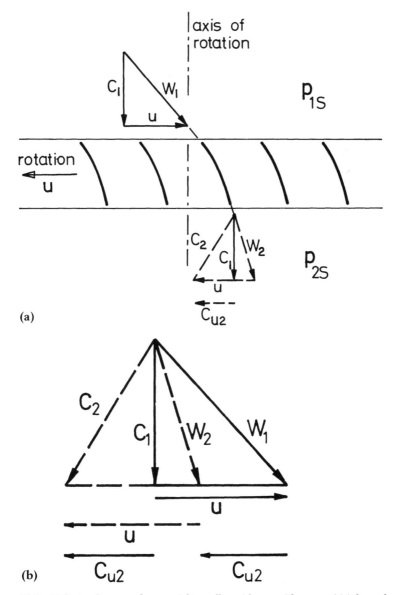

Figure 10.9 Velocity diagrams for an axial impeller without guide vanes: (a) inlet and outlet velocity diagrams; (b) inlet and outlet velocity diagrams combined.

For simplicity each blade is shown as a curved vane rather than an aerofoil section.

The air approaches the moving impeller axially at a velocity, C_1. At the optimum design point, the C_1 vector combines with the blade velocity vector, u, to produce a velocity relative to the blade, W_1, and which is tangential to the leading edge of the blade. At the trailing edge, the air leaves at a relative velocity, W_2, which also combines with the blade velocity to produce the outlet absolute velocity, C_2. This has a rotational component, C_{u2}, imparted by curvature of the blade. The initial axial velocity, C_1, has remained unchanged as the impeller has no component of axial velocity in a fan.

Figure 10.9(b), drawn at a larger scale for clarity, shows how the inlet and outlet velocity diagrams can be combined. As W_1 and W_2 are both related to the same common blade velocity, u, the vector difference between the two, $W_1 - W_2$, must be equal to the final rotational component C_{u2}.

As the axial velocity, C_1, is the same at outlet as inlet, it follows that the increase in total pressure across the impeller is equal to the rise in static pressure:

$$p_{ft} = p_{2s} - p_{1s}$$

Now let us consider again the relative velocities, W_1 and W_2. Imagine for a moment that the impeller is standing still. It would impart no energy and Bernouilli's equation (2.16) tells us that the increase in static pressure must equal the decrease in velocity pressure (in the absence of frictional losses). Hence

$$p_{ft} = p_{2s} - p_{1s} = \rho \left(\frac{W_1^2}{2} - \frac{W_2^2}{2} \right) \quad \text{Pa} \tag{10.20}$$

Applying Pythagoras' theorem to Fig. 10.9(b) gives

$$W_1^2 = C_1^2 + u^2$$

and

$$W_2^2 = C_1^2 + (u - C_{u2})^2$$

resulting in

$$p_{ft} = \frac{\rho}{2} \left[u^2 - (u - C_{u2})^2 \right]$$

$$= \frac{\rho}{2} \left(2C_{u2}u - C_{u2}^2 \right)$$

i.e.

$$p_{ft} = \rho u C_{u2} - \rho \frac{C_{u2}^2}{2} \quad \text{Pa} \tag{10.21}$$

Comparison with equation (10.12) shows that this is the pressure given by Euler's equation, less the velocity head due to the final rotational velocity.

The vector diagram for an axial impeller with inlet guide vanes is given on Fig. 10.10. In order to retain subscripts 1 and 2 for the inlet and outlet sides of the moving impeller, subscript 0 is employed for the air entering the inlet guide vanes. At the optimum design point, the absolute velocities, C_0 and C_2 should be equal and axial, i.e. there should be no rotational components of velocity at either inlet or outlet of the guide vane–impeller combination. This means that any vortex action imparted

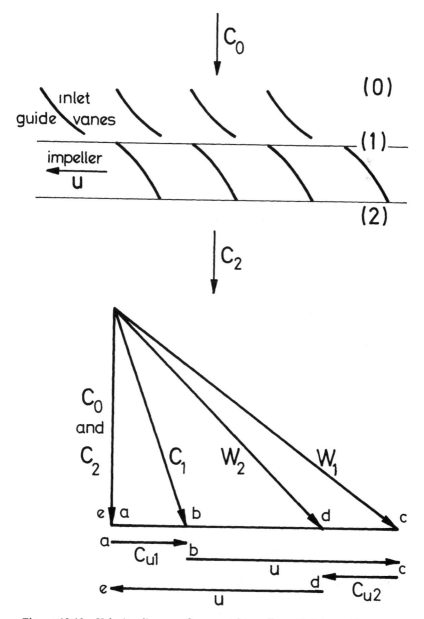

Figure 10.10   Velocity diagrams for an axial impeller with inlet guide vanes.

Fans

by the guide vanes must be removed by an equal but opposite vortex action imparted by the impeller.

We can follow the process on the vector diagram on Fig. 10.10 using the labelled points a, b, c, d, and e.

a. The air arrives at the entrance to the guide vanes with an axial velocity, C_0, and no rotational component (point a).
b. The turn on the inlet guide vanes gives a rotational component, C_{u1}, to the air. The axial component remains at C_0. Hence, the vector addition of the two results in the absolute velocity shown as C_1 (point b).
c. In order to determine the velocity of the air relative to the moving impeller at location 1, we must subtract the velocity vector of the impeller, u. This brings us to position c on the vector diagram and an air velocity of W_1 relative to the impeller.
d. The turn on the impeller imparts a rotational component C_{u2} to the air. The velocity of the air relative to the impeller is thus reduced to W_2 and we arrive at point d.
e. In order to determine the final absolute velocity of the air, we must add the impeller velocity, u. This will take us to point e on the vector diagram (coinciding with point a), with no remaining rotational component provided that $C_{u1} = C_{u2}$. It follows that the absolute velocities at inlet and outlet, C_0 and C_2, must be both axial and equal.

To determine the total theoretical pressure developed by the system, consider first the stationary inlet guide vanes (subscript g). From Bernoulli's equation with no potential energy term

$$p_g = \rho \frac{C_0^2 - C_1^2}{2}$$

Applying Pythagoras' theorem to the vector diagram of Fig. 10.10 gives this to be

$$p_g = -\rho \frac{C_{u1}^2}{2} \tag{10.22}$$

and represents a pressure loss caused by rotational acceleration across the guide vanes.

The gain in total pressure across the impeller (subscript i) is given as (see, also, equation (10.20))

$$p_i = \frac{\rho}{2}(W_1^2 - W_2^2) \tag{10.23}$$

Using Pythagoras on Fig 10.10 again gives this to be

$$p_i = \frac{\rho}{2}[C_0^2 + (u + C_{u2})^2 - (C_0^2 + u^2)]$$

$$= \frac{\rho}{2}(C_{u2}^2 + 2C_{u2}u)$$

However, as $C_{u1} = C_{u2}$, this can also be written as

$$p_i = \frac{\rho}{2}(C_{u1}^2 + 2C_{u2}u) \quad \text{Pa} \tag{10.24}$$

Now the total theoretical pressure developed by the system, p_{ft}, must be the combination $p_g + p_i$. Equations (10.22) and (10.24) give

$$p_{ft} = p_g + p_i = \rho C_{u2}u \quad \text{Pa} \tag{10.25}$$

Once again, we have found that the theoretical pressure developed by a fan is given by Euler's equation (10.12). Furthermore, comparison with equation (10.21) shows that elimination of the residual rotational component of velocity at outlet (by balancing C_{u1} and C_{u2}) results in an increased fan pressure when guide vanes are employed.

The reader might wish to repeat the analysis for outlet guide vanes and for a combination of inlet and outlet guide vanes. In these cases, Euler's equation is also found to remain true. Hence, wherever the guide vanes are located, the total pressure developed by an axial fan operating at its design point depends only on the rotational component imparted by the impeller, C_{u2}, and the peripheral velocity of the impeller, u.

It is obvious that the value of u will increase along the length of the blade from root to tip. In order to maintain a uniform pressure rise and to inhibit undesirable cross flows, the value of C_{u2} in equation (10.25) must balance the variation in u. This is the reason for the 'twist' that can be observed along the blades of a well-designed axial impeller.

Actual characteristic curves for an axial fan

The losses in an axial fan may be divided into **recoverable** and **non–recoverable** groups. The recoverable losses include the vortices or rotational components of velocity that exist in the airflow leaving the fan. We have seen that these losses can be recovered at the design point by the use of guide vanes. However, as we depart from the design point, swirling of the outlet air will build up.

The non-recoverable losses include friction at the bearings and drag on the fan casing, the hub of the impeller, supporting beams and the fan blades themselves. These losses result in a transfer from mechanical energy to heat which is irretrievably lost in its capacity for doing useful work.

Figure 10.11 is an example of the actual characteristic curves for an axial fan. The design point, C, coincides with the maximum efficiency. At this point the losses are at a minimum. In practice, the region A to B on the pressure curve would be acceptable. Operating at low resistance, i.e. to the right of point B, would not draw excessive power from the motor as the shaft power curve shows a non-overloading characteristic. However, the efficiency decreases rapidly in this region.

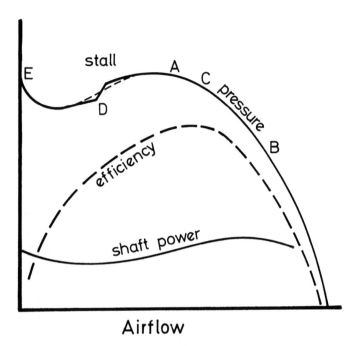

Figure 10.11 Typical characteristics curves for an axial fan.

The disadvantage of operating at too high a resistance, i.e. to the left of point A, is, again, a decreasing efficiency but, more importantly, the danger of approaching the stall point, D. There is a definite discontinuity in the pressure curve at the stall point although this is often displayed as a smoothed curve. Indeed, manufacturers' catalogues usually show characteristic curves to the right of point A only. In the region E to D, the flow is severely restricted. Boundary layer breakaway takes place on the blades (see the section on aerofoils) and centrifugal action occurs, producing recirculation around the blades.

A fixed-bladed axial fan of constant speed has a rather limited useful range and will maintain good efficiency only when the system resistance remains sensibly constant. This can seldom be guaranteed over the full life of a main mine fan. Fortunately, there are a number of ways in which the range of an axial fan can be extended.

1. The angle of the blades may be varied. Many modern axial fans allow blade angles to be changed, either when the rotor is stationary or while in motion. The latter is useful if the fan is to be incorporated into an automatic ventilation control system. Figure 10.12 gives an example of the characteristic curves for an axial fan of variable blade angle. The versatility of such fans gives them considerable advantage over centrifugal fans.
2. The angle of the inlet and/or outlet guide vanes may also be varied, with or without modification to the impeller blade angle. The effect is similar to that illustrated by Fig. 10.12.

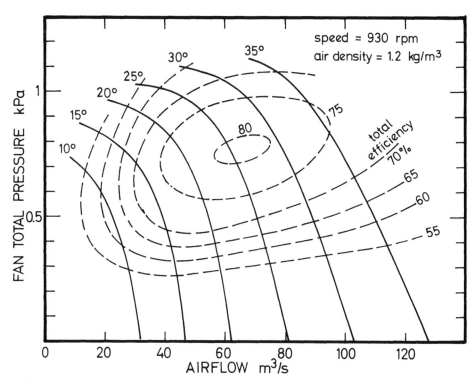

Figure 10.12 Example of a set of characteristic curves for an axial fan with variable blade angle.

3. The pitch of the impeller may be changed by adding or removing blades. The impeller must, of course, remain dynamically balanced. This technique can result in substantial savings in power during time periods of relatively light load.

4. The speed of the impeller may be changed either by employing a variable speed motor or by changing the gearing between the motor and the fan shaft. The majority of fans are driven by a.c. induction motors at a fixed speed. Variable speed motors are always more expensive although they may produce substantial saving in operating costs. Axial fans may be connected to the motor via flexible couplings which allow a limited degree of angular or linear misalignment. Speed control may be achieved by hydraulic couplings, or, in the case of smaller fans, by V-belt drives with a range of pulley sizes.

10.4 FAN LAWS

The performance of fans is normally shown as a series of pressure, efficiency and shaft power characteristic curves plotted against airflow for specified values of rotational speed, air density and fan dimensions. It is, however, convenient to be able to determine the operating characteristic of the fan at other speeds and air densities.

It is also useful to be able to use test results gained from smaller prototypes to predict the performance of larger fans that are geometrically similar.

Euler's equation and other relationships introduced in section 10.3 can be employed to establish a useful set of proportionalities known as the fan laws.

10.4.1   Derivation of the fan laws

Fan pressure

From Euler's equations (10.12) and (10.25)

$$p_{ft} = \rho u C_{u2}$$

However, both the peripheral speed, u, and the rotational component of outlet velocity, C_{u2} (Figs. 10.3 and 10.10) vary with the rotational speed, n, and the impeller diameter, d. Hence,

$$p_{ft} \propto \rho(nd)(nd)$$

or

$$p_{ft} \propto \rho n^2 d^2 \tag{10.26}$$

where $\propto$ means 'proportional to'.

Airflow

For a centrifugal fan, the radial flow at the impeller outlet is given as

$$Q = \text{area of flow at impeller outlet} \times C_{m2}$$
$$= \pi d \times \text{impeller width} \times C_{m2}$$

where d = diameter of impeller and C_{m2} = radial velocity at outlet (Fig. 10.3). However, for geometric similarity between any two fans, impeller width $\propto d$, giving

$$Q \propto d^2 C_{m2}$$

Again, for geometric similarity, all vectors are proportional to each other. Hence

$$C_{m2} \propto u \propto nd$$

giving

$$Q \propto nd^3 \tag{10.27}$$

A similar argument applies to the axial impeller (Fig. 10.10). At outlet,

$$Q = \pi d^2 \times C_2$$

where C_2 = axial velocity. As before, vectors are proportional to each other for geometric similarity,

$$C_2 \propto u \propto nd$$

giving, once again,

$$Q = \propto nd^3$$

Density

From Euler's equation (10.12) or (10.25) it is clear that fan pressure varies directly with air density:

$$p_{ft} \propto \rho \qquad (10.28)$$

However, we normally accept volume flow, Q, rather than mass flow as the basis of flow measurement in fans. In other words, if the density changes we still compare operating points at corresponding values of volume flow.

Air power

For incompressible flow, the mechanical power transmitted from the impeller to the air is given as (see equation (5.56))

$$P_{ow} = p_{ft} Q$$

Employing proportionalities (10.26) and (10.27) gives

$$P_{ow} \propto \rho n^3 d^5 \qquad (10.29)$$

10.4.2 Summary of fan laws

In the practical utilization or design of fans, we are normally interested in varying only one of the independent variables (speed, air density, impeller diameter) at any given time while keeping the other two constant. The fan laws may then be summarized as follows:

Variable n	*Variable* d	*Variable* ρ
$p \propto n^2$	$p \propto d^2$	$p \propto \rho$
$Q \propto n$	$Q \propto d^3$	Q fixed
$P_{ow} \propto n^3$	$P_{ow} \propto d^5$	$P_{ow} \propto \rho$

These laws are applicable to compare the performance of a given fan at changed speeds or densities, or to compare the performance of different sized fans provided that those two fans are geometrically similar.

If the two sets of operating conditions, or the two geometrically similar fans, are identified by subscripts a and b, then the fan laws may be written, more generally, as the following equations

$$\frac{p_{ft,a}}{p_{ft,b}} = \frac{n_a^2 \, d_a^2 \, \rho_a}{n_b^2 \, d_b^2 \, \rho_b} \tag{10.30}$$

$$\frac{Q_a}{Q_b} = \frac{n_a \, d_a^3}{n_b \, d_b^3} \tag{10.31}$$

$$\frac{P_{ow,a}}{P_{ow,b}} = \frac{n_a^3 \, d_a^5 \, \rho_a}{n_b^3 \, d_b^5 \, \rho_b} \tag{10.32}$$

where p_{ft} = fan pressure (applies for both total and static), Q = airflow, P_{ow} = air-power, n = rotational speed, d = impeller diameter and ρ = air density.

Example Characteristic curves are available for a fan running at 850 rpm and passing air of inlet density 1.2 kg/m³. Readings from the curves indicate that at an airflow of 150 m³/s, the fan pressure is 2.2 kPa and the shaft power is 440 kW. Assuming that the efficiency remains unchanged, calculate the corresponding points if the fan is run at 1100 rpm in air of density 1.1 kg/m³.

Solution As it is the same fan that is to be used in differing conditions, there is no change in impeller diameter,

Equation (10.30) gives the new pressure to be

$$p_f = 2.2 \times \left(\frac{1100}{850}\right)^2 \times \frac{1.1}{1.2}$$

$$= 3.377 \text{ kPa}$$

From equation (10.31), the new volume flow is

$$Q = 150 \times \frac{1100}{850} = 194.1 \text{ m}^3/\text{s}$$

Equation (10.32) refers to the airpower rather than the shaft power delivered to the impeller. However, if we assume that the impeller efficiency remains the same at corresponding points on the characteristic curves then we may also use that equation for shaft power:

$$P_{ow} = 440 \times \left(\frac{1100}{850}\right)^3 \times \frac{1.1}{1.2}$$

$$= 874 \text{ kW}$$

By treating a series of points from the original curves in this way, a second set of characteristic curves can be produced that is applicable to the new conditions.

10.5 FANS IN COMBINATION

Many mines or other subsurface facilities have main, and perhaps, booster fans sited at differing locations; for example, there may be two or more upcast shafts, each with its own surface exhausting fan. With the advent of simulation programs for ventilation network analysis (Chapters 7 and 9), the relevant pressure–volume characteristic data may be entered separately for each fan unit. The system resistance offered to each of those fans becomes a function not only of the network geometry but also the location and operating characteristics of other fans in the system. That resistance is sometimes termed the **effective resistance** 'seen' by the fan.

There are situations in which it is advantageous to combine fans either in series or in parallel at a single location. This enables a wide spectrum of pressure–volume duties to be attained with only a limited range of fan sizes. In general, fans may be connected in series in order to pass a given airflow against an increased resistance, while a parallel combination allows the flow to be increased for any given resistance. Although ventilation network programs can allow each fan to be entered separately,

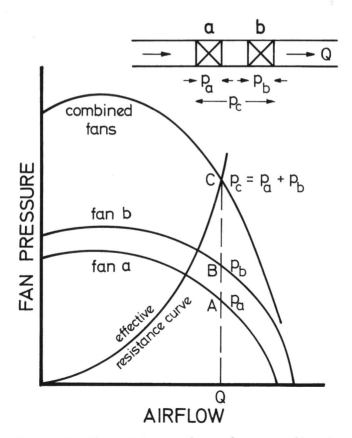

Figure 10.13 Characteristic curves for two fans connected in series.

it is sometimes more convenient to produce a pressure–volume characteristic curve that represents the combined unit.

10.5.1   Fan in series

Figure 10.13 shows two fans, a and b, located in series within a single duct or airway. The corresponding pressure–volume characteristics and the effective resistance curve are also shown. The characteristic curve for the combination is obtained simply by adding the individual fan pressures for each value of airflow.

The effective operating point is located at C, where the resistance curve intersects the combined characteristics. Fans a and b both pass the same airflow, Q, but develop pressures p_a and p_b respectively. The individual operating points are shown as A and B. For three or more fans, the process of adding fan pressures remains the same. However, if the change in density through the combination becomes significant then the fan laws (section 10.4.2) should be employed to correct the individual characteristic curves.

As shown in Fig. 10.13, the individual fans need not have identical characteristic curves. However, if one fan is considerably more powerful than the other, or if the system resistance falls to a low level, then the impeller of the weaker unit may be driven in turbine fashion by its stronger companion. The weaker fan then becomes an additional resistance on the system.

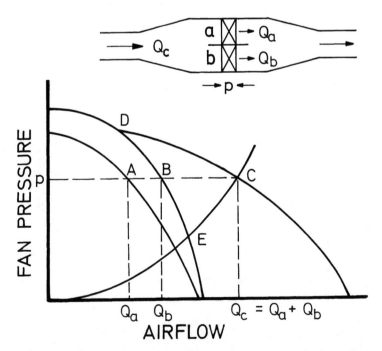

Figure 10.14 Characteristic curves for two fans connected in parallel.

10.5.2   Fans in parallel

For fans that are combined in parallel, the airflows are added for any given fan pressure in order to obtain the combined characteristic curve. As shown in Fig. 10.14, fans a and b pass airflows Q_a and Q_b, respectively, but at the same common pressure, p.

The operating point for the complete unit occurs at C with the individual operating points for fans a and b at A and B respectively. Here again, the fans need not necessarily be identical. However, particular care must be taken to ensure that the operating points A and B do not move too far up their respective curves. This is particularly important in the case of axial fans because of their pronounced stall characteristic. If the system resistance increases to such an extent that the operating point moves from C to D then, at that point, fan b will pass all of the airflow while the impeller of fan a is faced with an infinite effective resistance and, hence, passes no airflow. In practice, before this condition is reached, the fans may exhibit a noticeable 'hunting, effect. For these reasons, the maximum variation in system resistance that is likely to occur should be investigated before installing fans in parallel.

Here again, three or more fans may be combined in parallel, adding airflows to obtain the combined characteristic curve. Although Fig. 10.14 shows two differing fan characteristics, it is prudent to employ identical fans when connected in parallel. This will reduce the tendency for one of them to approach stall conditions before the other. However, differences in the immediate surroundings of ductwork or airway geometry often result in the fans operating against slightly different effective resistances. Hence, even when identical fans are employed, it is usual for measurements to indicate that they are producing slightly different pressure– volume duties. In the case of fans located in separate ducts or airways that are connected in parallel, the resistance of those ducts or airways may be taken into account by subtracting the frictional pressure losses in each branch from the corresponding fan pressures. In these circumstances, it is preferable to consider the fans as separate units for the purposes of network analysis.

An advantage of employing fans in parallel is that if one of them fails then the remaining fan(s) continue to supply a significant proportion of the original flow. In the example shown on Fig. 10.14, if fan a ceases to operate then the operating point for fan b will fall to position E, giving some 70% of the original airflow. The latter value depends on the number of fans employed, the shape of their pressure–volume characteristic curves and the provision of non-return baffles at the fan outlets.

Fans may be connected in any series–parallel combination, adding pressures and airflows respectively to obtain the combined characteristic curve. This is particularly useful for booster fan locations. A mine may maintain an inventory of standard fans, combining them in series–parallel configurations to achieve any desired operating characteristic.

10.6   FAN PERFORMANCE

Power is delivered to the drive shaft of a fan impeller from a motor (usually electric) and via a transmission assembly. Losses occur in both the motor and transmission.

For a properly maintained electrical motor and transmission, some 95% of the input electrical power may be expected to appear as mechanical energy in the impeller drive shaft. The impeller, in turn, converts most of that energy into useful airpower to produce both movement of the air and an increase in pressure. The remainder is consumed by irreversible losses across the impeller and in the fan casing (sections 10.3.1 and 10.3.2) and produces an additional increase in temperature of the air.

Impeller efficiency may be defined as

$$\frac{\text{airpower}}{\text{shaft power}} \tag{10.33}$$

while the **overall efficiency** of the complete motor–transmission–impeller unit is given as

$$\frac{\text{airpower}}{\text{motor input power}} \tag{10.34}$$

In the following subsection, we shall define other measures of fan efficiency. As there are several different measures of fan efficiency it is prudent, when perusing manufacturers' literature, to ascertain the basis of any quoted values of efficiencies. It is also important to use the same measure of efficiency when comparing one fan with another. However, the one parameter that really matters is the input power required to achieve the specified pressure–volume duty, as this is the factor that dictates the operating cost of the fan.

10.6.1 Compressibility, fan efficiency and fan testing

In our previous analyses, we have defined airpower as the product pQ on the basis of incompressible flow. That simplification will give an error of less than 1% when testing fans that develop pressures up to 2.8 kPa. Unfortunately, fan operating costs at many large mines are such that 1% may represent a significant expenditure. Furthermore, mine fan pressures exceeding 6 kPa are not uncommon. For these reasons, we should take air compressibility into account when measuring fan performance.

Applying the steady-flow energy equation (3.25) to a fan gives

$$\frac{u_1^2 - u_2^2}{2} + (Z_1 - Z_2)g + W = \int_1^2 V\, \mathrm{d}P + F_{12} = H_2 - H_1 - q_{12} \quad \frac{\text{J}}{\text{kg}} \tag{10.35}$$

where subscripts 1 and 2 refer to the fan inlet and outlet respectively, $W =$ impeller shaft work (J/(kg of air)), $V =$ specific volume of air (m³/kg), $P =$ absolute (barometric) pressure (Pa), $F_{12} =$ frictional losses (J/kg), $H =$ enthalpy and $q_{12} =$ heat added through fan casing (J/kg).

Now, for the purposes of this analysis we shall assume that the change in elevation through the fan is negligible, $Z_1 - Z_2 = 0$. Furthermore, we shall assume that the change in air velocity across the fan is also negligible compared with other terms, $u_1 = u_2$. This latter assumption implies that the fan total pressure, referred to here simply as p_f, is equal to the increase in barometric pressure across the fan, $P_2 - P_1$. Then

$$W = \int_1^2 V \, dP + F_{12} = H_2 - H_1 - q_{12} \quad \frac{J}{kg} \tag{10.36}$$

It is also reasonable to assume that the heat transferred from the surroundings through the fan casing is small compared with the shaft work. The steady-flow energy equation them simplifies to the adiabatic equation

$$W = \int_1^2 V \, dP + F_{12} = H_2 - H_1 \quad \frac{J}{kg} \tag{10.37}$$

In order to define a thermodynamic efficiency for the fan impeller, we must first designate a 'perfect' fan against which we can compare the performance of the real fan. In the perfect fan, there are no losses, $F_{12} = 0$. As we now have a frictionless adiabatic, i.e. an isentropic, compression we can write

$$W_{isen} = \int_1^2 V \, dP = H_{2, isen} - H_1 \quad \frac{J}{kg} \tag{10.38}$$

where the subscript isen denotes isentropic conditions.

The pressure–volume method

There are two methods of determining the efficiency of a fan. The technique that is accepted as standard by most authorities relies on measurement of the fan pressure, airflow and shaft power, and is known as the pressure–volume method.

The first step is to evaluate the integral $\int_1^2 V \, dp$. Here, we have a choice. We can use the isentropic relationship

$$PV^\gamma = \text{constant } C \quad J \tag{10.39}$$

where γ = the isentropic index C_p/C_v (1.4 for dry air). This will lead to the **isentropic efficiency** of the fan. Alternatively, we could employ the actual polytrope produced by the fan

$$PV^n = \text{constant}$$

and assume that it is a reversible (ideal) process. This will lead to the **polytropic efficiency** of the fan. Both measures of efficiency are acceptable provided that the choice is stated clearly. While the polytropic efficiency is closer to a true measure of output/input, and takes any heat transfer, q_{12}, into account, the isentropic efficiency has the advantage that the index γ is defined for any gas. For this reason, we shall continue the analysis on the basis of isentropic compression within the ideal fan. In

practice, polytropic and isentropic efficiencies are near equal for most fan installations.*

Substituting

$$V = (C/P)^{1/\gamma}$$

into equation (10.38) gives

$$W_{isen} = \int_1^2 \frac{C^{1/\gamma}}{P^{1/\gamma}} \, dp$$

$$= C^{1/\gamma} \frac{[P^{1-1/\gamma}]_1^2}{1 - 1/\gamma}$$

However, as $C^{1/\gamma} = P^{1/\gamma} V$,

$$W_{isen} = \frac{\gamma}{\gamma - 1} (P_2 V_2 - P_1 V_1)$$

$$= \frac{\gamma}{\gamma - 1} P_1 V_1 \left(\frac{P_2 V_2}{P_1 V_1} - 1 \right)$$

Now

$$\frac{V_2}{V_1} = \left(\frac{P_1}{P_2} \right)^{1/\gamma}$$

from equation (10.39), giving

$$W_{isen} = \frac{\gamma}{\gamma - 1} P_1 V_1 \left[\left(\frac{P_2}{P_1} \right)^{1 - 1/\gamma} - 1 \right] \quad \frac{J}{kg}$$

In order to convert the shaft work from J/kg to shaft power, P_{ow} (J/s or watts),
we must multiply by the mass flow of air

$$M = Q_1 \rho_1 = \frac{Q_1}{V_1} \quad \frac{kg}{s}$$

where $\rho = $ air density (kg/m^3). Then

$$P_{ow,isen} = \frac{\gamma}{\gamma - 1} P_1 Q_1 \left[\left(\frac{P_2}{P_1} \right)^{1 - 1/\gamma} - 1 \right] \quad W \qquad (10.40)$$

---

* Fan efficiencies can be defined in terms of areas on the Ts diagram of Fig. 3.7:

$$\eta_{isen} = \frac{ACBXY}{DBXZ}$$

$$\eta_{poly} = \frac{ADBXY}{BDXZ}$$

Now let us multiply both the numerator and denominator by fan pressure, p_f:

$$P_{ow,isen} = p_f Q_1 \frac{\gamma}{\gamma - 1} \frac{P_1}{p_f} \left[\left(\frac{P_2}{P_1} \right)^{1 - 1/\gamma} - 1 \right] \quad W$$

We can rewrite this equation as

$$P_{ow,isen} = p_f Q_1 K_p \quad W \tag{10.41}$$

where

$$K_p = \frac{\gamma}{\gamma - 1} \frac{P_1}{p_f} \left[\left(\frac{P_2}{P_1} \right)^{1 - 1/\gamma} - 1 \right] \tag{10.42}$$

and is known as the **compressibility coefficient**. By substituting $p_f = P_2 - P_1$, we

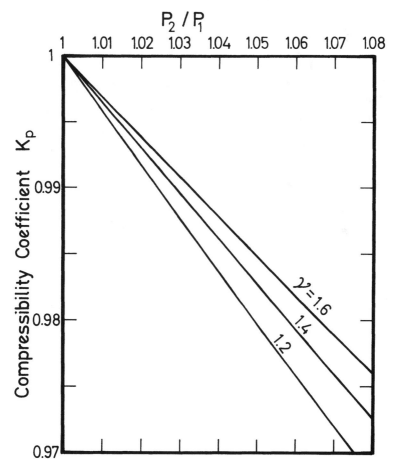

Figure 10.15 Variation of compressibility coefficient with respect to pressure ratio. $\gamma = 1.4$ for dry air.

obtain

$$K_p = \frac{\gamma}{\gamma - 1}\left[\frac{(P_2/P_1)^{(\gamma - 1)/\gamma} - 1}{P_2/P_1 - 1}\right] \tag{10.43}$$

We can now see that our earlier assumption of incompressible flow ($P_{ow} = p_f Q$) involved an error represented by the deviation of K_p from unity. Figure 10.15 shows the variation in compressibility coefficient with respect to the pressure ratio P_2/P_1. For unsaturated air the value $\gamma = 1.4$ may be used.

The isentropic efficiency of the fan can now be defined as

$$\eta_{isen} = \frac{P_{ow,isen}}{\text{shaft power}}$$

$$= \frac{p_f Q_1 K_p}{\text{shaft power}} \tag{10.44}$$

Employment of this equation is a standard technique of determining fan efficiency and is known commonly as the pressure–volume method.

Two other terms used frequently in the literature are static efficiency and total efficiency. These are obtained simply by using fan static pressure and fan total pressure, respectively, for p_f in equation (10.44).

All of these measures of efficiency are matters of definition rather than precision. However, care should be taken to ensure that the same measure is employed when comparing fan performances.

Equation (10.44) indicates that the pressure–volume method of fan testing requires the measurement of pressures, air volume flow and the impeller shaft power. It is frequently the case that a large fan installed at a mine site does not quite meet the specifications indicated by factory tests. Part of the problem may be the less than ideal inlet or outlet conditions that often exist in field installations. In particular, uneven distribution of airflow approaching the fan can produce a diminished performance and may even result in premature blade failure.

Another problem is that turbulence and an asymmetric velocity profile may make it difficult to obtain good accuracy in the measurement of airflow (section 6.2). Water droplets in the airstream can also result in erroneous readings from both anemometers and pitot tubes.

Impeller shaft power can be obtained accurately in the laboratory or factory test rig by means of torquemeters or, in the case of smaller fans, swinging carcass (dynamometer) motors. However, for *in-situ* tests, it may be necessary to resort to the measurement of input electrical power and to rely on manufacturer's data for the efficiences of the motor and transmission.

Despite these difficulties, it is always advisable to conduct an *in-situ* test on a new main fan in order to verify, or modify, the fan characteristic data that are to be used for subsequent ventilation network exercises.

Example   A fan passes an airflow of 300 m³/s at the inlet, and develops a pressure of 2.5 kPa. The pressure at the fan inlet is 97 kPa. The motor consumes an electrical power of 1100 kW. Assuming a combined motor–transmission efficiency of 95%, determine the isentropic efficiency of the impeller and, also, of the total unit.

Solution $P_1 = 97$ kPa, $P_2 = 97 + 2.5 = 99.5$ kPa. Using a value of $\gamma = 1.4$ for air, equation (10.40) gives the air power (or isentropic shaft power) as

$$P_{ow,isen} = \frac{\gamma}{\gamma - 1} P_1 Q_1 \left[\left(\frac{P_2}{P_1} \right)^{(\gamma - 1)/\gamma} - 1 \right]$$

$$= 3.5 \times 97\ 000 \times 300 \left[\left(\frac{99.5}{97} \right)^{0.286} - 1 \right]$$

$$= 743.9 \times 10^3\ W \quad \text{or} \quad 743.9\ kW$$

Actual shaft power,

$$P_{ow} = 1100 \times 0.95 = 1045\ kW$$

Isentropic efficiency of impeller

$$\eta_{isen} = \frac{P_{ow,isen}}{P_{ow}}$$

$$= \frac{743.9}{1045} = 0.712 \quad \text{or} \quad 71.2\%$$

The overall isentropic efficiency of the unit is

$$\eta_{isen}\ (\text{overall}) = \frac{743.9}{1100} = 0.676 \quad \text{or} \quad 67.6\%$$

Note that if compressibility had been ignored, the isentropic shaft power would have been

$$P_{ow,isen} = p_f Q$$

$$= 2.5 \times 300 = 750\ kW$$

involving an error 0.8% from the true value of 743.9 kW.

The thermometric method

The problems associated with *in-situ* pressure–volume tests on mine fans have led researchers to seek another method that did not require the measurement of either air-flow or shaft power. The earliest such work appears to have been carried out in the 1920s (Whitaker) although later work was required to make it a practical proposition (McPherson, 1971; Drummond, 1972).

The thermometric method of fan testing is based on the enthalpy terms in the steady flow energy equation. The shaft work for the ideal isentropic fan is given by equation (10.38):

$$W_{isen} = H_{2,isen} - H_1$$

If we assume that the air contains no free water droplets and that neither evaporation nor condensation occurs within the fan then

$$W_{isen} = H_{2,isen} - H_1 = C_p(T_{2,isen} - T_1) \quad J/kg \qquad (10.45)$$

where C_p = specific heat at constant pressure of the air (1005 J/(kg K) for dry air). Similarly, for the real fan (no secondary subscript)

$$W = C_p(T_2 - T_1) \quad J/kg \qquad (10.46)$$

The isentropic efficiency is then given as then ratio of the shaft work for the isentropic fan to that for the actual fan

$$\eta_{isen} = \frac{W_{isen}}{W} = \frac{T_{2,isen} - T_1}{T_2 - T_1} \qquad (10.47)$$

or

$$\eta_{isen} = \frac{\Delta T_{isen}}{\Delta T} \qquad (10.48)$$

where

$$\Delta T_{isen} = T_{2,isen} - T_1 \quad °C$$

and

$$\Delta T = T_2 - T_1 \quad °C$$

ΔT_{isen} is, therefore, the increase in dry bulb temperature that would occur as air passes through an isentropic fan, while ΔT is the temperature rise that actually occurs in the real fan. These two parameters are illustrated on the Ts diagram of Fig. 3.7. As ΔT can be measured, it remains to find an expression for the isentropic temperature rise. This was derived in Chapter 3 as equation (3.53):

$$\frac{T_{2,isen}}{T_1} = \left(\frac{P_2}{P_1}\right)^{(\gamma-1)/\gamma} \qquad (10.49)$$

where γ = ratio in specific heats, C_p/C_v (1.4 for dry air). Then

$$\Delta T_{isen} = T_{2,isen} - T_1 = T_1\left[\left(\frac{P_2}{P_1}\right)^{(\gamma-1)/\gamma} - 1\right] \quad °C \qquad (10.50)$$

giving

$$\eta_{isen} = \frac{\Delta T_{isen}}{\Delta T} = \frac{T_1}{\Delta T}\left[\left(\frac{P_2}{P_1}\right)^{(\gamma-1)/\gamma} - 1\right] \qquad (10.51)$$

This equation may be employed directly as all of its variables are measurable. It can, however, be simplified by employing the compressibility coefficient, K_p, from eqution (10.42):

$$\left(\frac{P_2}{P_1}\right)^{(\gamma-1)/\gamma} - 1 = K_p\frac{p_f}{P_1}\frac{\gamma-1}{\gamma}$$

Substituting in equation (10.51) gives

$$\eta_{\text{isen}} = \frac{\gamma - 1}{\gamma} \frac{p_f}{P_1} \frac{T_1}{\Delta T} K_p \tag{10.52}$$

Using the value of $\gamma = 1.4$ for dry air gives

$$\eta_{\text{isen}} = 0.286 \frac{p_f}{P_1} \frac{T_1}{\Delta T} K_p \tag{10.53}$$

Furthermore, for fan pressures not exceeding 2.8 kPa, the compressibility coefficient may be ignored for 1% accuracy, leaving the simple equation

$$\eta_{\text{isen}} = 0.286 \frac{p_f}{P_1} \frac{T_1}{\Delta T} \tag{10.54}$$

This analysis has assumed that the airflow contains no liquid droplets of water. The presence of water vapour has very little effect on the accuracy of equation (10.53) or (10.54) provided that the air remains unsaturated. (see equation (10A.15) in the appendix following this chapter.)

In many deep and hot mines, the reduction in temperature as the return air ascends the upcast shaft may cause condensation. Exhaust fans operating at the surface will then pass fogged air—a mixture of air, water vapour and droplets of liquid water. Equation (10.53) no longer holds because of the cooling effect of evaporation within the fan. A more complex analysis, taking into account the air, water vapour, liquid droplets and the phase change of evaporation produces the following differential equation which quantifies the temperature–pressure relationship for the isentropic behaviour of fogged air:

$$\frac{dT}{dP} = \frac{(R + R_v X_s)\dfrac{T}{P} + \dfrac{LX_s}{P - e_s}}{C_p + XC_w - BX_s + \dfrac{L^2 P}{P - e_s} \dfrac{X_s}{R_v T^2}} \quad °C/Pa \tag{10.55}$$

A derivation of the equation and the full definition of the symbols are given in the appendix at the end of this chapter.

Inserting the values of the constants gives

$$\frac{dT}{dP} = \frac{0.286\left[(1 + 1.6078 X_s)\dfrac{T}{P} + \dfrac{LX_s}{287.04(P - e_s)}\right]}{\left[1 + 4.1662 X - 2.3741 X_s + \dfrac{L^2 P X_s}{463.81 \times 10^3 (P - e_s) T^2}\right]} \quad °C/Pa \tag{10.56}$$

where $T =$ absolute temperature (K), $P =$ absolute pressure (Pa), $L =$ latent heat of evaporation (J/kg), $X =$ total moisture content (kg/(kg dry air)), $X_s =$ water vapour content at saturation (kg/(kg dry air)) and $e_s =$ saturation vapour pressure at temperature T (Pa). This equation can be programmed into a calculator or microprocessor for the rapid evaluation of dT/dP.

For the relatively small pressures and temperatures developed by a fan, we can write to a good approximation

$$\Delta T_{\text{isen}} = p_f \frac{\mathrm{d}T}{\mathrm{d}P} \quad °\text{C} \tag{10.57}$$

Furthermore, to improve accuracy, the values used for T and P should be the mean temperature and pressure of the air as it passes through the fan.

Although the thermometric method eliminates the need for airflow and shaft power in the determination of fan efficiency, it does introduce other practical difficulties. The temperature of the air may vary with both time and position over the measurement cross-sections because of vortex action and thermal stratification. The method of measuring the temperature rise across the fan must give an instantaneous reading of the difference between the mean temperatures at inlet and outlet measuring stations. This may be accomplished by thermocouples, connected in series, with the hot and cold junctions distributed over supporting grids at the outlet and inlet measuring stations respectively (Drummond).

Poorly designed evasees on exhausting centrifugal fans may suffer from a re-entry down-draught of outside air along one side. A test should be made for this condition before positioning the exit thermocouple heads.

Although the thermometric technique does not require an airflow in order to calculate efficiency, the airflow is nevertheless still needed if that efficiency is to be compared with a manufacturer's characteristic curve. If the site conditions are such that greater confidence can be placed in a determination of shaft power than airflow, then the latter may be computed as

$$Q = \frac{\text{shaft power}}{\rho_1(H_2 - H_1)} \quad \frac{\text{m}^3}{\text{s}} \tag{10.58}$$

where the shaft power is in watts, ρ_1 = air density at inlet (kg/m^3), and H = enthalpy (J/kg). If the air is unsaturated then

$$H_2 - H_1 = C_p(T_2 - T_1)$$

In the case of fogged air, the enthalpies must be determined from equation (10A.3) given in the appendix.

Example 1 A temperature rise of 5.96 °C is measured across a fan developing an increase in barometric pressure of 5 kPa and passing unsaturated air. The inlet temperature and pressure are 25.20 °C and 101.2 kPa respectively. Determine the isentropic temperature rise and, hence, the isentropic efficiency of the impeller.

Solution From equation (10.50)

$$\Delta T_{\text{isen}} = (273.15 + 25.2)\left[\left(\frac{101.2 + 5}{101.2}\right)^{0.286} - 1\right]$$

$$= 4.143\,°\text{C}$$

Then

$$\eta_{isen} = \frac{\Delta T_{isen}}{\Delta T}$$

$$= \frac{4.143}{5.96} = 0.695 \text{ or } 69.5\%$$

Ignoring the effects of compressibility allows equation (10.54) to be applied, giving

$$\eta_{isen} = 0.286 \times \frac{5}{101.2} \times \frac{273.15 + 25.2}{5.96}$$

$$= 0.707 \text{ or } 70.7\%$$

Hence, in this example, ignoring compressibility causes the fan efficiency to be overestimated by 1.2%.

Example 2 A surface exhausting fan passes fogged air of total moisture content $X = 0.025$ kg/(kg dry air). The pressure and temperature at the fan inlet are 98.606 kPa and 18.80 °C respectively. If the temperature rise across the fan is 2.83 °C when the increase in pressure is 5.8 kPa, calculate the isentropic efficiency of the impeller.

Solution The mean barometric pressure in the fan is

$$P = 98.606 + 5.8/2 = 101.506 \text{ kPa}$$

At inlet,

$$T_1 = 273.15 + 18.8 = 291.95 \text{ K}$$

and at outlet,

$$T_2 = 291.95 + 2.84 = 294.79 \text{ K}$$

Hence, the mean temperature is

$$T = 293.37 \text{ K} \quad \text{or} \quad t = 20.22 \,°\text{C}$$

The psychrometric equations in section 14.6 allow the following parameters to be calculated. For saturation conditions, the wet and dry bulb temperatures are, of course, equal.

1. *Saturation vapour pressure.*

$$e_s = 610.6 \exp\left(\frac{17.27t}{237.3 + t}\right)$$

$$= 610.6 \exp\left(\frac{17.27 \times 20.22}{237.3 + 20.22}\right) = 2369.5 \text{ Pa}$$

2. *Saturation vapour content.*

$$X_s = 0.622 \frac{e_s}{P - e_s}$$

$$= 0.622 \times \frac{2369.5}{101\,506 - 2369.5} = 0.014\,87 \text{ kg/kg}$$

3. *Latent heat of evaporation.*

$$L = (2502.5 - 2.386\,t)1000$$

$$= (2502.5 - 2.386 \times 20.22)1000 = 2454.26 \times 10^3 \text{ J/kg}$$

Substituting the known values into equation (10.56) gives

$$\frac{dT}{dP} = 0.286 \frac{0.002\,959 + 0.001\,282}{1 + 0.104\,16 - 0.035\,30 + 2.296\,93}$$

$$= 3.604 \times 10^{-4} \quad °\text{C/Pa}$$

This example shows that the term involving the total moisture content, X, is relatively weak (0.104 16). Hence, no stringent efforts need be made to measure this factor with high accuracy. Then (equation (10.57)) gives

$$\Delta T_{isen} = p_f \frac{dT}{dP}$$

$$= 5800 \times 3.604 \times 10^{-4} = 2.09\,°\text{C}$$

The isentropic efficiency is

$$\eta_{isen} = \frac{\Delta T_{isen}}{\Delta T} = \frac{2.09}{2.83}$$

$$= 0.739 \text{ or } 73.9\%$$

10.6.2 Fan specifications

The results of ventilation network planning exercises will produce a range of pressure–volume duties required of any new major fan that is to be installed. The process of finding and ordering the fan often commences by the ventilation engineer perusing the catalogues of fan characteristics produced by fan manufacturers. A number of those companies should be invited to submit tenders for the manufacture and, if required, installation of the fan. However, in order for those tenders to be complete, information in addition to the required pressure–volume range should be provided by the purchasing organization.

1. The mean temperature, barometric pressure, humidity and, hence, air density at the fan inlet should be given. This allows data based on standard density to be corrected to the psychrometric conditions expected in the field.

2. In many cases of both surface and underground fans, noise restrictions must be applied. These should be quantified in terms of noise level and, if necessary, with respect to direction.
3. A plan and sections of the site should be provided showing the proposed fan location and, in particular, highlighting any restrictions on space.
4. The request for a tender must identify and, whenever possible, quantify the concentrations and types of pollutants to be handled by the fan. These include dusts, gases, water vapour and liquid water droplets. In particular, any given agents of a corrosive nature should be stressed. The purchaser should further indicate any preference that may exist for the materials to be used in the manufacture of the fan impeller and casing. Specifications on paints or other protective coatings may also be given.
5. Any preference for the type of fan should be indicated. Otherwise, the manufacturer should be specifically invited to propose one or more fans that will meet the other specifications.
6. The scope of the required tender should be clearly defined. If the contractor is to be responsible for providing, installing and commissioning the new fan, the individual items should be specified for separate costing.
7. The motor, transmission, electrical switchgear and monitoring devices may be acquired and installed either by the provider of the fan or separately. In either case, the voltage and any restrictions on power availability should be stated.
8. Areas of responsibility for site preparation and the provision and installation of ducting should be identified.

APPENDIX: DERIVATION OF THE ISENTROPIC TEMPERATURE–PRESSURE RELATIONSHIP FOR A MIXTURE OF AIR, WATER VAPOUR AND LIQUID WATER DROPLETS

It is suggested that the reader delays working on this appendix until Chapter 14 has been studied.

During a flow process, the work done by expansion or compression of the air is given by $V\,dP$. In the case of fogged air containing X kg of water (vapour plus liquid) per kg of dry air, this becomes $V_s\,dP$ where the specific volume of the $1 + X_s$ kg of the air-vapour mixture is (see equation (14.14))

$$V_s = (R + R_v X_s)\frac{T}{P}$$

where R = gas constant for dry air (287.04 J/(kg K)) and R_v = gas constant for water vapour (461.5 J/(kg K)). X_s is the mass of water vapour associated with 1 kg of 'dry' air in the saturated space. The volume of the $X - X_s$ kg of liquid water is negligible compared with that of the gases. In a fogged airstream, the wet and dry bulb temperatures are equal.

The steady flow energy equation (10.35) for an isentropic process $(dF = dq = 0)$

then becomes

$$V_s \, dP = (R + R_v X_s) \frac{T}{P} dP = dH \quad \frac{J}{kg} \tag{10A.1}$$

The enthalpy of the $1 + X$ kg of air–vapour–liquid mix, based on a datum of $0\,°C$, is the combination of $C_p t$, the heat required to raise 1 kg of dry air through $t\,°C$, $XC_w t$, the heat required to raise X kg of liquid water through $t\,°C$, and LX_s, the heat required to evaporate X_s kg of water at $t\,°C$, where C_p = specific heat of dry air $(1005\,J/(kg\,K))$, C_w = specific heat of liquid water $(4187\,J/(kg\,K))$ and (see equation (14.6))

$$L = \text{latent heat of liquid water at temperature } t\,(°C)$$
$$= (2502.5 - 2.386t)\,1000 \quad J/kg \tag{10A.2}$$

Then

$$H = C_p t + XC_w t + LX_s \quad J/kg \tag{10A.3}$$

Differentiating,

$$dH = (C_p + XC_w)\,dt + X_s\,dL + L\,dX_s \quad J/kg \tag{10A.4}$$

It is assumed that no water is added or removed during the process. Hence, X remains constant. We must now seek to formulate this expression in terms of temperature and pressure. First, we evaluate dL and dX_s as functions of temperature and pressure.

1. *Change of latent heat of evaporation, dL.* From equation (10A.2),

$$dL = -B\,dt \tag{10A.5}$$

   where $B = 2386$.
2. *Change of vapour content, dX_s.* From equation (14.3) for saturation conditions

$$X_s = G \frac{e_s}{P - e_s} \tag{10A.6}$$

where $G = 0.622$, e_s = saturation vapour pressure (Pa) at $t\,°C$ and P = absolute (barometric) pressure (Pa). By partial differentiation,

$$dX_s = \frac{\partial X_s}{\partial e_s} de_s + \frac{\partial X_s}{\partial P} dP \tag{10A.7}$$

From equation (10A.6)

$$\frac{\partial X_s}{\partial e_s} = G \left[\frac{1}{P - e_s} + \frac{e_s}{(P - e_s)^2} \right]$$

$$= \frac{GP}{(P - e_s)^2} \tag{10A.8}$$

and

$$\frac{\partial X_s}{dP} = \frac{-Ge_s}{(P-e_s)^2} \qquad (10A.9)$$

Furthermore, the Calusius–Clapeyron equation (14.5) gives

$$de_s = \frac{Le_s}{R_v T^2} dt \quad \text{Pa} \qquad (10A.10)$$

where R_v = gas constant for water vapour (461.5 J/(kg K)) and $T = (273.15 + t)$ K, giving $dt = dT$.

Substituting from equations (10A.8), (10A.9) and (10A.10) into equation (10A.7) gives

$$dX_s = \frac{GP}{(P-e_s)^2} \frac{Le_s}{R_v T^2} dT - \frac{Ge_s}{(P-e_s)^2} dP \qquad (10A.11)$$

However, as

$$X_s = \frac{Ge_s}{P-e_s}$$

from equation (10A.6), equation (10A.11) becomes

$$dX_s = \frac{X_s P}{P-e_s} \frac{L}{R_v T^2} dt - \frac{X_s}{P-e_s} dP \qquad (10A.12)$$

Having evaluated dL (equation (10A.5)) and dX_s (equation (10A.12)), we can substitute these expressions into equation (10A.4) and, hence, (10A.1):

$$dH = (C_p + XC_w)\, dt - BX_s\, dt + \frac{X_s P}{P-e_s} \frac{L^2}{R_v T^2} dt - \frac{LX_s}{P-e_s} dP$$

$$= (R + R_v X_s)\frac{T}{P} dP$$

Collecting the dt and dP terms, and replacing dt by dT for consistency (the two are identical),

$$\frac{dT}{dP} = \frac{(R + R_v X_s)\dfrac{T}{P} + \dfrac{LX_s}{P-e_s}}{C_p + XC_w - BX_s + \dfrac{L^2 P}{(P-e_s)R_v T^2}} \quad \frac{°C}{Pa} \qquad (10A.13)$$

This is a **general isentropic relationship** describing the rate of change of temperature with respect to pressure for a saturated mixture of air, water vapour and liquid water. All of the variables X_s, L, and e_s depend only on T and P.

For unsaturated air involving no liquid water, X_s becomes X and there is no change of phase. Hence, the terms that include L become zero. Then⋆

$$\frac{dT}{dP} = \frac{R + R_v X}{C_p + (C_w - B)X} \frac{T}{P} \quad \frac{^{\circ}C}{Pa}$$ (10A.14)

Inserting values for the constants gives

$$\frac{dT}{dP} = \frac{R}{C_p} \frac{1 + 1.6078 X}{1 + 1.7920 X} \frac{T}{P}$$

Expansion of $(1 + 1.7920 \, X)^{-1}$ and ignoring terms of X^2 and smaller, leads to the approximation

$$\frac{dT}{dP} = \frac{R}{C_p}(1 - 0.2X)\frac{T}{P}$$

$$= 0.286 \, (1 - 0.2 \, X)\frac{T}{P}$$ (10A.15)

For dry air, $X = 0$, leaving

$$\frac{dT}{dP} = 0.286 \frac{T}{P}$$ (10A.16)

(This correlates with equation (10.54) when $dT = \Delta T$, $dP = p_f$ and $\eta_{isen} = 1$.)

Equation (10A.15) shows that if moist but unsaturated air is assumed to be dry then the fractional error involved is approximately $0.2X$. Hence, for an error of 1%,

$$0.2X = 0.01$$

or

$$X = 0.05 \, kg/(kg \, dry \, air)$$

This vapour content represents fully saturated conditions at $40\,^{\circ}C$ and $100\,kPa$. It is unlikely that such a moisture content will exist in mining circumstances without liquid water being present. Thus, for unsaturated air in the normal atmospheric range, assuming dry air will give an accuracy of within 1% for the temperature/pressure gradient.

REFERENCES

Drummond, J. A. (1972) Fan efficiency investigations on mines of Union Corp. Ltd. *J. Mine Vent. Soc. Afr.* **25**, 180–95.

McPherson, M. J. (1971) The isentropic compression of moist air in fans. *J. Mine Vent. Soc. S. Afr.* **24**, 74–89.

⋆ By considering an isothermal–isobaric change of phase, it can be shown that $C_w - B = C_{pv}$, where C_{pv} = specific heat at constant pressure of water vapour. This gives a value for C_{pv} to be $4187 - 2386 = 1801 \, J/(kg\,K)$. The 4.4% deviation from the normally accepted value of $1884 \, J/(kg\,K)$ indicates the uncertainly of C_{pv} within the atmospheric range of temperatures.

FURTHER READING

ASHRAE (1985) *Laboratory Methods of Testing Fans for Ratings*, ANSI/AMCA, Atlanta, GA.

British Standards Institution (1963) Methods of testing fans for general purposes. *British Standard 848*, Part 1, HMSO, London.

De La Harpe, J. H. (1982) Basic fan engineering. *Environmental Engineering in South African Mines*, Chapter 7, Mine Ventilation Society of South Africa.

McFarlane, D. (1966) *Ventilation Engineering*, Davidson & Co., Belfast.

Stepanoff, A. J. (1955) *Turboblowers*, Wiley, New York.

Whitaker, J. W. (1926) The efficiency of a fan. *Trans. Inst. Min. Eng.* **72**, 43.

Whitaker, J. W. (1928) The efficiency of a fan. *Trans. Inst. Min. Eng.* **74**, 93.

PART THREE

Gases in the Subsurface

11

Gases in subsurface openings

11.1 INTRODUCTION

When air enters any mine or other subsurface structure, it has a volume composition of approximately 78% nitrogen, 21% oxygen and 1% other gases on a moisture-free basis. A more precise analysis is given in Table 14.1. However, as the air progresses through the network of underground openings, that composition changes. There are two primary reasons for this. First, the mining of subsurface structures allows any gases that exist in the surrounding strata to escape into the ventilating airstream. Such **strata gases** have been produced over geological time and remain trapped within the pores or fracture networks of the rock. Methane and carbon dioxide are commonly occurring strata gases.

Secondly, a large number of chemical reactions may cause changes in the composition of mine air. Oxidation processes reduce the percentage of oxygen and will often cause the evolution of carbon dioxide or sulphur dioxide. The action of acid mine water on sulphide minerals may produce the characteristic odour of hydrogen sulphide while the burning of fuels or the use of explosives produce a range of gaseous pollutants. Most of the fatalities resulting from mine fires and explosions have been caused by the large volumes of toxic gases that are produced rapidly in such circumstances.

Several of the gases that appear in subsurface facilities are highly toxic and some are dangerously flammable when mixed with air. Their rate of production is seldom constant. Furthermore, their propagation through the multiple airways and often convoluted leakage paths of the ventilation system is further modified by the effects of gas density differences, diffusion and turbulent dispersion. These matters all influence the variations in concentrations of mine gases that may be found at any time and location in an underground mine.

A primary requirement of a mine ventilation system is to dilute and remove air-borne pollutants (section 1.3.1). It is, therefore, necessary that the subsurface environmental engineer should be familiar with the physical, chemical and physiological properties of mine gases, how they may be detected and preferred methods of control. In this chapter we shall classify and discuss those gases that may appear in subsurface ventilation systems.

11.2 CLASSIFICATION OF SUBSURFACE GASES

Table 11.1 lists the gases that are most commonly encountered in underground openings. Each of those gases is discussed in further detail within this section.

11.2.1 Threshold limit values

Threshold limit values (TLVs) of airborne substances refer to those concentrations within which personnel may be exposed without known adverse effects to their health or safety. The guideline TLVs given on Table 11.1 have been arrived at through a combination of industrial experience and from both animal and human studies. They are based primarily on recommendations of the American Conference of Governmental Industrial Hygienists (ACGIH) and the US National Institute for Occupational Safety and Health (NIOSH). The values quoted should be regarded simply as guidelines rather than clear demarcations between safe and dangerous concentrations. There are two reasons for this. First, wide variations in personal response to given substances occur between individuals, depending on one's state of health, exposure history and personal habits (e.g. smoking and use of alcohol). Secondly, TLVs are necessarily based on current scientific knowledge and are subject to revision as new evidence becomes available.

To ensure compliance with statutory requirements, environmental engineers and industrial hygienists should familiarize themselves with any mandatory TLVs that have been established within their particular country or state.

Three types of threshold limit values are expressed in Table 11.1. The **time-weighted average** (TWA) is the average concentration to which nearly all workers may be exposed over an 8 h shift and a 40 h work week without known adverse effects. However, many substances are sufficiently toxic that short-term exposures at higher concentrations may prove harmful or even fatal. The **short-term exposure limit** (STEL) is a time-weighted average concentration occurring over a period of not more than 15 min. That is, concentrations above the TWA and up to the STEL should not last for longer than 15 min. It is also recommended that such circumstances should not occur more than four times per day nor at intervening intervals of less than one hour.

The **ceiling limit** is the concentration that should not be exceeded at any time. This is relevant for the most toxic substances or those that produce in an immediate irritant effect.

11.2.2 Oxygen, O_2

Human beings and, indeed, the entire animal kingdom are completely dependent on the oxygen that constitutes some 21% of fresh atmospheric air. Oxygen diffuses

Table 11.1 Classification of gases most commonly found in subsurface openings

Name	Symbol	Molecular weight (based on C^{12})	Density at 20 °C and 100 kPa (kg/m^3)	Density relative to dry air	Gas constant ($J/(kg\,°C)$)	Primary sources in mines	Smell, colour, taste	Hazards	Guideline TLVs	Methods of detection	Flammability limits in air (%)
Dry air		28.966	1.1884	1	287.04	Air	None		—		
Oxygen	O_2	31.999	1.3129	1.1047	259.83	Air	None	Oxygen deficiency, may cause explosive mixtures with reactive gases	> 19.5	Electro-chemical, Para-magnetic, flame lamp	
Nitrogen	N_2	28.015	1.1493	0.9671	296.80	Air, strata	None	Inert		By difference	
Methane	CH_4	16.04	0.6581	0.554	518.35	Strata	None	Explosive, layering	1%: isolate electrical power 2%: remove personnel	Catalytic oxidation, flame lamp, thermal conductivity, optical, acoustic	5 to 15
Carbon dioxide	CO_2	44.00	1.805	1.519	188.96	Oxidation of carbon, fires, explosions, IC engines, blasting, respiration	Slight acid taste and smell	Promotes increased rate of respiration	TWA = 0.5% STEL = 3.0%	Optical, infrared	

Table 11.1 (*Contd.*)

Name	Symbol	Molecular weight (based on C¹²)	Density at 20°C and 100 kPa (kg/m³)	Density relative to dry air	Gas constant (J/(kg °C))	Primary sources in mines	Smell, colour, taste	Hazards	Guideline TLVs	Methods of detection	Flammability limits in air (%)
Carbon monoxide	CO	28.01	1.149	0.967	296.8	Fires, explosions, IC engines, blasting, spontaneous or incomplete combustion of carbon compounds	None	Highly toxic, explosive	TWA = 0.005% STEL = 0.04%	Electrochemical, catalytic oxidation, semiconductor, infrared	12.5 to 74.2
Sulphur dioxide	SO₂	64.06	2.628	2.212	129.8	Oxidation of sulphides, acid water on sulphide ores, IC engines	Acid taste, suffocating smell	Very toxic, irritant to eyes, throat and lungs	TWA = 2 ppm STEL = 5 ppm	Electrochemical, infrared	
Nitric oxide	NO	30.01	1.231	1.036	277.1	IC engines, blasting, fumes, welding	Irritant to eyes, nose and throat	Oxidizes rapidly to NO₂	TWA = 25 ppm	Electrochemical, infrared	
Nitrous oxide	N₂O	44.01	1.806	1.519	188.9		Sweet smell	Narcotic (laughing gas)	TWA = 50 ppm	Electrochemical	
Nitrogen dioxide	NO₂	46.01	1.888	1.588	180.7		Reddish brown,	Very toxic, throat and	TWA = 3 ppm ceiling: 5 ppm	Electrochemical chemical	

Name	Formula					Source	Odour	Physiological effects	Detection / limits	Range
							acidic smell and taste	lung irritant, pulmonary infections	infrared	
Hydrogen sulphide	H_2S	34.08	1.398	1.177	244.0	Acid water on sulphides, stagnant water, strata, decomposition of organic materials	Odour of bad eggs	Highly toxic, irritant to eyes and respiratory tracts, explosive	TWA = 10 ppm STEL = 15 ppm Electrochemical, semiconductor	4.3 to 45.5
Hydrogen	H_2	2.016	0.0827	0.0696	4124.6	Battery charging, strata, water on incandescent materials, explosions	None	Highly explosive	Catalytic oxidation	4 to 74.2
Radon	Rn	≈222	9.108	7.66	37.45	Uranium minerals in strata	None	Radioactive and decays to produce radioactive particles	Radiation detectors	1 working level (WL) and 4 WL-months per year
Water vapour	H_2O	18.016	0.739	0.622	461.5	Evaporation of water, IC engines, respiration, spontaneous combustion and other fires	None	Affects climatic environment	Psychrometers, dielectric effects	

[a] In addition to the methods listed, most of these gases can be detected by gas chromatography and stain tubes.

Table 11.2   Gas exchange during respiration (based on work by Forbes and Grove (1954))

Activity	Breaths/min	Inhalation rate (l/s)	Oxygen consumption (l/s)	Carbon dioxide produced (l/s)
At rest	12–18	0.08–0.2	≈ 0.005	≈ 0.004
Moderate work	30	0.8–1.0	≈ 0.03	≈ 0.027
Vigorous work	40	≈ 1.6	≈ 0.05	≈ 0.05

through the walls of the alveoli in the lungs to form oxyhaemoglobin in the blood-stream. This unstable substance breaks down quite readily to release the oxygen where required throughout the body.

As muscular activity increases, so also does the rate of respiration and the volume of air exchanged at each breath. However, the percentage of oxygen that is utilized decreases at heavier rates of breathing. For low levels of physical activity, exhaled air contains approximately 16% oxygen, 79% nitrogen and 5% carbon dioxide. Table 11.2 indicates typical rates of oxygen consumption and production of carbon dioxide. This table is helpful in estimating the effects of respiration on the gas concentrations in a confined area.

Example   Air is supplied at a rate of $5\,m^3/s$ to ten persons working at a moderate rate in a mine heading. The intake air contains 20.6% oxygen and 0.1% carbon dioxide. Determine the changes in concentrations of these gases caused by respiration.

Solution   For a moderate rate of activity, Table 11.2 indicates an individual consumption rate of some 0.03 l/s for oxygen and a production rate of 0.027 l/s for carbon dioxide. Hence, for ten persons,

$$\text{oxygen depletion} = 10 \times 0.03 \times 10^{-3} = 0.0003 \quad m^3/s$$

$$\text{carbon dioxide added} = 10 \times 0.027 \times 10^{-3} = 0.000\,27 \quad m^3/s$$

In the intake air,

$$\text{oxygen flow} = 5 \times 0.206 = 1.03 \quad m^3/s$$

$$\text{carbon dioxide flow} = 5 \times 0.001 = 0.005 \quad m^3/s$$

Therefore, in the exit air:

$$\text{oxygen flow} = 1.03 - 0.0003 = 1.0297 \quad m^3/s$$

$$\text{carbon dioxide flow} = 0.005 + 0.000\,27 = 0.005\,27 \quad m^3/s$$

$$\text{oxygen concentration} = \frac{1.0297}{5} \times 100 = 20.594\%$$

$$\text{carbon dioxide concentration} = \frac{0.005\,27}{5} \times 100 = 0.1054\%$$

This example illustrates the very limited effect of respiration on the concentrations of oxygen and carbon dioxide in a mine ventilation system. Indeed, the ventilating air necessary for breathing is negligible compared with other airflow requirements in the subsurface (section 9.3).

As air flows through an underground facility, it is probable that its oxygen content will decrease. This occurs not only because of respiration but, more importantly, from the oxidation of minerals (particularly coal and sulphide ores) and imported materials. The burning of fuels within internal combustion engines and open fires also consume oxygen. The primary danger with this gas is, therefore, that it may be depleted below a level that is necessary for the well-being of the work force. In Table 11.1, oxygen is the only gas whose concentration should be maintained above its recommended threshold limit value. The effects of oxygen depletion are as follows:

Per cent oxygen in air	Effects
19.0	Flame height on a flame safety lamp reduced by 50%
17	Noticeable increase in rate and depth of breathing—this effect will be further enhanced by an increased concentration of carbon dioxide
16	Flame lamp extinguished
15	Dizziness, increased heartbeat
13 → 9	Disorientation, fainting, nausea, headache, blue lips, coma
7	Coma, convulsions and probable death
Below 6	Fatal

Oxygen deficiency implies an increased concentration of one or more other gases. Hence, although not listed explicitly in Table 11.1, even non-toxic gases will endanger life by asphyxiation if they are present in sufficient concentration to cause a significant oxygen deficiency.

11.2.3 Nitrogen, N_2

Nitrogen constitutes approximately 78% of air and is, therefore, the most abundant gas in a ventilated system. It is fairly inert and occurs occasionally as a strata gas, usually mixed with other gases such as methane and carbon dioxide.

11.2.4 Methane, CH_4

Methane is produced by bacterial and chemical action on organic material. It is evolved during the formation of both coal and petroleum, and is one of the most common strata gases. Methane is not toxic but is particularly dangerous because it is flammable and can form an explosive mixture with air. This has resulted in the deaths of many thousands of miners over the past two centuries. A methane:air mixture is sometimes referred to as **firedamp**.

Although methane is especially associated with coal mines, it is often found in other types of subsurface openings that are underlain or overlain with carbonaceous or oil-bearing strata. The methane is retained within fractures, voids and pores in the rock either as a compressed gas or adsorbed on mineral (particularly carbon) surfaces. When the strata are pierced by boreholes or mined openings, then the gas pressure gradient that is created induces migration of the methane towards those openings through natural or mining-induced fracture patterns. The phenomena of methane retention within the rock and the mechanisms of its release are discussed in Chapter 12.

Although methane itself has no odour, it is often accompanied by traces of heavier hydrocarbon gases in the paraffin series that do have a characteristic oily smell. As indicated in Table 11.1, methane has a density that is a little over half that of air. This gives rise to a dangerous behaviour pattern—methane can form pools or layers along the roofs of underground openings. Any ignition of the gas can then propagate along those layers to other emanating sources (section 12.4.2). The buoyancy of methane can also create problems in inclined workings.

Methane burns in air with a pale blue flame. This can be observed over the lowered flame of a safety lamp at concentrations as small as $1\frac{1}{4}\%$. In an abundant supply of air, the gas burns to produce water vapour and carbon dioxide:

$$CH_4 + 2O_2 \rightarrow 2H_2O + CO_2 \tag{11.1}$$

Unfortunately, within the confines of mined openings and during fires or explosions, there may be insufficient oxygen to sustain full combustion, leading to formation of the highly poisonous carbon monoxide:

$$2CH_4 + 3O_2 \rightarrow 4H_2O + 2CO \tag{11.2}$$

The explosible range for methane in air is normally quoted as 5% to 15%, with the most explosive mixture occurring at 9.8%. While the lower limit remains fairly constant, the upper explosive limit reduces as the oxygen content of the air falls. The flame will propagate through the mixture within the range 5% to 14% (flammability limits). Figure 11.1 illustrates a well known diagram first produced by H. F. Coward in 1928. This can be used to track the flammability of air:methane mixtures as the composition varies. In zone A, the mixture is not flammable but is likely to become so if further methane is added or that part of the mine is sealed off. In zone B, the mixture is explosive and has a minimum nose value at 12.2% oxygen. Zones C and D illustrate mixtures that may exist in sealed areas. A mixture is zone C will become explosive if the seals are breached and the gases intermingle with incoming air. However, dilution of mixtures in zone D can be accomplished without passing through an explosive range.

Mining law specifies actions that must be taken when certain fractions of the

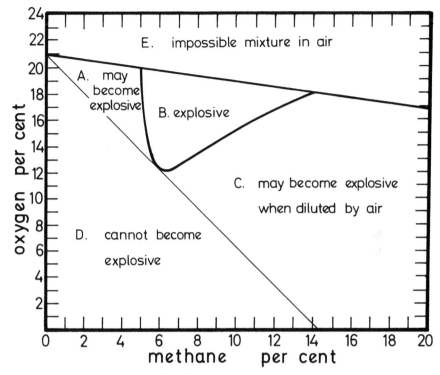

Figure 11.1 The Coward diagram for methane in air.

lower flammable limit have been reached. For exmple, electrical power must be switched off when the methane concentration exceeds 1% to $1\frac{1}{4}$%. Personnel other than those concerned with the improvement of ventilation should leave any area when the methane concentration exceeds 2% to $2\frac{1}{2}$%. The legislation of the relevant country or state must be consulted for the precise values of limiting concentrations and required actions. Other regulations specify the frequency at which measurements of methane concentration must be taken and threshold limit values to be applied in specified locations.

In many countries, underground mines are classified as either **gassy** or **non-gassy**. These legal terms relate to the potential for methane to be emitted into the workings. It is prudent that all underground coal mines should be designated as gassy. Any other mine may become legally gassy if (a) methane emissions from the roof, floor or sides of openings have been observed or the surrounding strata are deemed capable of producing such emissions, (b) a methane ignition has occurred in the past, or (c) the mine is connected underground to another mine that has already been classified as gassy. Imposing a gassy designation on a mine can result in a significant increase in capital and operating costs as all equipment and operating procedures must then be designed and maintained to minimize the risk of igniting a methane:air mixture.

11.2.5   Carbon dioxide, CO_2

Carbon dioxide appears in subsurface openings from a variety of sources including strata emissions, oxidation of carbonaceous materials, internal combustion engines, blasting, fires, explosions and respiration. Stagnant mixtures of air in sealed off areas often have an increased concentration of carbon dioxide and a decreased oxygen content. Such mixtures are sometimes called **blackdamp**.

Table 11.1 indicates that carbon dioxide is more than 50% heavier than air and will, therefore, tend to collect on the floors of mine workings. This was the reason for the small animals depicted in some of Agricola's woodcuts of sixteenth century mine workings. It is common to find emanations of blackdamp from the bottoms of seals behind which are abandoned workings, particularly during a period of falling barometric pressure. Particular care must be taken when holing through into a mined area that has not been ventilated for some time.

In addition to diluting oxygen in the air, carbon dioxide acts as a stimulant to the respiratory and central nervous systems. The solubility of carbon dioxide is about 20 times that of oxygen. Diffusion of the gas into the bloodstream is rapid and the effects on rate and depth of breathing are soon noticed. Cylinders of oxygen used for resuscitation often contain some 4% of carbon dioxide to act as a respiratory stimulant (carbogen gas).

The physiological effects of carbon dioxide have been listed by Strang and MacKenzie-Wood (1985) as follows:

Percentage in air	Effects
0.03	None, normal concentration of carbon dioxide in air
0.5	Lung ventilation increased by 5%
2.0	Lung ventilation increased by 50%
3.0	Lung ventilation doubled, panting on exertion
5 to 10	Violent panting leading to fatigue from exhaustion, headache
10 to 15	Intolerable panting, severe headache, rapid exhaustion and collapse

Fortunately, the administration of oxygen accompanied by warmth and an avoidance of exertion will usually lead to recovery with no known long-term effects.

11.2.6   Carbon monoxide, CO

The high toxicity of carbon monoxide coupled with its lack of smell, taste or colour make this one of the most dangerous and insidious of mine gases. It has a density very close to that of air and mixes readily into an airstream unless it has been heated by involvement in a fire, in which case it may layer with smoke along the roof.

Carbon monoxide is a product of the incomplete combustion of carbonaceous material. Although colourless, it has the traditional name of **whitedamp**. The great majority of fires and explosions in mines produce carbon monoxide. Indeed, most fatalities that have occurred during such incidents have been a result of carbon monoxide poisoning. The mixture of gases, including carbon monoxide, resulting from a mine explosion, is often referred to as **afterdamp**. Carbon monoxide is formed by internal combustion engines, blasting and spontaneous combustion in coal mines. It can also be generated as a component of water gas (carbon monoxide and hydrogen) when water is applied to incandescent coal during firefighting operations.

Carbon monoxide burns with a blue flame and is highly flammable, having a wide range of flammability, 12.5% to 74.2% in air, with the maximum explosibility at 29%.

In order to understand the physiological effects of carbon monoxide, let us recall from section 11.2.2 that oxygen passes through the walls of alveoli in the lungs and is absorbed by haemoglobin (red cells) in the blood to form the fairly unstable oxyhaemoglobin. Unfortunately, haemoglobin has an affinity for carbon monoxide that is about 300 times greater than that for oxygen. To compound the problem, the new substance formed in the bloodstream, **carboxyhaemoglobin** (CO·Hb), is relatively stable and does not readily decompose. The consequences are that very small concentrations of carbon monoxide cause the formation of carboxyhaemoglobin which accumulates within the bloodstream. This leaves a reduced number of red cells to carry oxygen molecules throughout the body. The physiological symptoms of carbon monoxide arise because of oxygen starvation to vital organs, particularly the brain and heart.

Physiological reactions to carbon monoxide depend on the concentration of the gas, the time of exposure and the rate of lung ventilation, the latter being governed primarily by physical activity. In order to relate the symptoms of carbon monoxide poisoning to a single parameter, the degree of saturation of the blood by carboxyhaemoglobin is employed. Although variations exist between individuals, the following list provides a guideline to the progressive symptoms.

Blood saturation per cent CO·Hb	Symptoms
5–10	Possible slight loss of concentration
10–20	Sensation of tightness across forehead, slight headache
20–30	Throbbing headache, judgment impaired
30–40	Severe headache, dizziness, disorientation, dimmed vision, nausea, possible collapse
40–60	Increased probability of collapse, rise in rates of pulse and respiration, convulsions
60–70	Coma, depressed pulse and respiration, possible death
70–80	Fatal

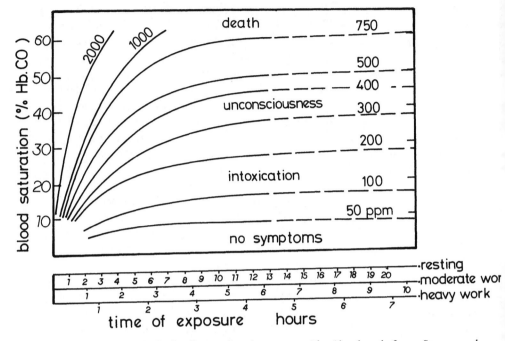

Figure 11.2 Physiological effects of carbon monoxide (developed from Strang and MacKenzie-Wood (1985)).

Figure 11.2 gives a more practical guide to physiological reactions to carbon monoxide and takes the level of physical activity and exposure time into account.

Because of the bright red colour of carboxyhaemoglobin, a victim of carbon monoxide poisoning may have a flushed appearance, even after death. At high blood saturation levels, deterioration of the body is often retarded after death owing to the absence of internal oxygen.

Persons suffering from carbon monoxide poisoning should be kept warm and removed from the polluted atmosphere, preferably, on a stretcher. It may take more than 24 h for blood saturation levels to return to normal. During this time, severe headaches may be experienced. However, the return to normal blood saturation levels can be accelerated significantly by the administration of pure oxygen. The rapidity with which carbon monoxide is absorbed into the bloodstream and the slowness of its expurgation can result in dangerous blood saturation levels occurring in firefighters who make short repeated expeditions into a polluted area.

In addition to the physical symptoms that have been listed for carbon monoxide poisoning, experience of personnel involved in mine fires has indicated significant psychological reactions that have had grave repercussions. Low levels of blood saturation can give an appearance of intoxication including impairment of judgment and an unsteady gait preceding collapse. Victims may become silent and morose, and may resist or fail to comprehend instructions that will lead them to safety. The sense of time may be affected, a particularly significant symptom when self-rescuer devices are being worn. However, all of these reactions vary considerably between individuals. In particular, total collapse will occur rapidly in high concentrations of carbon monoxide.

A small degree of acclimatization has been observed in persons who are repeatedly exposed to low levels of carbon monoxide as experienced, for example, by habitual smokers. This is thought to occur because of an increase in the number of red cells within the bloodstream.

11.2.7 Sulphur dioxide, SO_2

This is another highly toxic gas but one which, fortunately can be detected at very low concentrations both by its acidic taste and the intense burning sensation it causes to the eyes and respiratory tracts. The latter are a result of the high solubility of the gas in water to form sulphurous acid:

$$H_2O + SO_2 \rightarrow H_2SO_3$$

This, in turn, can oxidize to sulphuric acid, H_2SO_4.

When discussing very small concentrations, it is convenient to refer to parts per million (ppm) rather than percentages. The conversion between the two is accomplished simply by moving the decimal point four places in the appropriate direction. The following list of physiological reactions to sulphur dioxide employs parts per million.

Concentration of sulphur dioxide (ppm)	Effects
1	Acidic taste
3	Detectable by odour
20	Irritation of eyes and respiratory system
50	Severe burning sensation in eyes, nose and throat
400	Immediately dangerous to life

First aid for sulphur dioxide poisoning includes the administration of oxygen, immobility and warmth. The longer-term treatment is for acid corrosion of the eyes and respiratory system.

Sulphur dioxide is formed in internal combustion engines and by the oxidation of sulphide ores, for example, zinc blende

$$2ZnS + 3O_2 \rightarrow 2ZnO + 2SO_2 \tag{11.3}$$

or iron pyrites

$$4FeS_2 + 11O_2 \rightarrow 2Fe_2O_3 + 8SO_2 \tag{11.4}$$

These and other similar reactions can occur when sulphide ores are heated in a fire or by spontaneous combustion. Although sulphur dioxide is colourless, white fumes may be seen as a result of condensation of acidic water vapours or traces of sulphur trioxide, SO_3.

11.2.8   Oxides of nitrogen, NO_x

Three oxides of nitrogen are listed on Table 11.1. Nitric oxide, NO, nitrous oxide, N_2O, and nitrogen dioxide, NO_2, are formed in internal combustion engines and by blasting. The proportion of nitrous oxide is likely to be small. Furthermore, nitric oxide converts rapidly to nitrogen dioxide in the presence of air and water vapour:

$$2NO + O_2 \rightarrow 2NO_2 \tag{11.5}$$

As nitrogen dioxide is the most toxic of these oxides of nitrogen, it is sensible to concentrate on the physiological effect of this gas.

At the temperatures found in underground openings, it is probable that nitrogen dioxide will be mixed with a companion gas, nitrogen tetroxide, N_2O_4, having similar physiological effects. The brown fumes of nitrogen dioxide dissolve readily in water to form both nitrous (HNO_2) and nitric (HNO_3) acids:

$$2NO_2 + H_2O \rightarrow HNO_2 + HNO_3 \tag{11.6}$$

These acids cause irritation and, at higher concentrations, corrosive effects on the eyes and respiratory system. The progressive symptoms are as follows:

Concentration of nitrogen dioxide (ppm)	Effects
40	May be detected by smell
60	Minor throat irritation
100	Coughing may commence
150	Severe discomfort, may cause pneumonia later
200	Likely to be fatal

The immediate treatment for nitrogen dioxide poisoning is similar to that for sulphur dioxide, namely, the administration of oxygen, immobility and warmth. An insidious effect of nitrogen dioxide poisoning is that an apparent early recovery can be followed, soon afterwards, by the development of acute bronchopneumonia.

11.2.9   Hydrogen sulphide, H_2S

The presence of this highly toxic gas is readily detected by its characteristic smell of bad eggs. This has given rise to the colloquial name **stinkdamp**. Unfortunately, hydrogen sulphide has a narcotic effect on the nervous system including paralysis of the olfactory nerves. Hence, after a short exposure, the sense of smell can no longer be relied on.

Hydrogen sulphide is produced by acidic action or the effects of heating on sulphide ores. It is formed naturally by the bacterial or chemical decomposition of organic compounds and may often be detected close to stagnant pools of water in underground mines. Hydrogen sulphide may occur in natural gas or petroleum reserves and migrate through the strata in a weakly acidic water solution. It can also be generated in gob fires. In such cases, free sulphur may be deposited by partial oxidation of the gas:

$$2H_2S + O_2 \rightarrow 2H_2O + 2S \tag{11.7}$$

This can sometimes be seen as a yellow deposit in burned areas. However, in a plentiful supply of air, hydrogen sulphide will burn with a bright blue flame to produce sulphur dioxide:

$$2H_2S + 3O_2 \rightarrow 2SO_2 + 2H_2O \tag{11.8}$$

The physiological effects of hydrogen sulphide may be listed as follows.

Concentration of hydrogen sulphide (ppm)	Effects
0.1 to 1	Detectable by smell
5	Beginning of toxicity
50 to 100	Slight irritation to eyes and respiratory tract, headache, loss of odour after 15 min.
200	Intensified irritation of nose and throat
500	Serious inflammation of eyes, nasal secretions, coughing, palpitations, fainting
600	Chest pains due to corrosion of respiratory system, may be fatal
700	Depression, coma, probable death
1000	Paralysis of respiratory system, very rapid death

A victim who recovers from hydrogen sulphide poisoning may be left with longer term conjunctivitis and bronchitis.

11.2.10 Hydrogen, H_2

Although non-toxic, hydrogen is the most explosive of all the mine gases. It burns with a blue flame and has the wide flammable range of 4% to 74.2% in air. Hydrogen can be ignited at a temperature as low as 580 °C, and with an ignition energy about half of that required by methane.

Hydrogen occasionally appears as a strata gas and may be present in afterdamp at about the same concentrations as carbon monoxide. The action of water on hot coals can produce hydrogen as a constituent of water gas (section 11.2.6). Dangerous accumulations of hydrogen may occur at locations where battery charging is in progress. Hydrogen has a density only some 0.07 that of air. It will, therefore, tend to rise to the roof. Battery charging stations should be located in intake air with a duct at roof level that connects into a return airway.

11.2.11 Radon, Rn

This chemically inert gas is one of the elements formed during radioactive disinteg-ration of the uranium series. Although its presence is most serious in uranium mines, it may be found in many other types of underground openings. Indeed, seepages of radon from the ground into the basements of surface buildings have been known to create a serious health hazard.

Radon emanates from the rock matrix or from ground water that has passed over radioactive minerals. It has a half-life of 3.825 days and emits alpha radiation. The immediate products of the radioactive decay of radon are solids known as the radon daughters. These adhere to the surfaces of dust particles and emit alpha, beta and some gamma radiation.

The entry of radon and, especially, its daughters into the respiratory system is associated with a high incidence of lung cancer. This appears to be synergistic with the effects of smoking. Uranium miners, in particular, should be discouraged from smoking.

Radon and the problems of radiation in mines are discussed further in Chapter 13.

11.3 GAS MIXTURES

Although gases that occur most commonly underground have been discussed separately in section 11.2, it is more usual that several gaseous pollutants appear together as gas mixtures. Furthermore, the importation of an ever-widening range of materials into subsurface facilities also introduces the risk of additional gases being emitted into the mine environment.

11.3.1 Threshold limit values for gas mixtures

In order to widen the applicability of this section, the recommended threshold limit values of a number of other gases and vapours are given in Table 11.3 and grouped in terms of probable sources. Again, we are reminded that these TLVs are simply guidelines and that national or state-mandated limits should be consulted to ensure compliance with the relevant laws.

In any given atmosphere, if there are two or more airborne pollutants (gaseous or particulate) that have adverse effects on the same part of the body, then the threshold limit value should be assessed on the basis of their combined effect. This is calculated

Table 11.3 Table of threshold limit values for other gases and vapours that may be present underground

Substance	Guideline time-weighted average limit (ppm unless otherwise stated)
Cleaners and solvents	
Acetone	750 (STEL = 1000)
Ammonia	25 (STEL = 35)
Toluene	100 (STEL = 150)
Turpentine	100 (STEL = 150)
Refrigerants	
Ammonia	25 (STEL = 35)
R11	1000 (ceiling)
R12	1000
R22	1000
R112	500
Fuels	
Butane	800
Gasoline vapour	300 (STEL = 500)
Liquid petroleum gas	1000
Naphtha (coal tar)	100
Pentane	600 (STEL = 750)
Propane	1000
Welding and soldering	
General welding fumes	5 mg/m^3
Iron oxide fumes	5 mg/m^3
Lead fumes	0.15 mg/m^3
Ozone (arc welding)	0.1 ppm ceiling
Fluorides (fluxes)	2.5 mg/m^3
Heated plastics	
Carbon dioxide	0.5% (STEL = 3.0%)
Carbon monoxide	50 (STEL = 400)
Hydrogen chloride fumes	5 ppm ceiling
Hydrogen cyanide	10
Hydrogen fluoride	3 ppm ceiling
Phenol (absorbed by skin)	5
Explosives	
Carbon dioxide	0.5% (STEL = 3.0%)
Carbon monoxide	50 (STEL = 400)
Oxides of nitrogen	25
Ammonia	25 (STEL = 35)

(Contd.)

Table 11.3 *(Contd.)*

Substance	*Guideline time-weighted average limit (ppm unless otherwise stated)*
Sulphur dioxide	2 (STEL = 5)
Nitrous acid fumes	
Nitric acid fumes	2 (STEL = 4)
Others	
Chlorine (biocides)	0.5 (STEL = 1)
Cresol (wood preservative)	5
Mercury vapour	0.05 mg/m^3
Oil mist (mineral), vapour free	5 mg/m^3
Oil mist (vegetable), vapour free	10 mg/m^3
Sulphuric acid fumes (batteries)	1 mg/m^3

as the dimensionless sum

$$\frac{C_1}{T_1} + \frac{C_2}{T_2} + \cdots + \frac{C_n}{T_n} \tag{11.9}$$

where C = measured concentration and T = corresponding threshold limit value. If the sum of the series exceeds unity, then the threshold limit value of the mixture is deemed to be exceeded (ACGIH, 1989).

Example An analysis of air samples taken in a return airway indicates the following gas concentrations:

carbon dixoide	0.2%
hydrogen sulphide	2 ppm
carbon monoxide	10 ppm
sulphur dioxide	1 ppm

Determine the threshold limit values, TWA and STEL for the mixture.

Solution Sulphur dioxide and hydrogen sulphide are both irritants to the eyes and respiratory system. Furthermore, carbon dioxide is a stimulant to breathing, increasing the rate of ventilation of the lungs. Hence, it may be regarded as being synergistic with the sulphur dioxide and hydrogen sulphide. Carbon monoxide, however, affects the oxygen-carrying capacity of the bloodstream and need not be combined with the other gases in determining the threshold limit values for the mixture.

From Table 11.1, the following time-weighted averages (TWAs) and short-term exposure limits (STELs) are determined:

Component	Threshold limit values	
	TWA	*STEL*
Carbon dioxide	0.5%	3.0%
Hydrogen sulphide	10 ppm	15 ppm
Sulphur dioxide	2 ppm	5 ppm
Carbon monoxide	50 ppm	400 ppm

The carbon monoxide concentration measured at 10 ppm and treated separately is shown to be less than the TWA of 50 ppm.

The equivalent threshold limit value of the remainder of the mixture is assessed from equation (11.9) as

$$\text{TWA} \quad \underset{\substack{\text{carbon} \\ \text{dioxide}}}{\frac{0.2}{0.5}} + \underset{\substack{\text{hydrogen} \\ \text{sulphide}}}{\frac{2}{10}} + \underset{\substack{\text{sulphur} \\ \text{dioxide}}}{\frac{1}{2}} = 1.1$$

$$\text{STEL} \quad \frac{0.2}{3.0} + \frac{2}{15} + \frac{1}{5} = 0.4$$

The dimensionless TWA is greater than 1. Hence, the time-weighted average limit is exceeded. However, the short-term exposure limit is less than unity. The implication is that personnel may safely spend short periods of time in this atmosphere, but not a complete 8 hour shift.

In some cases of gas mixtures, it may be practicable to monitor quantitatively for one pollutant only, even though it is known that other gases or particulates are present. In such circumstances, a pragmatic approach is to reduce the threshold limit value of the measured substance by a factor that is assessed from the number, toxicity and estimated concentration of other contaminants known to be present.

11.3.2 Diesel emissions

The flexibility and reliability of diesel engines have resulted in a proliferation in their use for all types of underground mines. This is, however, tempered by the emissions of exhaust gases, heat and humidity that result from the employment of diesels. Guidance on the airflow requirements for diesel equipment is given in section 9.3.2.

The substances that are emitted from diesel exhausts include

1. nitrogen,
2. carbon monoxide,

3. carbon dioxide,
4. oxides of nitrogen,
5. sulphur dioxide,
6. diesel particulate matter, and
7. water vapour (sections 15.3.2 and 16.2.3).

The actual magnitude and composition of diesel exhausts are governed by the engine design, quality of maintenance, exhaust treatment units, rating for altitude and skill of the operator.

Catalytic converters cause the exhaust gases to filter through granulated oxidizing agents. These can successfully convert up to 90% of the carbon monoxide and 50% of unburned fuel to carbon dioxide and water vapour. Sulphur dioxide may be partially converted to sulphur trioxide and appear as vapour of sulphuric acid.

Water scrubbers give an improved removal of sulphur dioxide and particulate matter but do little for carbon monoxide. Engine gas recirculation (EGR) systems help to reduce the oxides of nitrogen.

The diesel particulate matter is a combination of soot, unburned fuel and aldehydes, and is regarded as being the component of diesel exhaust that is most hazardous to health. There are two reasons for this. First, the particles are typically less than 1 μm in diameter. They are inhaled deep into the lungs and have a high probability of being retained within the walls of the alveoli. Secondly, the porous or fibrous nature of the soot particles enables them to adsorb a range of polynuclear and aromatic hydrocarbons, giving diesel smoke its characteristic greasy feel. There is increasing evidence that these have carcinogenic properties (Waytulonis, 1988) although the relationship between exposure to diesel exhaust emissions and the incidence of cancer remains unclear (French, 1984). Exhaust filters assist greatly in the removal of particulate matter.

The operating advantages of diesel equipment are in direct conflict with the potentially hazardous nature of exhaust emissions. This resulted in a considerable research effort into the use of diesels in mining during the 1980s (e.g. World Mining, 1982; Daniel, 1989; Haney and Stoltz, 1989; Gunderson, 1990).

11.3.3 Fires

The predominant cause of loss of life associated with underground fires has been the gaseous products of combustion. The particular gases depend on the type of fire and the materials that are being burned. Although carbon monoxide has resulted in the majority of such fatalities, the importation of a widening variety of manufactured materials into mines has produced the potential for other toxic gases to be emitted during a mine fire.

Coal mine fires that involve burning of the coal itself are likely to produce an atmosphere that is deficient in oxygen and may contain carbon dioxide, methane, carbon monoxide, water vapour and smaller amounts of sulphur dioxide, hydrogen sulphide and hydrogen. Fires involving timber emit the same gases. The actual concentration of each gas depends primarily on the oxygen content of the air within the fire zone (Chapter 21).

Metal mine fires arising from the spontaneous combustion of sulphide ores will emit products of combustion that are rich in sulphur dioxide and, possibly, sulphuric acid vapour.

Fires involving diesel equipment will produce gases from burning diesel oil and also from any plastic or rubber components that may become heated. The burning oil itself will emit carbon monoxide, carbon dioxide, sulphur dioxide, oxides of nitrogen, hydrogen and hydrogen sulphide.

Plastic materials are employed for an increasing variety of purposes in underground workings, including brattice cloths, machine components, electrical cables and fittings, thermal insulation on pipes or airway linings, the pipes themselves, instrumentation, containers and meshing. Some plastic polymers will begin to disintegrate at about 250 °C and before actual combustion commences. However, the evolution of gases increases rapidly when the material burns. While all heated plastics give off carbon monoxide and carbon dioxide, the most dangerous are the polyurethanes, nylon and polyvinyl chloride (PVC) which may also produce hydrogen cyanide, hydrochloric acid fumes and hydrofluoric acid fumes. The phenolic plastics evolve the same gases but require a higher temperature for pyrolysis. Rubber-based materials may also emit hydrogen sulphide when heated. Local or national regulations may prohibit the use of certain plastics.

In addition to gases, fires will normally also produce copious amounts of the respirable particulates that appear as visible smoke. These particles themselves may be toxic and are irritating to the eyes and respiratory tracts.

11.3.4 Explosives

The gases that may be produced by blasting are listed in Table 11.3. The concentrations of gases depend on the type, quality and weight of explosive used, the means of detonation and the psychrometric condition of the air. The degree of confinement also affects the concentrations and time distribution of the gas emission — firing 'on the solid' produces a sharper and higher peak of blasting fumes than if a free face were available. Some of the gases and fumes may be adsorbed onto mineral faces within the fragmented rock and be emitted over a longer time period (Rossi, 1971).

The oxides of nitrogen are formed mainly as nitric oxide during the detonation. The rate at which this oxidizes to nitrogen dioxide depends on the degree to which the blasting fumes are diluted by a ventilating air current. Fumes that are purged from a mine during a re-entry period following blasting may still contain a significant proportion of nitric oxide when the pulse of fumes exits the mine.

11.3.5 Welding

The constituents of welding fumes can vary widely depending on the metals or alloys involved, the welding process being utilized and any fluxes that may be used. The fumes contain particles of amorphous slags and oxides of the metals being welded.

Arc welding in a local inert atmosphere such as argon will inhibit the formation of welding fumes but can produce ozone. On the other hand, arc welding within the air can produce copious fumes and carbon monoxide. Where stainless steel is involved, compounds of chromium and nickel can appear as constituents of the welding fumes. The employment of welding fluxes may also result in fluorides as well as a range of oxides being generated.

If no toxic compounds are present in the metals, welding rods or metal coatings, and the conditions do not favour the formation of toxic gases, then the threshold limit of $5 \, mg/m^3$ given for general welding fumes in Table 11.3 may be employed.

11.4   GAS DETECTION AND MONITORING

11.4.1   Objectives and overview

The primary purpose of monitoring the concentrations of airborne pollutants in a mine is to ensure that the atmosphere provides a safe environment free from levels of toxicants that would create a hazard to health. There are essentially three matters to consider. The first is the threshold limit values deemed to be acceptable for each pollutant. Second is the choice of instrumentation best suited for the detection and measurement of particular gases. Thirdly, the question of where and how frequently measurements are required must be addressed. Having discussed threshold limit values in sections 11.2.1 and 11.3.1, we now turn our attention to the operating principles of instruments used to measure gas concentrations and methods of sampling.

11.4.2   Principles of gas detection

Advances in the fields of electronics, electrochemistry and micromanufacturing have resulted in significant improvements in the accuracy and reliability of instrumentation for the detection and measurement of gas concentrations. Equipment now available is capable of indicating fractions of a part per million for some toxic gases. The same principles of detection method may be applied to more than one gas. Hence, in this section we shall concentrate on those principles rather than discussing each gas in turn. The methods of gas detection and monitoring used most frequently for subsurface gases are listed in Table 11.1.

Filament and catalytic oxidation (pellistor) detectors

These devices are used primarily for the measurement of methane and other gases that will burn in air such as carbon monoxide, hydrogen or the higher gaseous hydro-carbons. If an electrical filament is heated to a sufficiently high temperature, then a combustible gas in the surrounding air will burn and, hence, elevate the temperature of the filament even further. This will change the electrical resistance of the filament. By arranging for the filament to act as one arm of a Wheatstone bridge circuit, its change of resistance can be sensed as a change in voltage drop or current that is proportional to the concentration of combustible gas. A second filament of the Wheatstone bridge is also exposed to the air but inhibits oxidation of the gas either by a lower operating temperature or catalytic poisoning. This acts as a balancing control against variations in the temperature, moisture content and barometric pressure of the ambient air. The device is, essentially, a resistance thermometer sensing the temperature of the active filament. Filament detectors have been available since at least the 1950s.

Platinum has been used as the filament coil material as it has good resistance–temperature characteristics at the 900 to 1000 °C temperatures required to promote the oxidation process in a stable manner. There are, however, severe drawbacks to platinum filament detectors. First, the response of the instrument is very sensitive to the geometry of the coil. Slight variations in the pitch of the coil caused during manufacture or due to mechanical shock produce significant deviations in the electrical output. Secondly, at the required high operating temperature, evaporation of the metal causes the cross-sectional area of the wire to be reduced and, hence, increases its resistance. This is reflected by a slow but significant increase in bridge output for any given gas concentration and a zero drift in fresh air.

During the 1960s, research was carried out in a number of countries to overcome the difficulties inherent in filament methanometers. The platinum filament was reduced to less than 1 mm in length and encased in a tiny bead of refractory material, such as alumina. The outside of the bead was coated by a thin layer of catalytic material that produces stable oxidation of methane at temperatures of some 500 °C. The catalysts are, typically, metal–salt combinations of palladium and thorium or platinum and rhodium. These beads are called **pellistors** or **pelements** (Baker and Firth, 1969; Richards and Jones, 1970) and overcome most of the problems of the earlier filament detectors. First, the encapsulation of the coil makes it almost immune to mechanical damage. Indeed, more recent developments of electronic materials have enabled the coil to be eliminated. Secondly, there is no evaporation of the coil and, hence, near elimination of zero drift. Furthermore, the life of the unit is greatly increased while the reduced operating temperature requires less battery power and improves the safety of the instrument. The balancing or reference pellistors to compensate for air temperature, humidity and pressure are identical units except that the catalyst is poisoned by dipping it in a hot solution of potassium hydroxide. The first commercial methanometers to utilize pellistors appear to have been produced by Mine Safety Appliances (MSA).

Modern pellistor methanometers have a high degree of reliability, and operate satisfactorily and continuously on mining machines within conditions of high humidity, dust and vibration. The bridge output signals may be used to indicate on a dial, activate audio-visual alarms, isolate electrical power to a machine, drive a recorder or transmit to distant devices through a telemetering system. However, they still retain some disadvantages.

First, pellistors rely on oxidation process and, hence, the availability of oxygen. A good quality, heavy duty, pellistor will indicate increasing concentrations of methane up to 9% or 10%. Beyond that, the decreasing concentration of oxygen will diminish the rate and temperature of catalytic oxidation giving a reduced and false reading. Pellistor methanometers are, in the main, manufactured for the range 0% to 5%. However, use in an oxygen–deficient and methane–rich atmosphere could, again, give a false reading—apparently within the 0% to 5% scale. In some modern instruments, this promotes a warning signal or is combined with another mode of detection for high concentrations of combustible gas. Another possible consequence of repeated exposures to high concentrations of combustible gases is that 'cracking' of a hydro-carbon gas can deposit carbon within the catalytic layer resulting in scaling and eventual destruction of the pellistor. The catalyst should contain an inhibitor to minimize the effects of cracking. The choice of refractory material that forms the bead also influences the degree of cracking.

A second disadvantage of pellistors is that they are subject to poisoning by some other gases and vapours. Vaporized products of silicon compounds (greases, electrical components) or phosphate esters (fluid couplings) will produce permanent poisoning of pellistors while halogens from refrigerants or heated plastics can give a temporary reduction of the instrument output. The readings of pellistor methanometers should be treated with caution when such instruments are used downstream from a mine fire. Most of these instruments allow the ambient air to reach the pellistor heads by diffusing through a layer of absorbent material such as activated charcoal. This removes most potential poisons and also permits the instrument to be near independent of the air velocities generally encountered underground.

A third feature of pellistor transducers is that they will react to any combination of combustible gases that pass through the absorbent filter. The instruments used in mines are primarily calibrated for methane as this is the most common of the combustible gases found in the subsurface. However, hydrogen, carbon monoxide and ethane will also pass through simple activated filters and produce a response from the instrument. At 1% concentration of each gas, a catalytic combustion methanometer will indicate the readings shown on Table 11.4. The readings remain proportional to those given by methane up to the lower explosive limit of each gas.

Fortunately, a pellistor methanometer is unlikely to be used in concentrations of

Table 11.4 Approximate readings on a catalytic combustion methanometer given by a 1% concentration of each gas

Gas	Methanometer reading (per cent)
Methane	1.00
Carbon monoxide	0.39
Hydrogen	1.24
Ethane	1.61
Propane	1.96

carbon monoxide that would be indicated on the instrument. Furthermore, hydrogen, ethane and propane all give a reading above that for methane. Hence, the error is on the side of safety.

Example A methanometer used in a mixture of air and ethane gives a reading of 3.4%. What is the actual concentration of ethane? Assume that there are no other combustible gases present.

Solution From Table 11.4, the correction factor for ethane is 1.61. Hence,

$$\text{actual concentration of ethane} = \frac{3.4}{1.61} = 2.1\%$$

A useful feature of oxidation methanometers is that any given reading indicates approximately the same fraction of the lower flammable limit of the combustible gas being monitored. Hence, a pellistor methanometer would indicate 5% (approximately) if any one gas from Table 11.4 was present in a concentration equal to its lower, flammable limit. The same is true of mixtures of the gases, enabling a pellistor methanometer to be used as an indicator of how close a gaseous mixture is to its lower flammable limit.

Flame safety lamps

These lamps were introduced early in the nineteenth century for the purposes of providing illumination from an oil flame without igniting a methane–air mixture (section 1.2). Their use for illumination disappeared with the development of electric battery lamps. However, the devices have been retained for the purposes of testing for methane and oxygen deficiency.

A blue halo of burning gas over the lowered flame of the lamp becomes visible at about $1\frac{1}{4}\%$ methane. The size of the halo increases with methane concentration. At $2\frac{1}{2}\%$, it forms a very clear equilateral triangle. At 5% the flame spirals upwards into the bonnet of the lamp and either continues burning or self-extinguishes in a contained explosion. In both cases, the flame is prevented from propagating into the surrounding atmosphere by the tightly woven wire gauzes (section 1.2). The use of the flame safety lamp for methane testing has now largely been replaced by pellistor methanometers as these are more accurate, reliable and safer.

Some coal mining industries have retained the safety lamp for its ability to indicate oxygen deficiency (section 11.2.2). The height of the flame reduces progressively with oxygen content and is extinguished at 16% oxygen. However, in a methane-rich atmosphere, the flame may remain lit down to an oxygen concentration of 13%.

Various attempts have been made to convert modified flame lamps into alarm and recording devices (for example, Pritchard and Phelps, 1961). However, some ignitions of methane have been attributed to flame safety lamps that have been damaged or inadequately maintained (Strang and MacKenzie-Wood, 1985). The flame safety lamp has an honourable place in the history of mine ventilation but its role is almost over.

Thermal conductivity and acoustic gas detectors

At 20 °C and at normal atmospheric pressures, the thermal conductivity of methane is 0.0328 W/(m °C) compared with 0.0257 W/(m °C) for air. This difference is utilized in some high range methanometers. Two heated sensors are employed, one exposed to the gas sample and the other retained as a reference within a sealed air-filled chamber. A sample of the ambient air is draw through the instrument at a constant rate. The sample sensor cools at a greater rate owing to the higher thermal conductivity of the methane. The change in resistance of the sample sensor is detected within an electrical bridge to give a deflection on the meter. Typical ranges for a thermal conductivity methanometer are 2% or 5% to 100%. The principle is sometimes used in conjunction with pellistors to provide a dual range instrument or to override the falsely low readings that may be given by catalytic combustion detectors at high concentration (discussed earlier in this section). To reduce interference from other gases, suitable filters should be employed. In particular, carbon dioxide gives about half the response of methane but in a negative direction. For use in mining, a soda lime filter is, therefore, advisable.

Acoustic gas detectors rely on changes in the velocity of sound as the composition of a sample varies. They are used, primarily, for high concentrations and have been employed in methane drainage systems.

Optical methods

These subdivide into three groups. **Interferometers** utilize the refraction of light that occurs when a parallel beam is split, one half passing through the sample and the other through a sealed chamber containing pure air. The two beams are recombined and deflected through a mirror or glass prism arrangement for viewing through a telescope. The optical interference between the two beams causes a striped fringe pattern to appear in the field of view. This typically takes the form of two black lines in the centre with red and green lines on both sides. The fringe moves along a scale in proportion to the amount of gas present. Rotational adjustments of one of the deflecting prisms can be read on a vernier and added to the optical scale in order to widen the range of the instrument. Interferometers are sensitive to the presence of other gases and appropriate filters should be used when necessary. Hydrogen gives a negative response with respect to methane. At equal concentrations of the two gases, the reading is near zero. Carbon dioxide and methane give similar responses. For air containing both carbon dioxide and methane two readings are taken, one with a soda lime filter to remove the carbon dioxide and gives the methane concentration only, and the second without that filter. The difference between the two readings is an indication of the carbon dioxide concentration. The effects of other gases render the interferometer unsuitable for situations where the composition of the sample is dramatically different from that of normal air, for example, afterdamp, downstream from fires or samples taken from behind seals.

The **non-dispersive infrared gas analyser** is one form of absorption spectro-
meter that is frequently used for mine gas analysis. Identical beams of slowly pulsating
infrared radiation pass sequentially through two parallel chambers, one containing a
gas that does not absorb infrared (typically nitrogen) and the other fed by a stream
of the sample. The pulsations of 3 to 4 Hz may be achieved electronically or by a
rotating chopper arrangement. Beyond the sample and reference chambers is a
two-compartment sealed container (detector unit) filled with a pure specimen of that
particular gas which is to be detected. Detector units can be interchanged to determine
the concentrations of different gases. The two compartments of the detector unit are
at the same nominal pressure and are separated by a flexible diaphragm. The pulses
of infrared radiation are directed sequentially into the two sides of the detector unit,
heat the contained gas and, hence, raise its pressure. However, the beam that has
passed through the sample chamber has already been partially absorbed at the relevant
wavelength by molecules of the sought gas. Hence, the pulses of pressure induced in
that corresponding side of the detector unit are weakened. The amplitude of the
vibrating diaphragm is sensed by an electrical capacitor and translated electronically
to an output signal. Infrared gas analysers are employed primarily at fixed monitoring
stations. However, portable versions are available.

Laser spectroscopy is another means of air analysis that has considerable potential
for subsurface application. There are two systems that can be employed. One is the
differential absorption unit (DIAL system) in which two similar lasers are used, one
tuned to the absorption wavelength of the gas to be detected, and the other to a slightly
different wavelength. The two laser beams pass through the sample gas stream and
are reflected back to a single receiver unit. The difference in the two signals is
processed to indicate the concentration of the gas being monitored. The employment
of a reference beam eliminates the effects of dust, humidity or other gases (Holmes
and Byers, 1990).

The second laser technique for gas analysis is the light detection and ranging
(LIDAR) method which depends on the Raman effect. When a gas is excited by
monochromatic radiation from a laser, a secondary scattered radiation is produced.
The spectrum of this scattered radiation can be analysed to indicate the concentrations
of the gases that caused it.

The attraction of laser techniques for mine air sampling is that the laser beams may
be directed across or along subsurface openings to give continuous mean analyses for
large volumes of air. Furthermore, the lasers can also be used to monitor air velocity.
At the present time, the high cost of the units precludes their use in other than isolated
cases.

Another technique that is most promising involves passing pulses of light along
fibre optics. This method may also be used to indicate variations in temperature
along the length of the fibre.

Electrochemical methods

Very small concentrations of many gases can be detected by their influence on the
output from an electrochemical cell. There are two primary types, both based on

oxidation or reduction of the gas within a galvanic cell. The cell has at least two electrodes and an intervening electrolyte. In the polarographic or voltametric analyser, a voltage from an external battery is applied across the electrodes in order to induce further polarization (or retardation) of the electrodes. The gas sample is supplied to the interface between the electrolyte and one of the electrodes (the 'sensing electrode'). This may be accomplished by diffusion of the sample gas through a hollowed and permeable sensing electrode. The electrochemical reaction at the electrode–electrolyte interface changes the rate at which free electrons are released to flow through the electrolyte and to be collected by the 'receiving electrode'. For example, in a sulphur dioxide electrochemical analyser, the oxidation is as follows:

$$SO_2 + 2H_2O \rightarrow SO_4^{-2} + 4H^+ + 2e^-$$

The resulting change in electrical current is proportional to the concentration of sulphur dioxide in the sample.

In amperometric cells, the gas reacts directly with the electrolyte and, hence, enhances or reduces the current produced. Both types of cell are subject to interference by other gases. This is minimized by appropriate selection of the materials employed for the electrodes and electrolyte, the polarizing voltage applied to the polarographic cells and a suitable choice of filters.

Electrochemical cells may indicate the partial pressure rather than the concentration of a gas. These require the zero to be reset when taken through a significant change in barometric pressure—as will occur when travelling to different levels in a mine. Another disadvantage of electrochemical analysers is that they can become temporarily saturated when exposed to high concentrations of gas. The recovery period may be several minutes.

Mass spectrometers

In these instruments, the gas sample passes through a field of free electrons emitted from a filament or other source. Collision of the electrons with the gas molecules produces ions, each with a mass/charge ratio specific to that gas. The ions are accelerated by electromagnets and then pass through a magnetic deflection field which separates them into discrete beams according to their mass/charge ratios. The complete mass spectrum can be scanned and displayed on an oscilloscope or the signals transmitted to recorders.

Paramagnetic analysers

Oxygen is one of the very few gases that is paramagnetic, i.e. it aligns itself as a magnetic dipole in the presence of an applied magnetic field and, hence, creates a local anomaly within that field. This property is utilized in a paramagnetic oxygen analyser. A weak permanent dumb-bell magnet is suspended against a light applied torque within a non–uniform magnetic field. One end of the dumb-bell is encapsulated within a bulb of nitrogen while the other end is exposed to the gas sample. A rotation of the magnet is induced as the oxygen content of the sample varies. The movement is amplified optically or electrically for display or recording.

Gas chromatography

Gas chromatographs are used widely for the laboratory analysis of sampled mixtures of gases. Portable units are also manufactured. An inert carrier gas is pumped continuously through one or more columns (or coils) which contain gas adsorbents. The latter may be granulated solids or liquids. A small pulse of the sample gas mixture is injected into the line upstream from the columns. The constituent gases are initially adsorbed by the column materials. However, the continued flow of the carrier gas causes subsequent desorption of each gas at a time and rate dependent on its particular adsorption characteristics. The result is that the gases leave the adsorbent columns as discrete and separated pulses. Their identification and measurement of concentration is carried out further downstream by one or more of the detection techniques described in this section.

Semiconductor detectors

One of the most recent techniques introduced for gas detection involves passing the sample over the surface of a semiconducting material which is maintained at a constant temperature. Adsorption of gas molecules on to the surface of the semiconductor modifies its electrical conductance. Selectivity of the gas to be detected may be achieved by the choice of semiconductor and the operating temperature. However, filters may be necessary to avoid interference or poisoning from other airborne pollutants. Coupled with the development of thin film technology, the semiconductor technique holds promise for increased future utilization.

Stain tubes

Stain tubes are used widely for a large variety of gases. They are simply glass phials containing a chemical compound that changes colour in the presence of a specified gas. The phials are sealed at both ends. To carry out a test, the ends are snipped open and a metered volume of sample air pulled through at a constant rate, usually by means of a simple hand pump. The concentration of gas is estimated either from the length of the stain or by comparing the colour of the stain with a chart. Despite the lack of precision, stain tubes are used extensively in practice because of their simplicity, portability and low cost.

11.4.3 Methods of sampling

The methods of sampling for subsurface gases vary from judicious positioning of the human nose to sophisticated telemetering systems. The techniques most commonly applied can be divided into two classes: (a) manual and (b) automatic or remote.

Manual methods

The locations and times at which hand-held equipment should be employed to measure methane concentrations in gassy mines are normally mandated by the governing legislation. These include pre-shift inspections by qualified persons prior to entry of the work force, and at intervals of some 20 min throughout the shift on a working face.

The instruments that are most widely used for manual detection and measurement are catalytic oxidation methanometers, stain tubes and, in some countries, flame safety lamps. Light gases, including methane, are most likely to collect at roof level while carbon dioxide will tend to pool in low lying areas. Hence, measurements should be taken in those locations in addition to within the general body of air. Extension probes provide a means of drawing a sample from a location that is unsafe or out of reach.

Grab samples are volumes of air or gas mixtures that are collected in sample containers underground for subsequent laboratory analysis. Typically, the sample is drawn into a metal or plastic container by means of a hand-operated pump or by water displacement. This is usually the method employed for retrieving samples from behind seals or stoppings. The sample pipe should extend far enough beyond the seal to prevent contamination by air that has 'breathed' through the seal during preceding periods of rising barometric pressure. Sufficient gas should be drawn from the pipe prior to sampling in order to ensure that the captured gas is representative of the atmosphere beyond the seal. Care should be taken to prevent loss or pollution of the sample as the container is subjected to pressure changes during the journey to the surface. The gas seals on the containers should be well maintained and samples should be transported to the laboratory for expeditious analysis.

Personal samplers are devices that are worn continuously by personnel while at work. The simplest are badges that change colour in the presence of selected gases or radiation. More sophisticated samplers for dust or gases draw power from either an internal battery or a caplamp battery. Personal gas samplers emit audio-visual alarm signals when a preselected concentration of gas is exceeded.

Automatic and remote monitors

Permanent environmental monitors may be sited at strategic locations throughout a subsurface ventilation system. The gases most frequently subject to this type of monitoring are methane, carbon monoxide and carbon dioxide. Other gases may be monitored in workshops, repositories and for special applications such as hydrogen at the roofs of battery charging stations. Environmental monitors in an underground repository for nuclear waste must also be provided to detect airborne radionuclide contamination.

Permanent monitors should be mounted at strategic locations chosen on a site-specific basis. However, it is prudent to site transducers at the intake and return ends of working areas and at intervals along return routes. Carbon monoxide sensors mounted at intervals along conveyor roads provide an earlier warning of fire than temperature-sensitive devices.

In addition to providing local audio-visual alarms, an environmental monitoring system operates most efficiently when it is integrated into a telemetering network. The signals from gas, airflow, pressure and temperature transducers are transmitted to a remote computer and control station, usually on surface, where those signals are analysed for trends, recorded on magnetic media, operate alarms when appropriate and, possibly, generate feedback control signals to fans, doors or motorized regulators. (See, also, sections 9.6.3 and 9.6.4) Monitors at critical locations such as booster fans should be installed as dual or, even, triple units to safeguard against instrument failure. Rechargeable batteries can provide back-up in the event of a cut in electrical power. A system of planned maintenance should be implemented in order to ensure continuous and reliable operation. A computer-controlled telemetering system should be capable of detecting and identifying failed monitors.

Machine-mounted gas monitors should be provided on rock breaking equipment to test for methane in coal and other gassy mines. These devices must be particularly rugged in order to remain operational when subjected to dust, water sprays, vibration and impact. Machine-mounted methanometers should not only provide audio-visual warnings but also be connected such that the electrical power supply to the machine is isolated when the methane concentration reaches one percent (or other concentration as required by the relevant mining legislation). Here again, machine-mounted monitors may be integrated into a mine telemetering system.

Tube bundle systems are a slower alternative means of remote sampling. Air is drawn through plastic tubing from chosen locations in the mine to monitoring stations which may be located either underground or on surface. At a monitoring station, automatic valves on the tubes are operated in a cyclic manner in order to draw samples from the tubes sequentially, and to pass those samples into gas detection units. The latter are usually infrared analysers (section 11.4.2) connected in a series to monitor the concentrations of each of the required gases. A gas pump on each tube maintains continuous flow. However, the main disadvantage of the tube bundle system is the travel time of the gas between entering a tube and analysis. It is recommended that this should not exceed 1 h. While surface monitoring stations are preferred, workings distant from shaft bottoms may necessitate local underground stations. Tube bundle systems are not suitable for emergency situations that occur quickly such as an equipment fire. However, they have proved valuable for the identification of longer-term trends including incipient spontaneous combustion.

The tubes should be fitted with dust filters and water traps. Variations of temperature and pressure may cause condensation as the air passes through the tubes. Care must be taken during the installation of a tube bundle system, particularly at joints, as leaks are difficult to detect and locate. An important advantage of the system is that it requires no electrical power inbye the monitoring station. Hence, it remains operational when sampling from behind seals or when no electrical supply is available.

REFERENCES

ACGIH (1989) Threshold limit values and biological exposure indices for 1989–1990. *American Conf. Governmental Industrial Hygienists, Cincinnati, OH.*

Baker, A. R. (1969) A review of methanometry. *Min. Eng.* **128** (107), 643–53.

Daniel, J. H. (1989) Diesels—backbone of a changing mining industry. *Proc. 4th US Mine Ventilation Symp., Berkeley, CA*, 561–8.

Forbes, J. J. and Grove, G. W. (1954) Mine gases and methods for detecting them. *US Bureau of Mines, MC*, No. 33.

French, I. W. (1984) *Health Implications of Exposure of Underground Mine Workers to Diesel Exhaust Emissions—an Update*, Department of Energy, Mines and Resources, Ottawa, 607.

Gunderson, R. E. (1990) The advantages of electrically powered equipment for deep hot mines. *J. Mine Vent. Soc. S. Afr.* **43** (6), 102–111.

Haney, R. A. and Stolz, R. T. (1989) Evaluation of personal diesel particulate samplers. *Proc. 4th US Mine Ventilation Symp., Berkeley, CA*, 579–87.

Holmes, J. M. and Byers, C. H. (1990) *Countermeasures to Airborne Hazardous Chemicals*, Noyes, N.J.

Pritchard, F. W. and Phelps, B. A. (1961) A recorder of atmospheric methane concentration based on a butane flame lamp. *Colliery Eng.* **120** (February 9), 163–6.

Richards, G. O. and Jones, K. (1970) Some recent work on the detection and determination of firedamp. *Min. Eng.* **130** (121), 31–40.

Rossi, B. D. (1971) *Control of Noxious Gases in Blasting Work and New Methods of Testing Industrial Explosives*, NTIS (translated from Russian; Tech. Translation No. TT70-50163).

Strang, J. and MacKenzie-Wood, P. (1985) *A Manual on Mines Rescue and Gas Detection* Weston and Co., Kiama, Australia.

Waytulonis, R. W. (1988) Bureau of Mines diesel research highlights. *Proc. 4th Int. Mine Ventilation Congr., Brisbane*, 627–33.

World Mining (1982) *Diesels in Mining*, Special Issue, November, 6–54.

FURTHER READING

Baker, A. R. and Firth, J. G. (1969) The estimation of firedamp: applications and limitations of the pellistor. *Min. Eng.* **128** (100), 237–44.

Coward, H. F. (1928) Explosibility of atmospheres behind stoppings. *Trans. Inst. Min. Eng.* **77**, 94.

Coward, H. F. and Jones, G. W. (1952) Limits of flammability of gases and vapours. *Bull. US Bureau Mines*, No. 503.

NIOSH (1987) *Pocket Guide to Chemical Hazards*, National Institute for Occupational Safety and Health, US Department of Health and Human Services, Washington, DC.

US Bureau of Mines (1987) Diesels in underground mines. *Proc. Technology Transfer Semin.*, 1–165.

12

Methane

12.1 OVERVIEW AND ADDITIONAL PROPERTIES OF METHANE

The major properties and behavioural characteristics of methane were outlined in section 11.2.4. That section should be reviewed as an introduction to the more detailed treatment given in this chapter.

There are three primary reasons for giving particular attention to methane. First, it is the naturally occurring gas that most commonly appears in mined underground openings. Secondly, it has resulted in more explosions and related loss of life than any other cause throughout the recorded history of mining. The flammability characteristics of this gas have been studied since, at least, the time of Agricola in the sixteenth century. The number and severity of coal dust explosions initiated by methane ignitions declined after the development of electric caplamps and the replacement of shaft bottom furnaces with fans (section 1.2). However, the intensity and wider deployment of mechanized rock-breaking equipment resulted in a renewed incidence of frictional ignitions of methane following the 1960s (Richmond *et al.*, 1983). Fortunately, modern standards of ventilation and dust control prevent most of these from developing into the larger dust explosions and disastrous loss of life.

The third reason for giving special attention to methane concerns the continued development of methane drainage technology. Although the primary reason for extracting methane at high concentration from the strata around mines continues to be the reduction of methane emissions into these mines, a growing incentive has been the drainage of methane to provide a fuel source in its own right. Methane drainage may now be undertaken for this purpose from strata where there is no known intention of subsequent mining.

In this chapter we shall consider four broad areas; first, the manner in which methane is retained within the strata and the mechanisms of its release when the rock is disturbed by mine workings or boreholes. We shall also outline means of determining the gas content of carbonaceous strata.

Secondly, we shall examine the migration of the gas from its geologic sources towards workings or boreholes. This will encompass an analysis of strata permeability and flow through fracture networks.

The third consideration is the dynamic pattern of methane emission into active mine workings, varying from normal cyclic variations, through gas layering phenomena, to outburst activities.

Finally, we shall classify the major techniques of methane drainage. No one of these has universal application and the selection of drainage method must be made with great care according to the geology of the area, the methods and layout of mining (if any) and the natural or induced physical properties of the rocks.

First, however, let us list some further properties of methane in addition to those given in Table 11.1. The data are referred, wherever applicable, to standard temperature and pressure of $0°C$ and $101.3\,kPa$.

$$
\begin{array}{c}
H \\
| \\
H—C—H \\
| \\
H
\end{array}
$$

Molecular structure	(structure shown above)
melting point	$90.5\,K$
Boiling point	$111.3\,K$
Critical temperature and pressure	$190.5\,K$ and $4.63\,MPa$
Latent heat of vaporization (at 111.3 K)	$508.2\,kJ/kg$
Latent heat of fusion (at 90.5 K)	$58.8\,kJ/kg$
Specific heat C_p	$2184\,J/(kg\,K)$
Specific heat C_v	$1680\,J/(kg\,K)$
Isentropic index	1.300
Solubility in water	$55.6\,l$ per m^3 of water $11\,(33.1\,l/m^3$ at $20\,°C)$
Upper calorific value	$55.67\,MJ/kg$
Lower calorific value	$50.17\,MJ/kg$
Thermal conductivity	$0.0306\,W/(m\,°C)$ $(0.0328\,W/(m\,°C$ at $20\,°C)$
Dynamic viscosity (temperature $t = 0$ to $100\,°C$)	$(10.26 + 0.0305\,t) \times 10^{-6}\,N\,s/m^2$

12.2   THE RETENTION AND RELEASE OF METHANE IN COAL

In order to understand the manner in which methane is retained within coal and the mechanisms of its release, it is first necessary to comprehend the internal structures of coal. The existence of a large number of exceptionally small pores and the corresponding high internal surface area of coal has been recognized for many years and numerous conceptual models have been suggested. However, the availability of electron microscopes has revealed the actual structures that exist.

Coal is not a single material but a complex mix of fossilized organic compounds and minerals. The composition and structure depend on the nature of the original vegetation from which a given coal seam formed, the timing and turbidity of the

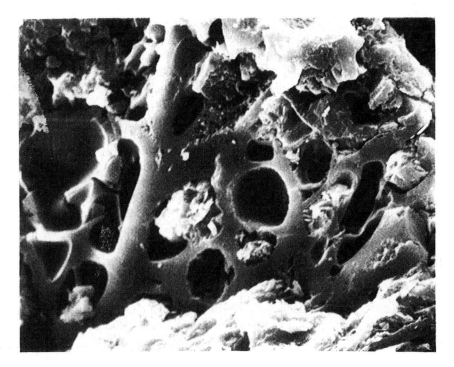

Figure 12.1 Electron micrograph of a bituminous coal (after Harpalani).

water flows involved in the sedimentary processes of deposition and the metamorphic effects of pressure, temperature and tectonic stressing over geological time. Electron micrographs have, indeed, confirmed the existence of pores, many of which may not be interconnected. An example is shown on Fig. 12.1. However, other areas may indicate amorphous, granular, sponge-like or fibrous structures, even within the same seam. Furthermore, the coal substance is intersected by a fracture network with apertures varying from those that are comparable with the pore diameters through to some large enough to be seen with the naked eye.

In this section we shall concentrate on the manner in which methane is held within the coal, the kinetics of its release when the geological equilibrium is disturbed by mining or drilling, and how the gas content of a seam may be determined.

12.2.1 Gas retention in coal

Methane exists within coal in two distinct forms, generally referred to as **free gas** and **adsorbed gas**. The free gas comprises molecules that are, indeed, free to move within the pores and fracture network. Porosities of coals have been reported from 1% to over 20%. However, those values depend on the chosen definition of porosity and the manner in which it is measured. **Absolute** (or **total**) **porosity** is the total internal voidage divided by the bulk volume of a sample and may be difficult to measure accurately. **Effective** (or **macroscopic**) **porosity** is the ratio of inter-connected void space to the bulk volume. The latter definition is more useful in the determination of recoverable gas from a seam. Unfortunately, the usual procedures of measuring porosity depend on saturating the internal voidage with some per-meating fluid. The probability of a gas molecule entering any pore or interconnection rises as its molecular diameter decreases or its mean free path increases. (The mean free path is the statistical average distance between collisions of the gaseous molecules.) Furthermore, any adsorptive bonding between the fluid and the solid may obstruct the narrower interconnections. The measured effective porosity is, therefore, depen-dent on the permeating fluid. In order to handle the extremely small pores that exist in coal, the maximum value of effective porosity is obtained by using helium as the permeating fluid. This gas has a small molecular diameter (some 0.27×10^{-9} m), a relatively large mean free path (about 270×10^{-9} m at $20\,^{\circ}$C and atmospheric pressure) and is non-adsorptive with respect to coal. For comparison, pore diameters of eastern US coals vary from over 30×10^{-9} to less than 1×10^{-9} m (Gan *et al.*, 1972).

An attractive force exists between the surfaces of some solids and a variety of gases. Coal surfaces attract molecules of methane, carbon dioxide, nitrogen, water vapour and several other gases. Those molecules adhere or are **adsorbed** onto the coal surface. When the adsorptive bond exceeds the short-distance repulsive force between gas molecules (section 2.1.1), then the adsorbed molecules will become

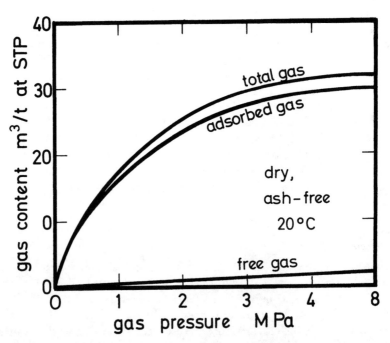

Figure 12.2 Examples of adsorbed, free and total gas isotherms for methane in coal.

packed together as a monomolecular layer on the surface. At very high gas pressures, a second layer will form with a weaker adsorptive bond (Jolly *et al.*, 1968).

Figure 12.2 illustrates the variation of both total and adsorbed methane with respect to gas pressure and at constant temperature. The curve illustrating adsorbed methane is known as the adsorption isotherm. Some 95% of the total gas will, typically, be in the adsorbed form, explaining the vast reserves of methane that are contained within many coal seams.

The most widely used mathematical relationship to describe adsorption isotherms was developed by Langmuir (1916, 1918) for a monomolecular layer. This may be expressed as

$$\frac{q}{q_{max}} = \frac{bP}{1 + bP} \tag{12.1}$$

where q = volume of gas adsorbed at any given pressure (m^3/t at NTP), q_{max} = the maximum amount of gas that can be adsorbed as a monomolecular layer at the prevailing temperature (m^3/t at NTP), P = gas pressure (usually expressed in MPa) and b = Langmuir's 'constant' (MPa^{-1}), a function of the adsorptive bond between the gas and the surface ($1/b$ is the pressure at which $q/q_{max} = 1/2$).

The values of Langmuir's constant depend on the carbon, moisture and ash contents of the coal as well as the type of gas and prevailing temperature. Boxho *et al.* (1980) report values of b decreasing from 1.2 MPa^{-1} at a volatile content of 5% to 0.5 MPa^{-1} at a volatile content of 40%. Figure 12.3(a) indicates the increased adsorptive capacities of the higher rank coals, these having greater values of carbon content. Values of q_{max} are shown to vary from about 14 to over 30 m^3/t. The effect of the type of gas is illustrated on Fig. 12.3(b).

Absorption isotherms are normally quoted on a dry, ash-free basis. The amount of methane adsorbed decreases markedly at small initial increases in moisture content of the coal. Most of this natural moisture is adsorbed on to the coal surfaces. However, adsorptive saturation of water molecules occurs at about 5% moisture above which there is little further decrease in methane content. A widely used approximation is given by Ettinger *et al.* (1958):

$$\frac{q_{moist}}{q_{dry}} = \frac{1}{1 + 0.31 h} \tag{12.2}$$

where q_{moist} = gas content of moist coal (m^3/t), q_{dry} = gas content of dried coal (m^3/t), and h = moisture content (percent) in the range 0% to 5% (assume 5% for greater moisture contents).

The mineral matter that constitutes the ash constituent of coal is essentially non-adsorbing. Hence the methane content decreases as the percentage of ash rises (Barker-Read and Radchenko, 1989). In order to express gas content on an ash-free basis, a simple correction may be applied.

$$q_{actual} = q_{ash\text{-}free}(1 - 0.01a) \tag{12.3}$$

where a = ash content (per cent).

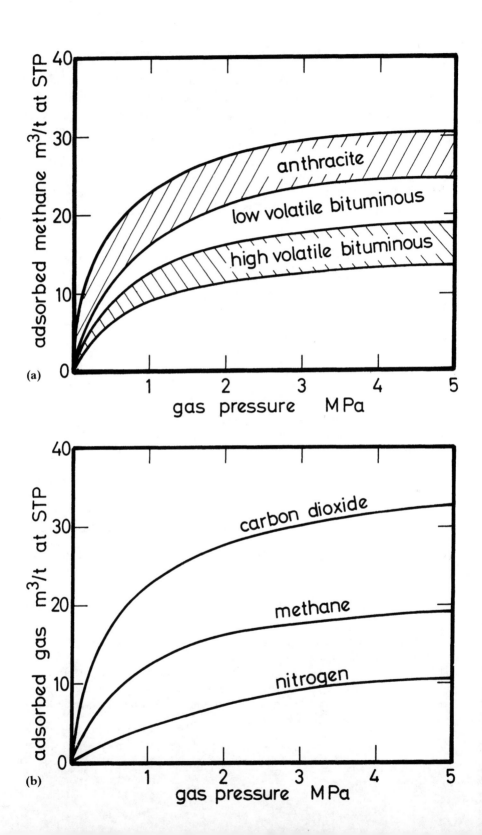

12.2.2 The release of methane from coal

In the undisturbed state, an equilibrium exists between free gas and adsorbed gas in the pores and fracture networks of coal. If, however, the coal seam is intersected by a borehole, or disturbed by mining, then the gas pressure gradient that is created will result in flow through natural or stress-induced fractures. The reduced gas pressure in the pores will promote desorption. The process will move from right to left along the appropriate adsorption isotherm. A glance at Fig. 12.2 or 12.3 indicates that the rate of desorption with respect to gas pressure increases as the pressure falls. The migration of methane from its original location is retarded by narrow and tortuous interconnections between pores and microfractures, obstruction by adsorbed molecules and the resistance offered by the fracture network.

Diffusion and Darcy flow

Current thinking is that two types of flow take place. Within the micropore structure, diffusion flow occurs while, in the fractures, laminar flow dominates. Let us first discuss diffusion flow. This occurs whenever a difference in concentration of molecules of a given gas occurs. The simplest means of quantifying diffusion flow is by Fick's law:

$$u_x = - D \frac{\partial C}{\partial x} \quad \text{m/s} \tag{12.4}$$

where u_x = velocity of diffusion in the x direction (m/s), D = coefficient of diffusion (m^2/s but often quoted in cm^2/s) and C = concentration of the specific gas (m^3/(m^3 of coal)). (The negative sign is necessary as movement occurs in the direction of decreasing concentration.)

Bulk diffusion is the normal gas to gas diffusion that occurs because of a concentration gradient in free space (see the appendix in Chapter 15, section 15A.4). However, within the confines of a micropore structure, two other forms of diffusion may become effective. **Surface diffusion** arises from lateral movement of the adsorbed layer of gas on the coal surfaces while **Knudson diffusion** occurs as a result of transient molecular interactions between the gas and the solid.

Laminar flow in the fracture network follows Darcy's (1856) law for permeable media:

$$u = - \frac{k}{\mu} \frac{\partial P}{\partial x} \quad \text{m/s} \tag{12.5}$$

where u = gas velocity (m/s), k = permeability (m^2), μ = dynamic viscosity (N s/m^2) and $\partial P/\partial x$ = pressure gradient (Pa/m). (Note the analogy between Darcy's

Figure 12.3 (a) Examples of adsorption isotherms. The amount of gas adsorbed increases with the carbon content of the coal. (b) Adsorption isotherms for carbon dioxide, methane and nitrogen in a bituminous coal at 25° C.

law and Fourier's law of heat conduction, equation (15.4)). Darcy's law is essentially empirical and assumes that the gas velocity at the surface is zero (section 2.3.3). For narrow passages, slippage at the walls (surface diffusion) may become significant and can be taken into account by adjusting the value of permeability (Klinkenberg effect, section 12.3.2).

Controversy has existed on whether diffusion flow or Darcy flow predominates in coal. Disagreements in the findings of differing researchers are probably a consequence of the wide variations that exist in coal structure. It would appear that diffusion flow governs the rate of degassing coals of low permeability while Darcy flow in the fracture network is the dominant effect in high permeability coals.

Desorption kinetics

A newly exposed coal surface will emit methane at a rate that decays with time. Figure 12.4(a) illustrates a desorption curve. This behaviour is analogous to the emission of heat (section 15.2.2). A number of equations have been suggested depending, primarily, on the conceptual model adopted for the structure of coal. For a spherical and porous particle with single sized and non-connecting capillaries, mathematical analyses leads to the series

$$\frac{q(t)}{q_{max}} = 1 - \frac{6}{\pi^2} \sum_{n=1}^{\infty} \frac{1}{n^2} \exp\left(-\frac{4Dn^2\pi^2 t}{d^2} \right) \tag{12.6}$$

where $q(t)$ = volume of gas (m^3) emitted after time t(s), q_{max} = total volume of gas in particle, D = coefficient of diffusion (m^2/s) and d = equivalent diameter (m) of particle (= 6 × volume/surface area). A more practicable approximation is given by Boxho *et al.* (1980) as follows:

$$\frac{q(t)}{q_{max}} \approx \sqrt{1 - \exp\left[-\frac{4\pi^2 Dt}{d^2} \right]} \tag{12.7}$$

Both of these equations approximate to

$$\frac{q(t)}{q_{max}} \approx \frac{12}{d} \sqrt{\frac{Dt}{\pi}} - \frac{12Dt}{d^2} \tag{12.8}$$

while for $q(t)/q_{max} < 0.25$, the relationship can be simplified further to

$$\frac{q(t)}{q_{max}} \approx \frac{12}{d} \sqrt{\frac{Dt}{\pi}} \tag{12.9}$$

This latter equation implies that during the early stages of desorption, the emission rate is proportional to $\sqrt{t}$. This is illustrated on Fig. 12.4(b). Furthermore, plotting the initial emission rate against $\sqrt{t}$ allows the corresponding value of the coefficient of diffusion, D, to be determined.

Unfortunately, there is a major problem associated with all of these equations.

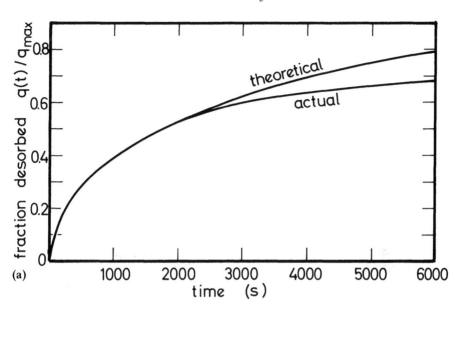

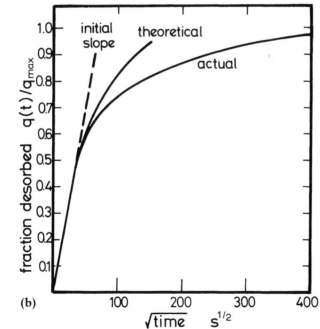

Figure 12.4 (a) Example of a methane desorption curve plotted against time. (b) A methane desorption curve plotted against $\sqrt{\text{time}}$.

The coefficient of diffusion, D, is not a constant for any given sample. It is dependent on

1. the size range of individual particles,
2. the distribution of pore sizes,
3. the number, sizes and tortuosity of fractures (this, in turn, depending on the stress history of the sample),
4. temperature, and
5. pressure and current gas content.

The latter factors are particularly troublesome as these, and hence D, will vary as the degassing proceeds. Lama and Nguyen (1987) have suggested the relationship

$$D = D_0 \frac{P}{P_0} \quad m^2/s \tag{12.10}$$

where P = current value of pressure (Pa) and D_0 = coefficient of diffusion at some original (seam) pressure, $P_0 (m^2/s)$. The variations in D contribute to the changes in apparent permeability of a coal sample when subjected to mechanical stress or variations in mean gas pressure (section 12.3.2). Considering the sensitivity of D to a significant number of variables, it is not altogether surprising that a wide range of values have been reported, from 1×10^{-8} to $1 \times 10^{-14} m^2/s$. In order to take the particle size into account, it has become common to express the diffusivity as the ratio D/a^2 (s^{-1}) where a = particle radius (m).

A further weakness is that the fundamental equation (12.6) is based on uniformly sized capillaries. This leads to significant deviations between predicted and observed desorption rates at the longer times $(q(t)/q_{max} > 0.6)$ as indicated on Fig. 12.4 (Smith and Williams, 1984a).

The difficulties encountered in the derivation and general applicability of analytical desorption equations led Airey (1968) to propose an empirical relationship:

$$\frac{q(t)}{q_{max}} \cdot 1 - \exp\left[-\left(\frac{t}{t_0}\right)^n \right] \tag{12.11}$$

where t_0 = 'time constant' = time for 63% of the gas to desorb and n = an index that varies from some 1/3 for bituminous coal to 1/2 for anthracites and depends, also on the degree of fracturing. Both t_0 and n are influenced by the range and magnitude of particle sizes and, hence, should be quoted with reference to a specific size spectrum. For example, values of t_0 for Welsh anthracite lie, typically, between 1500 and 2100 s for the size range 0.05 to 0.30 mm at 25 °C (Barker-Read and Radchenko, 1989). Smaller particles will give reduced values of both t_0 and n.

An improved desorption time parameter, τ, that is independent of the gas capacity of a sample has been proposed by Radchenko (1981) and Ettinger et al.(1986) (see also, Barker-Read and Radchenko, 1989) and may be related to Airey's constants:

$$\tau = 60\left\{ 1 - \exp\left[-\left(\frac{t_0}{60}\right)^{-n} \right] \right\}^{-2} \quad s \tag{12.12}$$

It can also be shown that

$$\tau = \frac{\pi a^2}{36D} \quad \text{s} \tag{12.13}$$

where a = radius of the particle (m).

The value of τ can be gained from the desorbed fraction vs. $\sqrt{\text{time}}$ curve (e.g. Fig. 12.4(b)):

$$\tau = 1/(\text{initial slope})^2 \quad \text{s} \tag{12.14}$$

Equations (12.13) and (12.14) provide a ready means of determinating D/a^2 from a desorption test.

12.2.3 Determination of gas content

There are two reasons for measuring the methane contents of coal seams and associated strata. First, such data are required in the assessment of methane emissions into mine workings and, hence, the airflows required to dilute those emissions to concentrations that are safe and within mandatory threshold limit values (sections 9.3.1 and 11.2.1). Secondly, the gas content of the strata is a required input for computer models or other computational procedures to determine the gas flows that may be obtained from methane drainage systems.

There are also two distinct approaches to the evaluation of gas content, the **indirect method** which employs adsorption isotherms and the **direct measurement method** which relies on observations of gas release from newly obtained samples. Let us consider each of these in turn.

Indirect method (adsorption isotherms)

In this technique, representative chippings are obtained from the full thickness of the seam, mixed and ground to a powder of known particle size range. The sample is dried at a temperature of not more than 80 °C and evacuated to remove the gases that remain in the pore structure. Unfortunately, the process of evacuation may alter the internal configuration of the pore structure owing to the liquefaction of tars (Harpalani, 1984). The phenomenon can be overcome by evacuating at low temperature—submerging the sample container into liquid nitrogen (-150 °C) prior to and during the evacuation.

After returning to the desired ambient temperature, methane is admitted in stages and the gas pressure within the sample container recorded at each increment. The volume of gas admitted may be monitored at inlet (volumetric method) or by measuring the increase in weight of the sample and container (gravimetric method). After correcting for free space in the sample container, a plot of cumulative methane admitted against pressure provides the total gas isotherm. If the porosity of the sample has been determined separately (section 12.2.1), then the results can be divided into free gas and adsorbed gas as illustrated on Fig. 12.2. The isotherms should, ideally, be determined at, or close to, the virgin rock temperature of the actual strata. If necessary, the curves can be corrected to seam temperature. At any given pressure, the quantity of gas adsorbed falls as the temperature increases. Starting at 26 °C, the gas adsorbed decreases by some 0.8% °C for bituminous coal and 0.6% °C for anthracite (Boxho *et al.*, 1980). The curves should also be corrected to the actual moisture content of the seam using Ettinger's formula, equation (12.2).

In order to utilize the corrected gas isotherm, a borehole is drilled into the seam, either from the surface or from an underground location. In the latter case, the hole must be sufficiently long (10 to 20 m) in order to penetrate beyond the zone of degassing into the mine openings. Seals are emplaced in the borehole to encapsulate a representative length within the seam. A tube from the encapsulated length of borehole is attached to a pressure gauge and the rise in pressure is monitored. The rate of pressure rise will be greater for coals of higher permeability. The initial rate of pressure rise may be used in conjunction with open hole flowrates to determine *in situ* permeability. However, for the purposes of assessing seam gas content, it is the maximum (or equilibrium) pressure that is required. Using this pressure, the gas content of the seam can be read from the corrected isotherm curve.

An advantage of the indirect method is that it gives the total gas content of the seam. However, this is not indicative of the actual gas that may be emitted into mine workings or recoverable by methane drainage. Furthermore, the measured *in situ* gas pressure will be influenced not only by methane but also by other gases that may be present and, particularly, by the presence of water.

Direct measurement method

This technique involves taking a sample of the seam from a borehole and placing it immediately into a hermetically sealed container. The gas is bled off to atmosphere in stages and its volume measured. The process is continued until further gas emissions are negligible.

An early method of direct measurement of gas content was developed in France (Bertrard *et al.*, 1970). This utilized samples of small chippings from the borehole. Further research carried out by the US Bureau of Mines during the 1970s led to a procedure that used complete cores (Diamond and Levine, 1981). Although developed primarily for surface boreholes, the technique is applicable also to horizontal holes drilled from mine workings. The sampling personnel must be present at the time the hole is drilled into the seam. A stop–watch is used to maintain an accurate record of the elapsed times between which the sample length of strata is penetrated, start of core retrieval, arrival of the sample at the mouth of the borehole and confinement within the sealed container.

Each container should be capable of holding about 2 kg of core and some 35 to 40 cm in length. Longer cores should be subdivided as a precaution against major loss of data should one container suffer from leakage. The seals on containers must be capable of holding a gas pressure of 350 kPa without leaking. A pressure gauge should be fitted to each container.

Gas is bled off from the container at intervals of time commencing at 15 min (or less if a rapid rise in container pressure is observed). The volume of gas emitted at each stage is measured, usually by water displacement, in a burette. The time intervals are increased progressively and the process allowed to continue until an average of not more than 10 cm^3 per day has been maintained for one week.

There is, inevitably, some gas lost from the core between the time of seam penetration and its confinement in a sample container. The volume of lost gas may be assessed by plotting the cumulative gas emitted from the container against $\sqrt{\text{time}}$. The initial straight line may be extrapolated backwards through the recorded elapsed

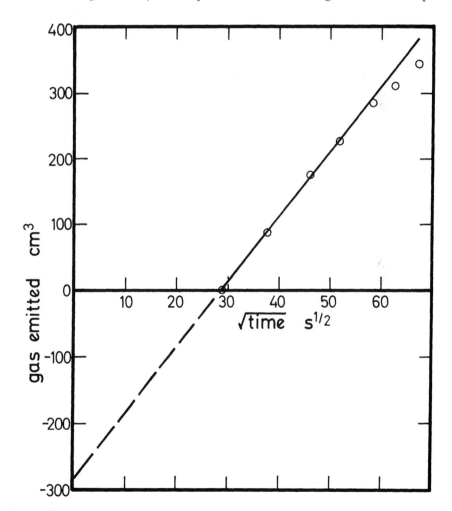

Figure 12.5 Assessment of gas lost during core retrieval. In this example, elapsed time before containment, $t = 840\,\text{s}$ (i.e. $\sqrt{t} = 29\,\text{s}^{1/2}$). Backward extrapolation of the initial straight line gives the lost gas as 285 cm^3.

time of core retrieval in order to quantify the lost gas. The retrieval time can be minimized by using wireline drilling. The technique is illustrated on Fig. 12.5. The analytical background to the procedure is embodied in equation (12.9). (Should subsequent tests show that the rate of desorption follows at t^n law where n deviates significantly from 0.5 (equation (12.11)), then the appropriate plot may be constructed as gas evolved vs. t^n in order to determine the lost gas.)

Another method of estimating the lost gas, developed by Smith and Williams (1984b), makes separate allowance for the elapsed times of drilling through the coal, core retrieval and period spent at the mouth of the borehole before containment. This more sophisticated technique is based on an analysis that takes account of the lack of uniformity in pore sizes (Smith and Williams, 1984a; Close and Erwin, 1989).

The combination of measured gas and lost gas gives an estimate of the maximum amount of gas that may be emitted into mine workings or captured by methane drainage. (It may be much less than this.) However, following the termination of gas evolution measurements, the core will still contain some residual gas. This can be quantified, if required, by crushing the coal and continuing to measure the additional gas that is then liberated. The crushing process should take place in a separate sealed ball mill within a nitrogen atmosphere. However, steel balls in the sample container allow the crushing to take place without removing the sample from the container. Another method is to activate a steel hammer by electromagnetic means within the container in order to crush a sample through a fixed grid. In either case, the sample and evolved gas should be cooled to the original ambient temperature before the gas is bled off.

A variation of the direct measurement method of gas content is the 'desorbmeter' (Hucka and Lisner, 1983) in which desorbing methane from coal chippings in a sample vessel pushes a small plug of fluid along a transparent spiral tube. An electronic version employing the same principle has been developed in Germany (Janas, 1980).

12.3   MIGRATION OF METHANE

Following release from its long-term geologic home, methane will migrate through rock under the influence, primarily, of a gas pressure gradient. That movement will occur through the coal seam and if the gas pressure gradient is transverse to the seam, also through adjoining permeable strata. We assume that the flow paths and velocities are sufficiently small that laminar flow exists. Hence, Darcy's law applies. This was introduced in its simplest form as equation (12.5). In this section, we shall consider the further ramifications of Darcy's law.

12.3.1   Fluid flow through a permeable medium

Incompressible flow

Let us consider, first, an incompressible fluid (liquid) passing through a permeable medium in the direction x as illustrated on Fig. 12.6. An elemental volume of the fluid

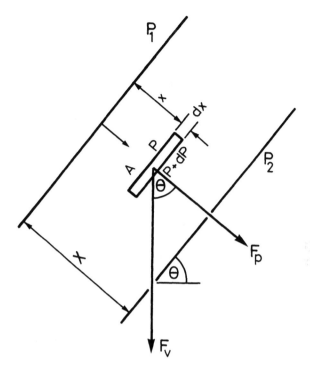

Figure 12.6 Forces on an elemental volume of fluid, $A\,dx$, within a permeable medium. Total force in the x direction, $F_t = F_p + F_v \cos\theta$.

within the medium has a thickness, dx, and an orthogonal area A. Two forces act upon the element.

First, a force in the direction of flow is exerted because of the pressure differential, dP, across its two faces. Let us call this F_p, where

$$F_p = -A\,dP \quad \text{N} \tag{12.15}$$

(negative because P decreases in the x direction). Secondly, a vertical downward force, F_v, exists as a result of gravitational pull:

$$F_v = mg$$

where the mass of the element (density, ρ, × volume, $A\,dx$)

$$m = \rho A\,dx$$

giving

$$F_v = \rho g A\,dx$$

If the direction of flow is at an angle θ to the vertical then the component of F_v in the x direction is

$$F_v \cos \theta = \rho g A \cos \theta \, dx \quad \text{N} \tag{12.16}$$

(Note that θ is also the inclination of the permeable medium to the horizontal.)
The total force on the element in the direction of flow is then

$$F_t = F_p + F_v \cos \theta$$
$$= -A \, dP + \rho g A \cos \theta \, dx \quad \text{N} \tag{12.17}$$

This can be expressed as force per unit volume

$$\frac{F_t}{A \, dx} = -\frac{dP}{dx} + \rho g \cos \theta \quad \frac{\text{N}}{\text{m}^3} \tag{12.18}$$

This is a fuller version of the pressure gradient $-\partial P/\partial x$ in Darcy's law, equation (12.5).

Hence, we can rewrite Darcy's law for an incompressible fluid in a non-horizontal flow field as

$$u_x = -\frac{k}{\mu}\left(\frac{dP}{dx} - \rho g \cos \theta\right) \quad \frac{\text{m}}{\text{s}} \tag{12.19}$$

where u_x = fluid velocity in the x direction (m), k = permeability of medium (m²), μ = dynamic viscosity of fluid (N s/m²) and dP is negative in the x direction. In the case of horizontal flow $(\cos \theta = 0)$ or where the fluid density is small, then equation (12.19) reverts to the simple form of Darcy's law:

$$u = -\frac{k \, dP}{\mu \, dx} \quad \frac{\text{m}}{\text{s}} \tag{12.20}$$

As the fluid is incompressible, then u remains constant and we can integrate directly between the bounding walls shown on Fig. 12.6, giving

$$u = -\frac{k}{\mu}\frac{P_2 - P_1}{X} = \frac{k}{\mu}\frac{P_1 - P_2}{X} \quad \frac{\text{m}}{\text{s}} \tag{12.21}$$

or, flowrate, Q, across a given orthogonal area, A, becomes

$$Q = uA = \frac{k}{\mu}A\frac{P_1 - P_2}{X} \quad \frac{\text{m}^3}{\text{s}} \tag{12.22}$$

Compressible flow

For gases the fluid density is, indeed, low and the gravitational term may be neglected even for non-horizontal flow. However, the gas will expand as it progresses along the flowpaths because of the reduction in pressure. Hence the flowrate, Q, and gas velocity, u, will both increase:

$$Q = uA = \frac{k}{\mu}A\frac{dP}{dx} \quad \frac{\text{m}^3}{\text{s}} \tag{12.23}$$

where A = given orthogonal area across which the flow occurs (m^2). As Q is a variable, we can no longer integrate directly. However, if we write the equation in terms of a steady-state (constant) **mass flow**, M (kg/s), then

$$M = Q\rho = -\frac{k}{\mu}A\rho\frac{dP}{dx} \quad \frac{kg}{s} \tag{12.24}$$

However, density

$$\rho = \frac{P}{RT} \quad \frac{kg}{m^3}$$

(general gas law, equation (3.11)) where R = gas constant (J/(kg K)) and T = absolute temperature (K), giving

$$M = -\frac{k}{\mu}\frac{A}{RT}\frac{P dP}{dx} \quad \frac{kg}{s} \tag{12.25}$$

As M is constant, we can integrate across the full thickness of the medium (Fig. 12.6) to give

$$M = \frac{k}{\mu}\frac{A}{RT}\frac{P_1^2 - P_2^2}{2X} \quad \frac{kg}{s} \tag{12.26}$$

This can be converted to a volume flow, Q, at any given density. In particular, at the density, ρ_m, corresponding to the arithmetic mean pressure, the flow becomes

$$Q_m = \frac{M}{\rho_m} \quad \frac{m^3}{s} \tag{12.27}$$

However, equation (12.26) can be written as

$$M = \frac{k}{\mu}\frac{A}{RT}\frac{(P_1 + P_2)}{2}\frac{(P_1 - P_2)}{X} \quad \frac{kg}{s} \tag{12.28}$$

$$= \frac{k}{\mu}A\frac{P_m}{RT}\frac{P_1 - P_2}{X}$$

$$= \frac{k}{\mu}A\rho_m\frac{P_1 - P_2}{X} \quad \frac{kg}{s} \tag{12.29}$$

Combining equations (12.27) and (12.29) gives

$$Q_m = \frac{k}{\mu}A\frac{P_1 - P_2}{X} \quad \frac{m^3}{s} \tag{12.30}$$

Comparing this with equation (12.22) shows that the volume flow of a gas at the position of mean pressure is given by the same expression as for an incompressible fluid.

At any other density, ρ, the volume flowrate, Q, is given by

$$Q = Q_m \frac{\rho_m}{\rho} \quad \frac{m^3}{s} \tag{12.31}$$

If the flow is isothermal (constant temperature) then this correction can be expressed in terms of pressure:

$$Q = Q_m \frac{P_m}{P} \quad \frac{m^3}{s} \tag{12.32}$$

$$= Q_m \frac{P_1 + P_2}{2P} \quad \frac{m^3}{s} \tag{12.33}$$

Radial flow of gas

Figure 12.7 illustrates a borehole of radius r_b intersecting a gas-bearing horizon of thickness h. The gas pressure in the borehole at the seam is P_b while, at some greater radius r_s into the seam, the gas pressure is P_s.

Consider the elemental cylinder of radius r and thickness dr, at which the gas

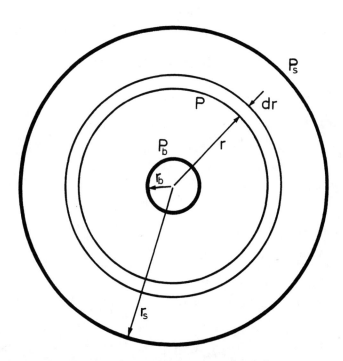

Figure 12.7 Radial flow into a borehole.

pressure is P. Then, from equation (12.25),

$$M = -\frac{k}{\mu RT}\frac{A}{}P\frac{dP}{dr}$$

where $A = 2\pi rh \ m^2$, giving

$$M = -\frac{k \ 2\pi h}{\mu RT}P \ dP\frac{r}{dr}$$

or

$$M\frac{dr}{r} = -\frac{k \ 2\pi h}{\mu RT}P \ dP \quad \frac{kg}{s} \tag{12.34}$$

Integrating between r_s and r_b and dropping the negative sign to denote flow towards the borehole as positive, gives

$$M \ln(r_s/r_b) = \frac{k \ 2\pi h}{\mu \ RT}\frac{P_s^2 - P_b^2}{2}$$

or

$$M = \frac{k \ 2\pi h}{\mu RT}\frac{(P_s + P_b)}{2}\frac{(P_s - P_b)}{\ln(r_s/r_b)} \quad \frac{kg}{s} \tag{12.35}$$

However, as

$$\frac{P_s + P_b}{2RT} = \frac{P_m}{RT} = \rho_m$$

$$Q_m = \frac{M}{\rho_m} = \frac{k}{\mu}2\pi h\frac{P_s - P_b}{\ln(r_s/r_b)} \quad \frac{kg}{s} \tag{12.36}$$

Equations (12.31) and (12.33) can again be used to give the flowrate at any other density or pressure.

Transient radial flow

The previous section assumed a steady-state distribution in gas pressure throughout the medium. This is not, of course, the real situation in practice. As a given source bed is drained, the gas pressure will decline with time. An analysis analogous to that given for heat flow (section 15.2.5) leads to the time transient equation for a non-adsorbing medium:

$$\frac{k}{\mu}\frac{P}{\phi}\left(\frac{\partial^2 P}{\partial r^2} + \frac{1}{r}\frac{\partial P}{\partial r}\right) = \frac{\partial P}{\partial t} \quad \frac{Pa}{s} \tag{12.37}$$

where ϕ = rock porosity (dimensionless) and t = time(s). For methane in coal, gas will be generated by desorption as the gas pressure falls (Fig. 12.2) and, hence, retard the rate of pressure decline very significantly. The gas desorption equations of section 12.2.2 must be coupled with equation (12.37) to track the combined effects of drainage and desorption.

12.3.2 The permeability of coal

In the previous section, it was assumed that the permeability, k, of the rock remained constant. Unfortunately, in some cases including coal, this no longer holds. The anisotropy of the material causes the natural permeability to vary with direction. Furthermore, the permeability changes with respect to mechanical stress, gas pressure and the presence of liquids. We shall examine each of these three effects in turn. First, however, let us clarify the concept of permeability and its dimensions in the SI system of units.

Permeability, k, may be construed as the **conductance** of a given porous medium to a fluid of known viscosity, μ. That conductance must depend only on the geometry of the internal flowpaths. In a rational system of units, the permeability must, therefore, be expressed in terms of the length dimension. This can be illustrated by re-expressing Darcy's law, equation (12.5) as

$$k = -u\mu \frac{\partial x}{\partial P} \quad \frac{\text{m}\,\text{N}\,\text{s}}{\text{s}\,\text{m}^2}\,\text{m}\,\frac{\text{m}^2}{\text{N}} = \text{m}^2$$

The units of permeability in the SI system are, therefore, m^2. However, the older units of darcies (or millidarcies, md) remain in common use. The darcy was defined as 'the permeability of a medium that passes a single-phase fluid of dynamic viscosity 1 centipoise ($0.01\,\text{N}\,\text{s/m}^2$) in laminar flow at a rate of $1\,\text{cm}^3/\text{s}$ through each cm^2 of cross-sectional area and under a pressure gradient of 1 atmosphere ($101\,324\,\text{Pa}$) per cm'. Such definitions make us grateful for the simplicity of the SI system. The conversion between the unit systems is given as

$$1\,\text{md} = 0.986\,93 \times 10^{-15}\,\text{m}^2 \tag{12.38}$$

Effect of mechanical stress

Many researchers have reported tests on the response of coal permeability to applied loading (e.g. Somerton *et al.*, 1974; Gawuga, 1979; Harpalani, 1984). Such tests consist of delicate machining of coal into sample cylinders of 30 to 50 mm in diameter and a length/diameter ratio of about 2 (Obert and Duval, 1967), then placing a sample into a triaxial permeameter. This device allows a liquid or gas to be passed while the sample is subjected to axial loading (by a stiff compression machine) and radial stressing (by oil pressure exerted on a synthetic rubber sleeve around the sample).

Such tests on coal have revealed the following phenomena during non-destructive loading.

1. Permeability reduces during loading and recovers during unloading. However,

there may be a significant hysteresis effect with lower permeabilities during unloading, particularly when the axial and radial stresses are unequal.

2. Samples left under a constant applied stress exhibit a creep effect. The permeability
reduces with time down to a limiting value.

3. Following a loading test, a sample may not recover its initial permeability.
Furthermore, repeated loading tests will give progressively reduced permeabilities.

These observations suggest that coals exhibit a combination of elasticity and non-

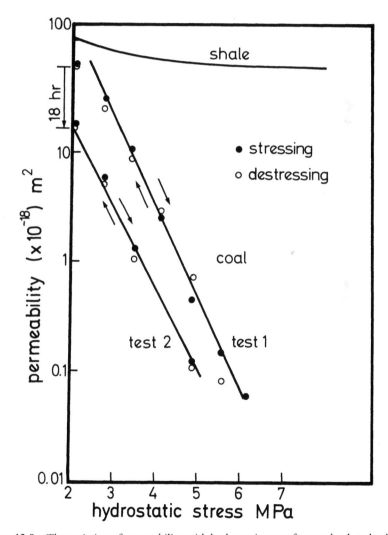

Figure 12.8 The variation of permeability with hydrostatic stress for two load–unload tests
of a bituminous coal. After the completion of test 1, the sample was left lightly loaded for
18 h before commencing test 2.

recoverable strain. The latter is thought to be caused by permanent damage to the weaker bridges between pores.

Figure 12.8 shows the results of hydrostatic loading (axial and radial stresses remaining equal) on the permeability of a bituminous coal. Hydrostatic stressing appears to minimize the hysteresis effect. Two load–unload cycles are shown. After completion of the first test, the sample was left for 18 h in the triaxial cell at a holding stress of slightly more than 2 MPa. The creep reduction in permeability before commencement of the second test shows clearly.

It appears from such tests that the indicated permeability of a coal sample measured during a laboratory test depends on the stress history of the sample; that is, the cycles of loading and unloading caused by mining near the original location of the sample, and the method of recovering that sample. However, for any given laboratory test, the coal permeability falls logarithmically with respect to applied hydrostatic stress, i.e. the relationship is of the form

$$k = A \exp(B\sigma) \quad \text{m}^2 \tag{12.39}$$

where σ = effective stress (MPa) = applied hydrostatic stress − pore (mean gas) pressure, A = constant (m²) and B = constant (1/MPa). The value of A is the theoretical permeability at zero stress and depends on not only the coal structure but also the stress history of the sample. Values of B from -1.5 to $-2.5/\text{MPa}$ have been reported for bituminous coal.

For comparison, Fig. 12.8 includes a similar test on shale. The reduction in permeability with applied stress is much less marked than that for coal.

Effect of gas pressure

For most rocks, the permeability increases as the pore pressure exerted by a saturating gas decreases. The effect was investigated by Klinkenberg (1941) who proposed the relationship

$$k = k_{\text{liq}}\left(1 + \frac{a}{P}\right) \quad \text{m}^2 \tag{12.40}$$

where k_{liq} = permeability of the medium to a single phase liquid, or permeability at a very high gas pressure (m²) and a = constant depending on the gas (Pa). The values of k_{liq} and a may be determined by plotting experimentally determined values of k against $1/P$.

It is thought that the Klinkenberg effect is caused by surface diffusion, i.e. slippage of gas molecules along the internal surfaces of the medium (section 12.2.2). However, in the case of coal, a straight-line relationship is not found when measured values of k are plotted against $1/P$. Hence, there is no longer any advantage in plotting the variables in that way.

Researchers have reported both rising and falling permeabilities with respect to mean gas pressure during laboratory tests on coal. Figure 12.9 illustrates a typical set

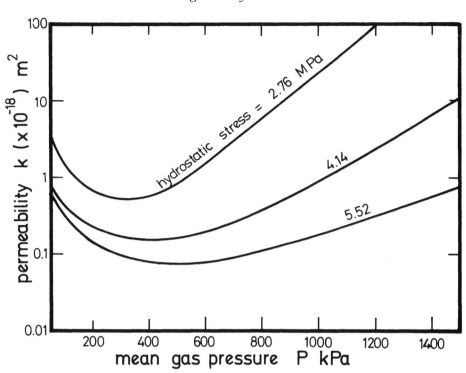

Figure 12.9 In addition to the effects of mechanical stress, the permeability of coal is also a function in the internal gas pressure.

of results and shows the effects of gas pressure and applied hydrostatic stress. Curve-fitting exercises have shown that such curves take the form

$$k = C_1 + \frac{C_2}{P} + \exp(C_3 P - C_4) \quad m^2 \qquad (12.41)$$

$$\underbrace{\qquad\qquad}_{\text{Klinkenberg effect}} \quad \underbrace{\qquad\qquad}_{\substack{\text{dilation of} \\ \text{flowpaths}}}$$

where C_1, C_2, C_3 and C_4 are constants for the curve. It will be observed that $C_1 + C_2/P$ has the form of the Klinkenberg effect and is dominant at low gas pressures. As the gas pressure rises, surface diffusion reduces and, furthermore, adsorbed molecules begin to obstruct the flowpaths. Hence, the permeability falls. However, with further increases in gas pressure, the exponential term in equation (12.41) becomes dominant causing the permeability to rise again. One hypothesis is that a sufficiently high pore pressure results in compression of the coal substance and unconnected pores while the flowpaths become dilated.

Two-phase flow

The pores and fracture networks of strata are often occupied by a mixture of fluids. In petroleum reservoirs, three-phase flow may occur with oil, gas and water as the occupying fluids. In most other cases, including coal, two-phase flow takes place as a mixture of gas and water.

The presence of water greatly inhibits the flow of gas and vice versa and, hence, reduces the permeability of the rock to both phases. The effect is described quantitatively in terms of the **relative permeabilities** $k_{r,g}$ and $k_{r,w}$ for gas and water respectively:

$$k_{r,g} = \frac{k_g}{k_{sg}} \quad \text{(dimensionless)} \tag{12.42}$$

and

$$k_{r,w} = \frac{k_w}{k_{sw}} \quad \text{(dimensionless)} \tag{12.43}$$

where k_g and k_w are the effective permeabilities of the rock to gas and water respectively (m^2) (dependent on degree of saturation), k_{sg} = permeability of the rock to gas when saturated by gas (m^2) (no water present) and k_{sw} = permeability of the rock

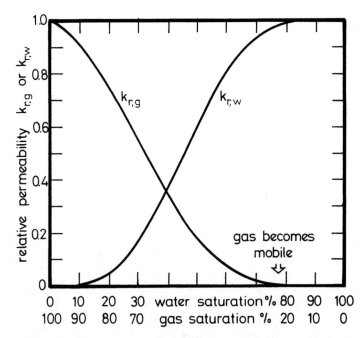

Figure 12.10 The relative permeabilities to gas $k_{r,g}$ and water $k_{r,w}$ depend on the degree of rock saturation by the two fluids.

to water when saturated by water (m^2) (no gas present). In the petroleum industry, it is often assumed that $k_{sg} = k_{sw}$ for the sandstones and limestones that are typical of petroleum reservoir rocks. However, the effects of adsorption and gas pressure indicate that this may not hold for coal.

Figure 12.10 illustrates a typical behaviour of relative permeabilities with respect to saturations of water and gas. The actual loci of the relative permeability curves depend on whether the coal substance is wetted preferentially by the water or the gas. This, in turn, varies with the proportion of coal constituents, vitrain and clarain tending to prefer the gas while durain and fusain are more easily wetted by water.

The curves on Fig. 12.10 suggest a net hydrophobic coal, i.e. the gas is the preferred wetting phase. The water will, therefore, tend to reside in the larger openings within the matrix and inhibit migration of the gas which exists in the smaller interstices. Hence, as indicated on Fig. 12.10, the gas will not become mobile until the water saturation has fallen significantly below 100%. This explains why considerable volumes of water may be produced from a borehole before gas flows appear.

12.4 EMISSION PATTERNS INTO MINE WORKINGS

The rates at which methane are emitted into mine workings vary from near steady state, through cycles that mimic rates of mineral production, to the dangerous phenomena of gas outbursts or 'sudden large emissions'. In general, the rate of emission from any given source depends on

1. initial gas content of the coal,
2. degree of prior degassing by methane drainage or mine workings,
3. method of mining,
4. thickness of the worked seam and the proximity of other seams,
5. the natural permeability of the strata and, in particular the dynamic variations in permeability caused by mining, and
6. comminution of the coal.

12.4.1 Sources of methane in coal mines

Variations in methane emissions into mines are influenced strongly by the dominant sources of the gas. In room and pillar workings, gas will be produced from the ribsides and the pillars of the seam being worked. While exposed pillars may be degassed fairly quickly (dependent upon the coal permeability), ribside gas may continue to be troublesome for considerable periods of time. In such cases, it is preferable for ribsides that border on virgin coal to be ventilated by return air (Fig. 4.7(a)). Peaks of gas emission will, in general, occur at the faces of the rooms because of the high rate of comminution caused by mechanized coal winning. This will be moderated by the degree of earlier degassing, either by methane drainage or by gas migration towards the workings. The latter is enhanced by a high coal permeability and a low rate of advance.

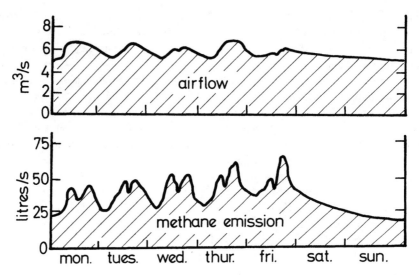

Figure 12.11 Recordings of airflow and methane flow in an airway returning from a longwall panel in a coal mine.

In addition to ribsides, the major sources of methane in longwall mines are the working faces and roof and floor strata. Ribside gas tends to be a slow and near constant source. It is convenient to classify the other sources of methane in a longwall mine into **face** (or **coal front**) **gas** and **gob gas.**

A peak emission occurs at the coal-winning machine owing to sudden fragmentation of the coal. It is this emission that gives rise to frictional ignitions at the pick point (section 12.1). The peak emission moves along the face with the machine. However, the freshly exposed coal front will also emit methane—rapidly at first and decaying with time until the machine passes that point again. Coal front gas is an immediate and direct load on the district ventilation system. Figure 12.11 is an example of a smoothed record of methane concentration in air returning from a longwall face. In this example, coal was produced on two out of three shifts for five days per week. The correlation of methane make with face activity shows clearly, decaying down to a background level over the weekend.

The gas flow into the gob area behind a longwall face originates from any roof or floor coal that has not been mined from the worked seam, but more particularly from source beds in the roof or floor strata. Any coal seams or carbonaceous bands within a range of some 200 m above to 100 m below the working horizon are liable to release methane that will migrate through the relaxed strata into the gob area. If methane drainage is not practiced, then that methane will subsequently be emitted into the mine ventilation system. Figure 12.12 illustrates the variations in stress-induced permeability of roof and floor strata that create enhanced migration paths for the gas.

As fragmented coal is transported out of the mine, it will continue to emit

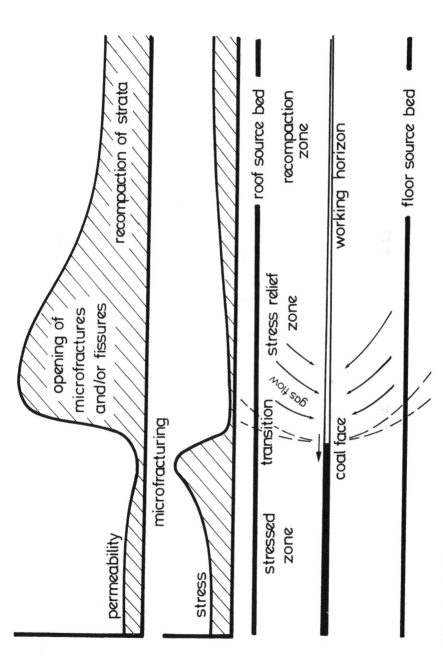

Figure 12.12 The migration of gas from roof and floor source beds towards the working horizon occurs because of the large increase in permeability of adjacent strata in the stress relief zone immediately behind a longwall face.

methane. The gas make depends on the degree of fragmentation, gas content on leaving the face, the tonnage and the time spent during transportation. The desorption equations given in section 12.2.2 allow estimates of the transport gas to be made. Precautions should continue to be taken against accumulations of methane after the coal has left the mine. Gas explosions have occurred in surface hoppers and in the holds of coal transport ships.

12.4.2 Methane layering

Methane emitted from the strata into a mine opening will often be at concentrations in excess of 90%. While being diluted down to safe general body concentrations, the methane will, inevitably, pass through the 5% to 15% range during which time it is explosive. It is, therefore, important that the time and space in which the explosive mixture exists are kept as small as possible. This can be achieved by good mixing of the methane and air at the points of emission. Unfortunately, the buoyancy of methane with respect to air (specific gravity 0.554) produces a tendency for concentrated methane to collect in roof cavities and to layer along the roofs of airways or working faces.

In level and ascentionally ventilated airways, as illustrated on Fig. 12.13(a), the layer will stream along the roof in the direction of airflow, increasing in thickness and decreasing in concentration as it proceeds. Multiple feeders of gas will, of course, tend to maintain the concentration at a high level close to the roof.

There are two main hazards associated with methane layers. First, they extend greatly the zones within which ignitions of the gas can occur. Secondly, when such an ignition has taken place, a methane layer acts very effectively as a fuse along which the flame can propagate—perhaps leading to much larger accumulations in roof cavities or gob areas.

Figure 12.13(b) indicates that, in a descentionally ventilated airway, the buoyant methane layer may stream uphill close to the roof and against the direction of the airflow. However, at the fringe between gas and air, viscous drag and eddy action will cause the gas–air mixture to turn in the same direction as the airflow. The result is that explosive mixtures may be drawn down into the airway upstream from points of emission.

Although the layering phenomena of methane in mines was all too obvious during the Industrial Revolution (section 1.2), systematic research on the topic seems not to have been well organized until the 1930s (Coward, 1937). A combination of analytical and experimental work at the Safety in Mines Research Establishment, England, in the early 1960s led to a quantification of the important parameters (Bakke and Leach, 1962). These were

1. velocity of the ventilating airstream, u(m/s),
2. rate of gas emission, Q_g (m^3/s),
3. width of airway W (m),

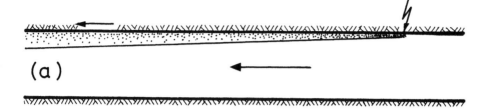

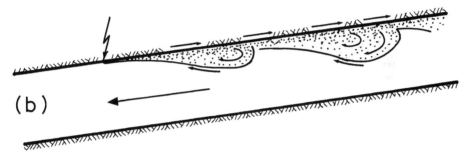

Figure 12.13 Methane layering in (a) a level airway and (b) a descentionally ventilated airway.

4. inclination of airway,
5. relative densities of the air and gas, and
6. roughness of the roof above the layer.

Although a rough lining will promote better mixing than a smooth one (except for free-streaming layers), the effect of roughness is fairly weak (Raine, 1960). Bakke and Leach found that the characteristic behaviour of a gas layer was proportional to the dimensionless group

$$\frac{u}{\left(g\,\dfrac{\Delta\rho}{\rho}\,\dfrac{Q_{\mathrm g}}{W}\right)^{1/3}} \tag{12.44}$$

where $\Delta\rho/\rho$ is the difference in relative densities of the two gases ($1 - 0.554 = 0.446$ for air and methane). Using a value of $g = 9.81$ m/s^2 gives the dimensionless number of methane layers in air to be

$$L = \frac{u}{(9.81 \times 0.446 \times Q_{\mathrm g}/W)^{1/3}}$$

or

$$L = \frac{u}{1.64}\left(\frac{W}{Q_g}\right)^{1/3} \tag{12.45}$$

The dimensionless group, L, is known as the **layering number** and is of fundamental significance in the behaviour of methane layers. Examination of equation (12.45) indicates that the air velocity is the most sensitive parameter in governing the layering number and, hence, the length and mixing characteristics of the layer. Although u is, theoretically, the air velocity immediately under the layer, the mean value in the upper third of the airway may be used. For non-rectangular airways, W may be taken as some three-quarters of the roadway width. The power of 1/3 in equation (12.45) reduces the effects of errors in estimated values of W and Q_g.

Experimental data from Bakke and Leach have been employed to produce Fig. 12.14 for level airways. For any given gas make and roadway width, the horizontal axis may be scaled in terms of air velocity. It can be seen from this graph that at low

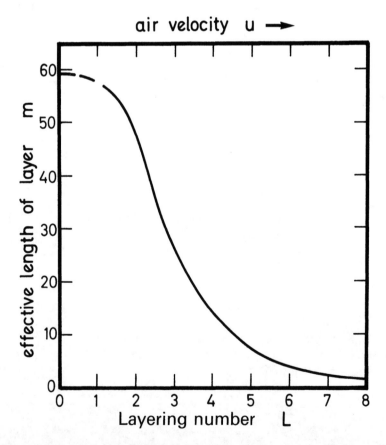

Figure 12.14 Variation of layer length with layering number for level

Table.12.1 Recommended minimum layering numbers at various roadway inclinations

Angle to horizontal (deg)	0	5	10	15	20	25	30	35	40	45
Ascentional	5	5.7	6.2	6.6	6.9	7.2	7.4	7.6	7.7	7.8
Descentional		3.3	3.7	4.1	4.3	4.5	4.6	4.7	4.8	4.9

layering numbers (and, hence, low velocity) the layer is affected very little by small additional increases in velocity. Indeed, in an ascentional airway, the layer will actually lengthen as the airspeed increases from zero to the free-streaming velocity of the methane. The mixing process is primarily due to turbulent eddies. The efficiency of mixing increases as the relative velocity between the methane and air rises. However, as the air velocity continues to increase giving layering numbers of over 1.5, then the layer shortens rapidly. On the basis of such results, it can be recommended that layering numbers should be not less than 5 in level airways.

Similar experiments in inclined airways have led to the recommended minimum layering numbers shown in Table 12.1 in order to inhibit the formation of methane layers.

Methane layers can be detected by taking methanometer readings or siting monitors at roof level. The most probable locations are in bleeder airways or return roadways close to a longwall face. On detecting a methane layer, an immediate temporary remedy is to erect a **hurdle cloth**, that is, a brattice cloth attached to the sides and floor but leaving a gap at the top. The size of the opening should be such that the increased air velocity near the roof disperses the layer. The hurdle cloth may need to extend some three-quarters of the height of the airway. However, anemometer readings should be taken to ensure that the overall volume flow of air through the area is not reduced significantly. Compressed air venturis or other forms of air movers may also be employed to disperse methane layers provided that they are earthed against electrostatic sparking.

The longer-term solutions are (a) to increase the airflow and, hence, the air velocities through the affected panel, (b) to reduce the rate of methane emission, or a combination of the two. Ventilation network analyses should be employed to investigate means of increasing the airflow (Chapter 7). These may include adjusting the settings of regulators or fans, and controlled partial recirculation (section 4.5). The methane emissions can most effectively be reduced by installing a system of methane drainage (section 12.5).

While methane layering was once a common occurrence in gassy coal mines, well-designed systems of ventilation and gas drainage are capable of eliminating this hazard from modern mines.

12.4.3 Gas outbursts

The most dramatic mode of gas emission into mine workings is the release of an abnormally large volume of gas from the strata in a short period of time. Such incidents have caused considerable loss of life. In many cases, the rate of emission has been explosive in its violence, fracturing the strata and ejecting large quantities of solid material into the workings. Gas outbursts are quite different from rock bursts that are caused by high strata loadings. However, the probability of disruptive gas release from adjacent strata is enhanced in areas of abnormally high stress such as a pillar edge in overlaying or underlying workings.

There are two distinct types of gas outbursts, **in-seam** bursts and sudden large emissions from **roof and floor**. Each of these will be discussed in turn.

In-seam outbursts

As the name implies, these are outbursts of gas and solids from the seam that is currently being mined. They have occurred in many countries, particularly in coal and salt (or potash) mines. The geologic conditions that lead to in-seam outbursts appear to be quite varied. The one common feature is the existence of mechanically weakened pockets of mineral within the seam and which also contain gas at high pressure. Methane, carbon dioxide and mixtures of the two have been reported from in-seam outbursts in coal mines while nitrogen may be the major component in potash mines (Robinson *et al.*, 1981).

The genesis and mechanisms of in-seam gas outbursts have been a matter of some controversy. A current hypothesis is that the structure of the material within outburst pockets has been altered by tectonic stressing over geological time. Seam outbursts in coal mines have been known to eject up to 5000 t of dust, much of it composed of particles less than 10 μm in diameter (Evans and Brown, 1973). This can be accompanied by several hundred thousand cubic metres of gas (Campoli *et al.*, 1985). It is thought that the pulverization of the coal has been caused by very high shear forces. Laboratory tests have shown that a sample of normal anthracite can be reduced to the fragile 'outburst' anthracite by such means. Coal mines located in areas that have been subject to thrust faulting are more prone to in-seam outbursts. A sample of 'outburst' coal is, typically, severely slickensided and friable to the extent that it may crumble to dust particles by squeezing it in the hand. The seam is often heavily contorted in the vicinity of the outburst pocket—again, evidence of the excessive tectonic stressing to which it has been subjected.

If the overlying caprock had maintained a low permeability during and since the pulverization of the coal, then the entrained methane will remain within the zone. However, the small size of the particles ensures that desorption of the gas can occur very quickly if the pressure is relieved (equation (12.9)).

When a heading or face approaches an outburst zone that contains highly compressed coal dust and gas, stress increases on the narrowing barrier of normal coal that lies between the free face and the hidden outburst pocket (Sheng and Otuonye, 1988). At some critical stage, the barrier will begin to fracture. This causes audible noise that has been variously described as 'cracking, popping' or 'like a two-stroke engine'. When the force exerted by the gas pressure in the outburst pocket exceeds the resistance of the failing barrier, the coal front bursts outwards explosively. The blast of expanding gas may initiate shock waves throughout the ventilation system of the mine. A wave of decompression also passes back through the pulverized coal. The expansion of the gas, reinforced by rapid desorption of large volumes of methane, expels the mass of dust into the airway. Although the outburst may last only a few seconds, the desorbing gas causes the dust to behave as a fluidized bed. This may flow down the airway, almost filling it and engulfing equipment and personnel for a distance that can exceed 100 m (Williams and Morris, 1972). The cavities that remain in the seam following an outburst may, themselves, indicate slickensides. Several cavities that become interconnected can contribute to a single outburst. The volume of dust subsequently removed from the airway or face has often appeared too large for the cavity from which it was apparently expelled.

The dangers associated with gas outbursts in mines are, first, the asphyxiation of miners by both gas and dust. Compressed air 'lifelines' may be maintained on, or close to, faces that are prone to in-seam outbursts. These are racks of flexible tubes connected to a compressed air pipe, and which open valves automatically when picked up. Such devices can save the lives of miners who become engulfed in outburst dust. A second hazard is that the violence of the outburst may damage equipment and cause sparking that can ignite the highly inflammable gas–dust–air mixture. Spontaneous ignitions of methane during outbursts in a salt mine have also been reported (Schatzel and Dunsbier, 1989). Thirdly, the sudden expansion of a large volume of gas can cause disruption of the ventilation system of a mine.

Precautionary measures against in-seam gas outbursts include forward drilling of exploratory boreholes. However, pre-drainage of outburst pockets has met with very limited success owing to rapid blockage of the boreholes. Similarly, testing samples of coal for strength or structure is uncertain as normal coal may exist very close to an outburst pocket. Monitoring for unusual microseismic activity is a preferred warning technique (Campoli *et al.*, 1985). When it is suspected that a face or heading is approaching an outburst pocket, then machine mining should be replaced by drill and blast methods (volley firing), clearing the mine of personnel before each blast. This should be continued until an outburst is induced or the dangerous area has been mined through.

Outbursts from roof and floor

These are most likely to occur in longwall mines. As discussed in section 12.4.1 and illustrated on Fig. 12.12, methane will migrate from higher or lower source beds towards the working horizon behind the working face. Strata in the stress relief zone normally exhibit a substantial increase in permeability owing to relaxation of the fracture network. The migration of gas then proceeds through the overlying and underlying strata at a rate that follows cyclic face operations and, under normal conditions, is quite controlled. However, a band of strong and low permeability rock (**caprock**) existing in the strata sequence between the source beds and the working horizon may inhibit the passage of gas. This can result in a reservoir of pressurized gas accumulating beyond the caprock. Any sudden failure of this retention band will then produce a large and rapid inundation of gas into the working horizon (Wolstenholme *et al.*, 1969).

Figure 12.15 illustrates the development of a potential outburst from the floor. In Fig. 12.15(a), relaxation of the strata allows gas to be evolved from the source bed and migrate upwards. An intervening bed of low natural permeability contains induced fractures that allow the gas to pass through. However, Fig. 12.15(b) illustrates a condition in which the low permeability bed has failed to fracture. Gas accumulates under this bed which is then subjected to an intensified gas pressure gardient. This is the potential outburst situation. The rate at which gas accumulates in the reservoir rock depends on the relative rates of gas make from the source seam and gas leakage through the caprock. Either an increase in the gas make from the source seam or a decrease in flow through the caprock can produce the potential outburst condition.

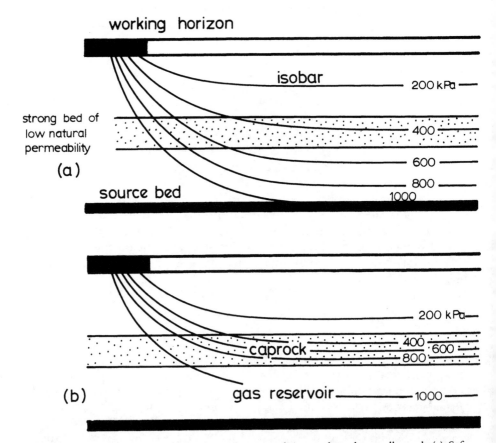

Figure 12.15 Development of a gas outburst condition under a longwall panel. (a) Safe condition: isobars distributed; strong bed sufficiently permeable to allow gas to migrate through in a controlled manner. (b) Potential outburst condition: increased pressure gradient across the low permeability bed with high pressure gas accumulations under this bed.

Resistance to gas flow through the fracture network of the caprock may occur by increases in its strength or thickness. Furthermore, a porous bed such as a fine-grained sandstone may suffer a large decrease in relative permeability to gas if it becomes partially saturated by water (Fig. 12.10).

There are considerable differences between the sudden large emissions that occur from roof and floor strata. Strata above the worked seam are subjected to larger vertical movement because of subsidence of the rock mass. This creates greater voidage in which gas can accumulate. Bed separation also promotes flowpaths parallel to the strata. Very large gas reservoirs may, therefore, develop in the overlying rocks. However, fracturing of low permeability beds is more probable and high pressures are less likely to develop in the gas reservoir rocks. For these reasons, sudden emissions from the roof have usually been characterized by large flows that may last from a few hours to several months (Morris, 1974). However, the initiation of the emission is unlikely to be accompanied by violent dislocation of the immediate roof strata.

In contrast, gas outbursts from the floor are usually of shorter duration but more violent—even to the extent of rupturing the floor strata upwards with ejection of solid material. The reduced deformation in the floor sequence enables a caprock to resist induced enlargement of fractures. This leads to high pressure accumulations of gas. However, the extent of the gas reservoir is likely to be much less than one in the roof strata. Morris (1974) reported methane emissions as large as $140\,000\,\mathrm{m}^3$ from floor outbursts, but over $8 \times 10^6\,\mathrm{m}^3$ from sudden roof emissions.

Gas outbursts from roof or floor are often preceded by smaller increases in general body gas concentrations, often intermittent in nature, during the hours before the major flow occurs. This is probably caused by increasing strain within the caprocks and interconnection of bed separation voidages before the main failure. Continuous gas monitors can detect such warnings. Immediately before the burst, severe weighting on the roof supports and caving in the gob may occur. Roof and floor outbursts are more likely to occur along the planes of maximum shear stress in the strata, i.e. in bleeders or airways bordering gob areas, or on the longwall face itself. Roof outbursts may occur because of the first failure of an overlying caprock within some two face lengths of the initial starting-off line of the longwall. Similarly, passing under or over old pillar edges in other seams may promote sudden failure of caprocks. However, previous workings may have partially degassed source beds. Most floor outbursts have occurred when no previous mining has taken place at a lower level.

The most effective means of preventing roof and floor outbursts is regular drilling of methane drainage holes wherever any potential caprock may exist. The holes should be angled over or under the caved area and should penetrate beyond the caprock to the source bed(s). The spacing between holes should be dictated by local conditions but may be as little as 10 m. It is particularly important that the drilling pattern be maintained close up to the face. All holes should be connected into a methane drainage pipework system (section 12.5.5). In outburst-prone areas, it is essential that drilling takes place through a stuffing box in order that the flow can be diverted immediately into the pipe range should a high pressure accumulation be penetrated.

It is often the case that routine floor boreholes produce very little gas. However, in areas liable to floor outbursts, these holes should continue to be drilled as a precautionary measure. Tests should be made of open-hole flow rates and rate of pressure build-up on the closed hole. When both of these tests show higher than normal values, then it is probable that an outburst condition is developing (Oldroyd *et al.*, 1971). In the event of any indication that a gas reservoir is accumulating in roof or floor strata, then additional methane drainage holes should be drilled immediately to relieve that pressure and to capture the gas into the drainage pipe system.

12.5 METHANE DRAINAGE

The organized extraction of methane from carboniferous strata may be practiced in order (a) to produce a gaseous fuel, (b) to reduce methane emissions into mine workings or (c) to carry out a combination of the two. If the intent is to provide a fuel for sale or local consumption, then it is important that the drained gas remain within prescribed ranges of purity and flowrates.

There is no single preferred technique of methane drainage. The major parameters that influence the choice of method include

1. the natural or induced permeability of the source seam(s) and associated strata,
2. the reason for draining the gas, and
3. the method of mining (if any).

In this section, we shall outline current methods of methane drainage, the infrastructure of pipe ranges and ancillary equipment, methods of predicting gas flows and conclude with a summary of the procedure for planning a methane drainage system.

12.5.1 In-seam drainage

Drainage of methane by means of boreholes drilled into a coal seam is successful only if the coal has a sufficiently high natural permeability or a fracture network can be induced in the seam by artificial methods. Hence, for example, while in-seam drainage can be practiced in some North American coalfields, it has met with very limited success in the low permeability coals of the United Kingdom or Western Europe. A knowledge of coalbed permeability is necessary before in-seam drainage can be contemplated.

The advances made in drilling technology have increased the performance potential of in-seam gas drainage (Schwoebel, 1987). Using down-the-hole motors and steering mechanisms, boreholes may be drilled to lengths of 1000 m within the seam. Provided that the coal permeability is sufficiently high, methane flows into mine workings can be reduced very significantly by pre-draining the seam to be worked. Gas capture efficiencies up to 50% are not uncommon, where

$$\text{gas capture efficiency} = \frac{\text{gas captured by methane drainage} \times 100\%}{\text{gas captured} + \text{gas emitted into ventilation}} \quad (12.46)$$

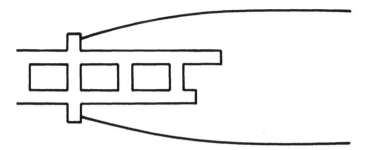

Figure 12.16 In-seam drainage boreholes to reduce methane flow into advancing headings.

Figure 12.16 illustrates flanking boreholes used to drain gas from the coal ahead of headings that are advancing into a virgin area. In gassy and permeable seams, ribsides bordering on solid coal are prolific sources of gas. On the other hand, previously driven headings or workings may have degassed the area to a very considerable extent.

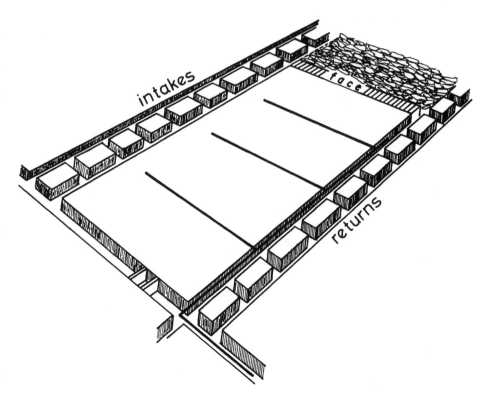

Figure 12.17 Horizontal boreholes draining methane from a coal seam in a two-entry retreating longwall.

In-seam gas drainage can also be effective in permeable seams that are worked by the retreating longwall systems (Mills and Stevenson, 1989; Ely and Bethard, 1989). Figure 12.17 illustrates the layout. Boreholes are drilled into the seam from the return airway and connect into the methane drainage pipe system. The preferred spacing of the holes depends on the permeability of the seam and may vary from 10 to over 80 m. The distance from the end of each borehole and the opposite airway should be about half the spacing between holes. The application of suction on the boreholes is often unnecessary but may be required for coals of marginal permeability or to increase the zone of influence of each borehole.

The time allowed for drainage should be at least six months and, preferably, over one year. Hence, the holes should be drilled during the development of what will become the tailgate of the longwall.

Spalling of coal into the borehole can be a problem, especially in the more friable coals. This may be reduced by employing smooth drill rods. Drill chippings are removed by means of a continuous water flush. Additionally, augers may be used to remove spalled coal from the boreholes. The internal diameter of some sections of finished borehole may be considerably larger than the drill bit. Perforated plastic liners can be inserted in order to maintain the holes open, subject to the governing legislation. The first 5 to 10 m of each borehole are drilled at, typically, 100 mm diameter. A standpipe is cemented into place and connected through a stuffing box into the methane drainage pipeline (section 12.5.5). The remainder of the hole is drilled through the standpipe at a diameter of some 75 mm.

The flowrate of gas from a gas drainage borehole will vary with time. Figure 12.18 illustrates a typical life cycle for an in-seam borehole. A high initial flow occurs from the expansion and desorption of gas in the immediate vicinity of the hole. This may diminish fairly rapidly but then increase again as the zone of influence in dewatered, hence, increasing the relative permeability of the coal to gas (Fig. 12.10). This, in turn, is followed by a decay as the zone of influence is depleted of gas.

In-seam boreholes drilled into outcrops or from surface mines may produce little methane. The gas content of coal seams tends to increase with depth. It is probable that seams near the surface have lost most of their mobile gas. However, multiple boreholes drilled with directional control from a surface drill rig can be diverted to follow a coal seam for distances that produce acceptable flowrates.

Vertical holes drilled from the surface to intersect coal seams are likely to produce very little gas because of the short length of hole exposed to any given seam. However, **hydraulic stimulation** or **hydrofracturing** may be used to enhance the flowrates. This involves injecting water or foam containing sand particles into the seam. Other sections of the borehole are cased. The objective is to dilate the fracture network of the seam by hydraulic pressure. The sand particles are intended to maintain the flowpaths open when injection ceases. The success of hydrofracturing depends on the natural fracture network that exists within the seam and the absence of clays that swell when wetted. Friable coals are more likely to respond well to hydrofracturing. On the other hand, in stronger coals, the induced fractures may be concentrated along discrete bedding planes and give a poor recovery of gas. In the

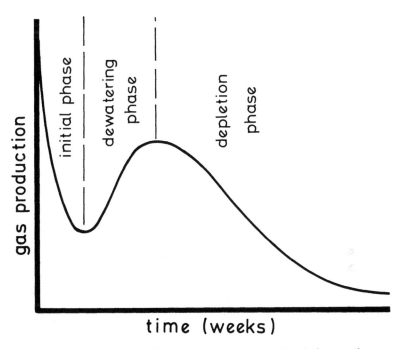

Figure 12.18 Typical life cycle of a gas drainage borehole in coal.

majority of cases, water must be pumped from surface boreholes before gas flows can be attained.

Another means of in-seam gas drainage is to sink a small diameter shaft to intersect the coal bed (US Bureau of Mines, 1980). Long multiple boreholes are drilled radially outwards from the shaft into the seam and connected into a methane drainage line that rises to the surface. This method may be considered when a shaft is to be sunk at a later date at that location for mine ventilation or service access. This enables degassing of an area for several years prior to mining.

12.5.2 Gob drainage by surface boreholes

The relaxation of strata above and below the caved zone in a longwall panel creates voidage within which methane can accumulate at high concentration, particularly when other coal beds exist within that strata. If this gas is not removed, then it will migrate towards the working horizon and become a load on the ventilation system of the mine (Fig. 12.12). Capture of this 'gob gas' may be accomplished either by underground by cross-measures drainage (section 12.5.3) or by drilling boreholes from the surface.

Figure 12.19 depicts methane drainage from the gob of a longwall panel by surface boreholes. This is a method that is favoured in the United States. Typically, three or four holes are drilled from surface rigs at intervals of 500 to 600 m along the

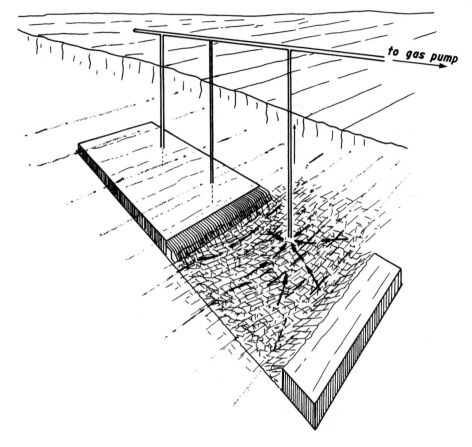

to gas pump

Figure 12.19 Gob drainage of a longwall panel. Each hole becomes productive after passage of the faceline.

centre-line of the panel and ahead of the coal face. The holes may be 200 to 250 mm in diameter and drilled to within some 8 to 10 m of the top of the coal seam (Mills and Stevenson, 1989). The holes should be cased from the surface to a depth that is dictated by the local geology and, in particular, to extend below any beds that are likely to act as bridging caprocks. A perforated liner can be employed in the rest of the hole to inhibit closure from lateral shear.

The initial gas made from the surface holes is likely to be small. However, as the face passes under each borehole, the methane that accumulates in the caved area will be drawn towards that borehole. Bed separation assists drainage from across the complete gob area. The first borehole should be located far enough from the face start line (typically about 150 m) to ensure that it connects into the caved zone. When a hole becomes active, the rate of gas production increases sharply and may yield over 50 000 m³/day of commercial quality methane for a period of several months, depending on the rate of mining. The more gradual decay is a result of reconsolidation of the caved strata. Gas drainage pumps located on surface ensure that the gas flow remains in the correct direction and may be employed to control both the rate of flow and gas purity. If the applied suction is too great, then ventilating air will be drawn into the gob and may cause excessive dilution of the drained methane.

In addition to longwall panels, gob drainage by surface boreholes can be utilized in pillar extraction areas. In both cases, the technique can result in very significant reductions in emissions of methane into mine workings.

12.5.3   Cross-measures methane drainage

Where the depth of coal workings mitigates against the drilling of methane drainage holes from surface, the extraction of methane from relaxed strata can be accomplished by drilling from underground airways. The principle of cross-measures drainage is illustrated on Fig. 12.20. Capture efficiencies in the range 20% to 70% have been reported for this technique, depending on the design and control of the system. Boreholes are drilled into the roof and, if necessary, also the floor strata. The holes are normally drilled parallel to the plane of the coal face but inclined over or under the waste. This is the dominant method of methane drainage in Europe and is particularly applicable to advancing longwall panels. However, the technique can also be employed for multiple-entry retreat systems.

The angle, length and spacing of boreholes must be decided on a site-specific basis. In general, the holes should intersect the major gas-emitting horizons. The spacing between holes should be such that their zones of influence overlap slightly. Significant increases in general body methane concentration along the airway is indicative of too great a distance between boreholes. Conversely, if closing a borehole causes a rapid increase in the flowrate from neighbouring holes, then the boreholes may be too close together. The hardness of the strata or other difficulties of drilling will also influence the spacing. Typical spacings lie in the range 10 to 25 m between boreholes.

Computer modelling procedures have been developed to assist in planning an optimum layout of cross-measures boreholes (e.g. O'Shaughnessy, 1980). However, a practical study of the local geology, followed by a period of experimentation, remains the most pragmatic means of establishing a successful drilling pattern.

The boreholes are normally drilled from a return airway. However, in particularly gassy situations, boreholes may be required on both sides of the panel. Although cross-measures methane drainage is usually carried out from the current working horizon, a variation is to drill from airways that exist in either overlying or underlying strata. Dewatering is facilitated by upward holes.

Routine underground drilling necessitates boreholes of smaller diameter than surface holes. Cross-measures drainage holes are normally in the range 50 to 100 mm diameter. However, it is necessary to employ standpipes in order to prevent excessive amounts of air being drawn into the system. For any given degree of suction applied to a borehole, the purity of the extracted gas will vary with the length of the standpipe. Hence, to ensure a high concentration, the standpipe may need to be

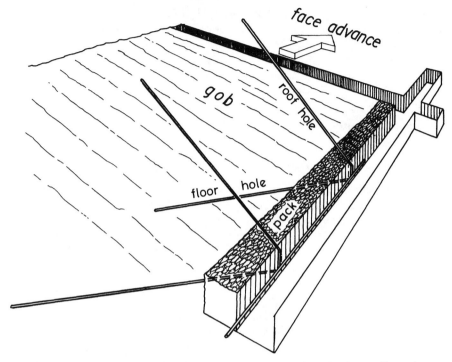

Figure 12.20   Cross-measures methane drainage in an advancing longwall panel.

extended to within a few metres of the main gas-producing horizon. Standpipes may be grouted into place with cement or resin, while rubber and sealing compounds have also been used (Boxho *et al.*, 1980).

The suction applied to gob drainage holes, whether drilled from the surface or underground, is a sensitive means of controlling both the flowrate and the purity of the drained gas. The suction to be applied on each individual borehole is a function of the gas pump duties, the geometry of the network of pipes and settings of borehole valves. In general, increasing the suction applied to an active borehole will increase the flow but decrease the purity. Borehole tests may be employed to determine the optimum suction when the mass flow of methane attains a maximum value. If the suction is too high, then the concentration of methane in the pipelines may fall below 15% when it becomes explosive. It is prudent to decrease the suction automatically when the concentration falls to 30%. Much higher cut-off concentrations may be imposed by local regulations or where the gas is utilized as a fuel. If the applied suction is too low, then excessive emissions of methane can occur into the mine ventilation system. The optimum suction required increases with respect to the rate of gas make from the source beds. A pragmatic approach is to ensure that the interface between the air and methane remains in the strata above the level of the airways. Although a borehole suction of some 10 kPa is fairly typical of a cross-measures drainage system, large variations occur in practice. Indeed, as indicated in section 12.4.3, gas may exit the borehole at a positive pressure.

12.5.4   Drainage from worked-out areas

Although the rate of gas emission from any given source decays with time, an abandoned area of a coal mine is likely to be surrounded by a large envelope of fragmented and relaxed strata. The total make of gas may be considerable and continue for an extended period. In cases where fragmentation has occurred across coal beds and between multiple seam workings, such emissions of gas may go on for several years. In order to reduce the pollution of a current ventilation system, methane drainage pipes can be inserted through stoppings and gas drained from the old workings. The pipes should extend beyond the stoppings to a point that is not affected by air penetration during a period of rising barometer. The objective is to extract the gas at approximately the same rate as it is produced. However, additional suction may be applied when the barometer is falling. Pressure balance chambers assist in reducing leakage across stoppings (section 21.5.5).

Gas drainage can also be practiced when a gassy mine is abandoned. Again, pipes are left in the seals on shafts or adits. This type of drainage may prevent seepages of methane escaping to the surface and, possibly, creating a hazard in the basements of buildings. Drainage pipes that are vented to the surface atmosphere should be elevated above ground level, well fenced against unauthorised approach and provided with notices giving warning of inflammable gas.

12.5.5   Components of a methane drainage system

The methane that is produced from drainage boreholes must be transported safely to the required point of delivery at surface level. The infrastructure that is necessary comprises five sets of components:

1. pipe ranges
2. monitors
3. safety devices
4. controls
5. extractor pumps (if required)

Pipe ranges

The standpipe of each borehole connects into a **branch pipe** via a flexible hose, a valve and a water trap. In the case of methane drainage in an underground mine, the branch pipes from each district or panel connect into **main** or **trunk** gas lines. Each main leads to the base of a vertical pipe that carries gas to the surface. Devices for monitoring, safety and control are located at strategic positions throughout the system.

Both steel and high density polyethylene have been used as pipe materials for methane drainage pipes. Steel has the advantage of mechanical strength. However, it must be galvanized or otherwise protected against corrosion on both the inner and outer surfaces. Where permitted by legislation, heavy duty plastic piping may be preferred because of its lighter weight and ease of installation (Thakur and Dahl, 1982). Polyethylene tubing up to some 75 mm in diameter may be brought into the mine on reels, eliminating most of the joints that are required every 5 or 6 m with steel pipes. Pipes are sized according to the gas flows expected and vary from 75 mm to 600 mm in diameter. All pipes should be colour coded for easy recognition and in correspondence to the relevant mine plan.

Pipes that are installed in underground airways should be suspended from the roof to reduce the impact of falling objects. Where steel pipes are employed, then the joints should be capable of flexing in order to accommodate airway convergence or other strata movements. Methane drainage pipes should not be located in any airway where electrical sparking might occur such as a trolley locomotive road. In any event, pipes should be earthed against a build-up of electrostatic charges. Following the installation of a new length of pipe, it should be pressure tested either by suction or by compressed air at a pressure of some 5000 kPa. All pipework should be inspected regularly for corrosion and damage.

Pipes should be installed with a uniform gradient and water traps installed at unavoidable low points including the base of the vertical surface connecting branch. Steel is preferred for the main vertical pipe. This may be located within an existing shaft. It is prudent to choose an upcast shaft and this may, indeed, be enforced by law. Shaft pipes should be fitted with telescopic joints to accommodate expansion and contraction. However, an alternative is to position the vertical pipe in a dedicated surface-connecting borehole. This method is attractive for the shallower mines and is the common technique within the United States. Surface pipes should be protected against frost damage, if necessary, either by thermal insulation or by heated cladding.

Monitors

There are, essentially, three types of parameters that should be monitored in a gas drainage system; pressure, flow and gas concentration. Pressure measurements are made at the mouths of boreholes, across orifice plates to indicate flow and at strategic locations within the pipe network. For boreholes that penetrate relaxed strata (Sections 12.5.2 and 12.5.3) there may be a sensitive relationship between borehole pressure, gas flowrate and concentration of the drained gas, particularly where extractor pumps are used to maintain a negative gauge pressure at the mouth of each borehole. Pressure differential indicators may vary from simple U-tubes (section 2.2.4) containing water or mercury to sophisticated capsule or diaphragm gauges (Boxho *et al.*, 1980). The latter may be capable of transmitting signals to remote recorders, alarms or control centres. A simpler arrangement employs mechanical regulators to maintain set pressures at each chosen location.

A variety of devices are used to indicate the rate of flow in gas pipes, including both swinging vane and rotating vane anemometers, and pitot tubes. However, the most popular method is to monitor the pressure drop across an orifice plate (Fig. 5A.8).

The gas flowrate, Q (m^3/s at pipe pressure), and the pressure drop across the orifice, p, are related as shown by equation (2.50), i.e.

$$p = R_t \rho Q^2 \quad \text{(Pa)}$$

where R_t = rational turbulent resistance of the orifice (m^{-4}) and ρ = density of the gas (kg/m^3). Furthermore, the rational turbulent resistance is given by equation (5.18) as

$$R_t = \frac{X}{2A^2}$$

where X = shock loss factor for the orifice (dimensionless) and A = cross-sectional area of the pipe (m^2). Combining these two equations gives

$$Q = \frac{A}{\sqrt{X}} \sqrt{\frac{2p}{\rho}} \quad \text{m}^3/\text{s} \qquad (12.47)$$

where the pressure drop, p, is measured in pascals. An estimate for the value of X may be made from Fig. 5A.8 (appendix to Chapter 5) for a clean sharp-edged orifice plate, or from the corresponding equations:

$$X = \frac{1}{C_c^2}\left[\left(\frac{D}{d}\right)^4 - 1\right] \qquad (12.48)$$

where D = diameter of pipe, d = diameter of orifice and the orifice coefficient C_c is given by

$$C_c = 0.48\,(d/D)^{4.25} + 0.6 \qquad (12.49)$$

For greater precision, further multiplying factors may be incorporated into equation (12.47) to account for variations in Reynolds' number, gas viscosity and, in particular, the roughness of the edge of the orifice.

Spot measurements of gas concentrations in methane drainage networks may be made by inserting probes through ports into the gas line and using hand-held methanometers. Permanent transducers are used to transmit signals of methane concentration at fixed locations to a remote monitoring or control centre.

While the majority of methanometers described in section 11.4.2 are intended for measurements made in the subsurface airflow system, those employed in methane drainage pipes must be capable of indicating gas concentrations up to 100%. The most popular instruments used for this purpose are based on the variation of thermal conductivity with respect to gas concentration. Interferometers and acoustic methanometers are also employed while infrared gas analysers may be used at permanent monitoring stations (section 11.4.2).

Controls and safety devices

The primary means of controlling an installed methane drainage system is by valves located in all main and branch lines as well as at the mouth of each borehole. Manually operated isolating valves are usually of the gate type in order to minimize pressure losses when fully open. These are employed to facilitate extension of the system and pressure tests on specific lengths of pipeline.

Diaphragm or other types of **activated valves** may be employed to react to control or alarm signals. These might be used to vary the gas pressure at the mouth of each borehole in response to changes in pipeline gas concentration. Abnormally high (or rapidly increasing) signals from a general body methane monitor located in an airway that carries a gas pipeline may be used to cut off gas flow into that branch. A simple but effective technique that is popular in the United States is to attach a length of mechanically weak plastic tubing along the upper surface of each gas pipe. The tubing is pressurized by compressed air or nitrogen which also holds open one or more mechanically activated values. Any fall of roof or other incident that may damage the pipe is likely to fracture the monitoring tube. The resulting loss of air or nitrogen pressure causes immediate closure of the corresponding valves.

The gas mixture that flows through methane drainage pipes is usually saturated with water vapour. Liquid water is introduced into the system from the strata, by condensation and, perhaps, following the connection of a borehole that has been drilled using a water flush. In order not to impede the flow of gas, it is necessary to locate a **water trap** in the connection between each borehole and the branch pipe, and also at all low points throughout the network. Water traps vary from simple U-tube arrangements to automated devices. The latter consist essentially of a compressed air or electric pump, activated by a float within a water reservoir. The size of the reservoir and the settings of the high and low water levels are chosen according to the make of water expected.

It is prudent to locate a lightening conductor close to any surface point of gas discharge. Furthermore, **flame traps** should be inserted into the surface pipeline to prevent any ignition from propagating into the subsurface system. Flame traps are devices that allow rapid absorption or dissipation of heat without introducing an unacceptable pressure drop. Alternative designs include closely spaced parallel plates, corrugated metal sheets or a series of wire gauze discs. A device that is popular in Germany consists of a vessel containing glass spheres (Boxho *et al.*, 1980). Flame traps are liable to be affected by dust deposits and should be designed for easy inspection and maintenance. It is useful to locate a pressure differential transducer across the flame trap in order to monitor for an unduly high resistance. If burning continues for any length of time at one end of a flame trap, then the device may loss its effectiveness. **Flame extinguishers** can be used in conjunction with flame traps. These are pressurized containers that pulse an extinguishing powder into the pipeline when activated by a temperature sensor.

The methane drainage network of a mine can be incorporated into an electronic environmental surveillance and control system (section 9.6.3). This has a number of significant advantages. The characteristic behaviour of the system can be investigated very readily from a control centre by varying valve openings at chosen boreholes and observing the effect on gas flow and concentrations. Optimum settings can be determined and site-specific control policies established for automatic control under the supervision of a computer. A fully monitored system also allows the rapid detection and location of problem areas such as blockages, abnormal flows, fractured pipes or faulty components.

Extractor pumps

If in-seam gas drainage is practiced (section 12.5.1) and the coal bed has a high natural permeability, then a satisfactory degree of degassification of the coal may be attained from the *in situ* gas pressure and without the use of extractor pumps. However, for gob or cross-measures drainage (Sections 12.5.2 and 12.5.3), it is usually necessary to apply controlled suction at the boreholes in order to achieve a satisfatory balance between flowrate, gas concentration and methane emissions into the mine.

There are two major types of gas extractor pumps used in methane drainage systems. **Water seal extractors** take the form of a curved, forward-bladed centrifugal impeller located either eccentrically within a circular casing or centrally within an elliptical casing. In both cases, one, or two, zones occur around the impeller where the casing approaches closely, but does not touch, the impeller. In the remaining zones the casing widens away from the impeller. During rotation, the water is thrown radially outwards by centrifugal action in the wide zones,drawing in a pocket of gas from a central inlet port; then, as the impeller continues to turn, the water is forced inwards by the converging casing, causing the pocket of trapped gas to be discharged into a conveniently located central discharge port. Figure 12.21 illustrates a typical pressure–volume characteristic curve for a water seal extractor.

The major advantage of water seal extractors is that a gas seal is maintained without any contact between moving and stationary components. Hence, there is little risk of igniting the gas or propagating a flame. Water seal extractors have proved to be reliable and robust in practice, and may be employed in banks of parallel machines to increase capacity. The temperature rise across each unit is small, enabling the machines to be used in continuous or automatic operation. However, periodic maintenance is required to remove any scaling or to guard against corrosive action of the water.

The second family of gas pumps are the **dry extractors**. These may be reciprocating machines or take the form of two lemniscate (dumb-bell) shaped cams that rotate against each other within an elliptical casing. This action encapsulates pockets of gas from the inlet port and transports them across to the discharge port. Dry extractors are compact and,for any given speed, produce a flowrate that is near independent of pressure differential. However, these devices are subject to wear, create a substantial temperature rise in the gas and may produce considerable noise.

In the majority of cases, extractor pumps are located on the surface of a mine and, indeed, this may be mandated by legislation. In some situations where the amount of drained gas is small, it may be possible to site the extractor pumps underground and to release the gas into a well-ventilated return airway (if permitted by law or special exemption). A section of the airway around the point of emission should be caged in and free from equipment to ensure that the gas has been diluted to permissible limits before it enters any areas of activity.

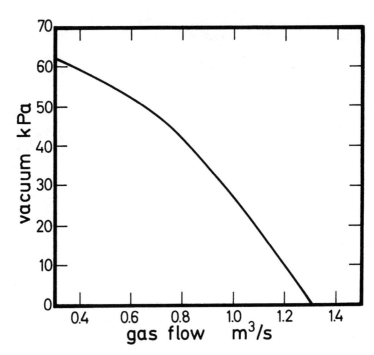

Figure 12.21 A typical pressure–volume characteristic curve for a wet-seal gas extractor pump.

12.5.6 Planning a methane drainage system

The detailed design and planning of a methane drainage system consists of two major phases. First, a variety of data must be collected and analysed in order to estimate the gas flows that will pass through the network of pipes. These flows provide the necessary input for the second phase, the design of the actual network itself, including the dimensions of pipes and the duties and locations of all ancillary components.

Data collection and estimation of gas capture

The data requirements and corresponding analyses for in-seam methane drainage (section 12.5.1) differ from those involved in gob drainage or cross-measures systems (sections 12.5.2 and 12.5.3). However, a common first requirement is to obtain samples of coal from all source beds and to determine the corresponding gas contents (section 12.2.3).

If particularly heavy emissions of methane occur in a mine from the worked horizon in headings, faces and ribsides in virgin areas, then in-seam drainage may be practicable. In this case, it is vital that permeability tests be carried out. Laboratory investigations are helpful in tracking the variations in permeability with respect to stress, gas pressure and two-phase flow (section 12.3.2). However, *in situ* tests provide the most reliable estimates of the actual permeabilities that will be encountered. These involve one or more boreholes with packers that encapsulate a length within the seam of interest. **Pressure build-up and draw-down tests** indicate the transient increase in pressure when the hole is shut in and the subsequent flow behaviour when it is re-opened. Analysis of the time transient data allows the *in situ* permeability to be determined (section 12.3.1). A simpler method that can be employed in beds of higher permeability is the **slug test** (GRI, 1988). In this technique, a surface borehole is drilled to intersect the coal bed. A packer is inserted above the seam through which passes a column into which is introduced a head of water. A pressure transducer monitors the time transient variation in pressure and, hence, water flowrate as the water enters the seam and the level in the column falls. Again, the transient radial flow equation given in section 12.3.1 is used as the basis of a computer algorithm to indicate *in situ* permeability.

The coal samples should also be used to conduct adsorption and desorption tests (sections 12.2.3 and 12.2.2). The adsorption isotherm (Fig. 12.2) can be employed to determine the volume of gas that will be produced from a given area of coal reserves as the seam gas pressure declines from its initial value to an estimated final state. The desorption tests provide data on the variation of diffusion characteristics with respect to gas pressure and particle size.

Computer simulation programs have been developed that accept these various items of data and model the flow of methane through coal beds towards boreholes or mine workings (Mavor and Schwoebel, 1991). The models are based on interactions between the relationships for gas release and migration given in sections 12.2 and 12.3. Such programs are most helpful in estimating the amount of gas that may be captured with differing sizes and locations of in-seam boreholes.

Where a methane drainage system is to be designed for an existing mine, the emission patterns into the mine workings should be established for all working areas, bleeder airways, development headings and old workings. This is particularly important where the method is to be gob or cross-measure drainage. Let us now turn to data collection and estimation of gas capture that are pertinent to these systems.

Historical records of monitored data are very helpful but should be related to the corresponding stratigraphy, mining layout and methods of working. The gas captured by gob and cross-measures methane drainage is released primarily because of the stress relaxation and fragmentation of source beds above and below the working horizon (Fig. 12.12, 12.19 and 12.20). The natural permeability of those seams prior to mining may, therefore, have little relevance to the subsequent rates of gas emission. However, the permeability–stress tests described in section 12.3.2 give an indication of the increased permeabilities induced by stress relaxation. These, coupled with gas content data, desorption kinetics and the gas migration relationships, allow predictions to be made of future gas makes. Here again, computer models have been developed to assist in such analyses (King and Ertekin, 1988). Several methods have also been produced that allow hand calculations of predicted methane emissions from relaxed roof and floor source beds (Dunmore, 1980).

In predicting gas production from future workings, it is important to take into account any changes in the mine layout and the method and rate of coal extraction. In multiseam areas, degassification by earlier workings within some 200 m above or below the current horizon is likely to result in a significant decrease in the gas available for drainage. Strata sequences should also be examined for gas outburst potential (section 12.4.3).

Design of the drainage network

The first task in designing the hardware configuration of a methane drainage system is to draw a line schematic of the layout. The network should be shown at its future maximum extent. The locations of boreholes at that time and the corresponding predicted flowrates should be indicated for all branches of the pipe network.

First estimates of pipe diameters can be made at this time (section 12.5.5). Gas velocities in the range 8 to 12 m/s are suggested but, in any event, should not be greater than 15 m/s. Pipes that are undersized for a future enhanced system will reduce the amount of gas that can be drained or necessitate higher duties of the extractor pumps.

A number of methods are available for assessing the pressure distribution throughout the network. Slide rules to assist in pipe network design are available (Boxho et al., 1980). A number of computer programs have been produced for this purpose varying from single branch assessment to full network analysis employing compressible flow relationships (section 7.2.2). These programs can also be used for compressed air networks and allow the consequences of changing pipe sizes, network layout or other design parameters to be assessed very rapidly.

Significant changes in pressure and, hence, density occur as gas proceeds through the pipes of a methane drainage system. It is, indeed necessary to employ compressible flow relationships for all but the simplest systems in order to determine frictional pressure drops. A methodology for hand calculations can be derived from equations (2.50) and (2.51):

$$p = r_t \rho Q^2 \quad \text{Pa} \qquad \text{(equation (2.50))}$$

where p = frictional pressure drop (Pa), ρ = gas density (kg/m^3), Q = flowrate (m^3/s) and r_t = rational turbulent resistance of the pipe (m^{-4}). (We are using lower case r_t for rational resistance here in order to avoid confusion with gas constants.) Now

$$r_t = \frac{fL \, \text{per}}{2A^3} \quad (\text{m}^{-4}) \qquad \text{(equation (2.51))}$$

$$= \frac{fL\pi d}{2(\pi d^2/4)^3} = \frac{64 fL}{2\pi^2 d^5} \qquad (12.50)$$

where f = coefficient of friction (dimensionless), L = length (m) and d = diameter of pipe.

In order to take compressibility into account, let us replace volume flow, $Q(m^3/s)$, by mass flow $M = Q\rho\,(kg/s)$ in equation (2.50), giving

$$p = r_t \frac{M^2}{\rho} \quad \text{Pa} \tag{12.51}$$

However,

$$\rho = \frac{P}{RT} \quad \text{kg/m}^3 \tag{12.52}$$

where P = absolute pressure (Pa), R = gas constant $(J/(kg\,K))$ and T = absolute temperature (K) of the gas, giving

$$Pp = r_t M^2 RT \quad (Pa)^2 \tag{12.53}$$

If we consider a very short length of horizontal pipe, dL, having a resistance (dr_t) and pressure drop dp, then equation (12.53) becomes

$$P\,dp = M^2 RT(dr_t) \quad Pa^2 \tag{12.54}$$

However, for a horizontal pipe, $dp = -\,dP$, and we may rewrite equation (12.54) as

$$-\,P\,dP = M^2 RT(dr_t) \quad Pa^2 \tag{12.55}$$

Integrating over the complete length of pipe (stations 1 and 2) and assuming isothermal conditions $(T = \text{constant})$, gives

$$\frac{P_1^2 - P_2^2}{2} = M^2 RTr_t \quad Pa^2 \tag{12.56}$$

Substituting for r_t from equation (12.50),

$$P_1^2 - P_2^2 = \frac{64 fL}{\pi^2 d^5} M^2 RT$$

or

$$P_1^2 - P_2^2 = 6.485 \frac{f}{d^5} LM^2 RT \quad Pa^2 \tag{12.57}$$

Hence, if the pressure at one end of the pipe is known, then the pressure at the other end can be determined.

Example A methane–air mixture containing 45% of methane enters an extractor pump at a pressure of $P_2 = 50$ kPa. The connecting pipe is 1000 m long, 300 mm in diameter and has a coefficient of friction, $f = 0.004$. If the volume flowrate is 1 m^3/s at standard atmospheric pressure and temperature while the actual mean temperature of the gas is 25 °C, determine the gas pressure, P_1, entering the pipe.

Solution   Let us, first, establish the values of the parameters required by equation (12.57):

$$P_2 = 50\,000\,\text{Pa}$$

$$f = 0.004$$

$$d = 0.3\,\text{m}$$

$$L = 1000\,\text{m}$$

$$T = 273 + 25 = 298\,\text{K}$$

To establish the gas constant and mass flow of the mixture, we utilize the fractional components of 0.45 (methane) and 0.55 (air):

$$\text{methane} \quad R_\text{m} = 518.35\,\text{J/(kg K)}$$

$$\text{air} \quad R_\text{a} = 287.04\,\text{J/(kg K)}$$

(from Table 11.1). Therefore, for the mixture,

$$R = (0.45 \times 518.35) + (0.55 \times 287.04)$$

$$= 391.13\,\text{J/kg}$$

The flowrate is given as $Q = 1\,\text{m}^3/\text{s}$ at standard conditions. As the density of air at those same conditions is $1.2\,\text{k/gm}^3$ and the relative density of methane is 0.554 (dimensionless, Table 11.1) then the corresponding density of the mixture is

$$\rho = (0.45 \times 0.554 + 0.55 \times 1)1.2$$

$$= 0.959\,\text{kg/m}^3$$

giving mass flow

$$M = Q\rho$$

$$= 1 \times 0.959 = 0.959\,\text{kg/s}$$

Substituting for the known values in equation (12.57) gives

$$P_1^2 - (50\,000)^2 = 6.485 \times \frac{0.004}{(0.3)^5} \times 1000 \times (0.959)^2 \times 391.13 \times 298$$

giving

$$P_1 = 60\,368\,\text{Pa or }60.37\,\text{kPa}$$

   This technique can be used, branch by branch, to determine the effective pressures (producing flow) throughout the network. If an extractor pump is used, then the starting point will be the absolute pressure at the pump inlet. On the other hand, the analysis may commence from a given borehole pressure and proceed through the network in the direction of flow. This indicates the suction pressure required to generate any desired flowrate. Iterative exercises can be carried out to determine optimum pipe sizes. Allowance must be made for the pressure drops that occur at all bends, valves, monitors and other fittings. Such assessments can normally be made from tables or charts provided by manufacturers.

The **buoyancy pressure**, p_{buoy}, that assists gas flow in vertical pipes may be estimated simply as

$$p_{buoy} = (\rho_a - \rho_g) g \Delta Z \quad \text{Pa} \tag{12.58}$$

where ρ_a = mean density of air in shafts (kg/m^3), ρ_g = mean density of gas mixture in the vertical pipe (kg/m^3), g = gravitational acceleration (m/s^2) and ΔZ = length of vertical pipe (m).

Where extractor pumps are employed, the buoyancy pressure is often ignored for design purposes. However, if no pumps are used, it becomes a significant factor affecting gas flowrate.

The selection of the network monitoring system, transducers and safety devices should be considered most carefully. Such choices must be made in accord with any national or state legislation appertaining to gas pipelines.

Finally, a colour-coded mine plan should be prepared showing the positions, dimensions and other specifications of all pipes, control points, monitors, safety devices and extractor pumps. This plan should be kept up to date as the system is extended.

12.5.7 Utilization of drained gas

Much of the methane produced from mines is passed into the surface atmosphere either via the mine ventilation system or by venting the gas directly to the atmosphere from methane drainage systems. Not only is this a waste of a valuable fuel resource but may have long-term global environmental repercussions.

Where coalbed methane drainage is practiced, the production of gas of a consistent quantity and quality may enable it to be employed in gas distribution facilities for sale to industrial or domestic users. For other methods of methane drainage, variations in purity are likely to restrict utilization of the gas to more local purposes. The simplest use of drained gas is for heating. Gas-fired boilers or furnaces can provide hot water, steam or heated air for low cost district heating schemes. Surface buildings on the mine and local communities can benefit from such systems. Mineral processing plants may utilize the heat for drying purposes or to facilitate chemical reactions. Variations in the purity of the gas fed to burners can be monitored continuously and automatic adjustment of air:fuel ratios employed to maintain the required heat output.

Gas-fired turbines can be employed to produce electricity. Units are available that will develop mine site electrical power from drained gas (Boxho *et al.*, 1980; Owen and Esbeck, 1988). These turbines can act as excellent co-generators producing not only several megawatts of electrical energy but also useful heat from the exhaust gases. Gas is normally delivered to the site of the engines at relatively low pressure but is compressed to a pressure of 1000 to 1400 kPa before injection into the combustion chambers. A supply of some $0.7 \, \text{m}^3/\text{s}$ (standard pressure and temperature) at 50% methane and 50% air can produce up to 4 MW of electrical power and 9 MW of useable heat from a mine site gas turbine. Dual units may accept either methane or fuel oil.

Other uses that have been found for drained methane include the firing of furnaces in local metallurgical plants, kilns and coke ovens and for the synthesis of chemicals. In very gassy mines, the methane concentration in the ventilating air returned to surface may be as high as 0.8%. Where furnaces are used on the mine site, then fuel savings may be achieved by feeding those furnaces with return air.

REFERENCES

Airey, E. M. (1968) Gas emission from broken coal: an experimental and theoretical investigation. *Int. J. Rock Mech. Min. Eng. Sci.* **5**, 475–94.

Bakke, P. and Leach, S. J. (1962) Principles of formation and dispersion of methane roof layers and some remedial measures. *Min. Eng.*, **121**(22), 645–58.

Barker-Read, G. R. and Radchenko, S. A. (1989) Methane emission from coal and associated strata samples. *Int. J. Min. Geol. Eng.*, **7**(2), 101–26.

Bertrard, C., Bruyet, B. and Gunther, J. (1970) Determination of desorbable gas concentration of coal (direct method). *Int. J. Rock Mech. Min. Sci.*, **7**, 43–65.

Boxho, J., *et al.* (1980) *Firedamp Drainage Handbook*, Coal Directorate of the Commission of the European Communities, 1–415.

Campoli, A. A., *et al.* (1985) An overview of coal and gas outbursts. *Proc. 2nd US Mine Ventilation Symp.*, *Reno, NV*, 345–51.

Close, J. C. and Erwin, T. M. (1989) Significance and determination of gas content data as related to coalbed methane reservoir evaluation and production implications. *Proc. 1989 Coalbed Methane Symp.*, *University of Alabama, Tuscaloosa, AL*, 19 pp.

Coward, H. L. (1937) Movement of firedamp in air. *Trans. Inst. Min. Eng.* **94**, 446.

Darcy, H. (1856) *Less Fontaines Publique de la Ville de Dijon*, Dalmante, Paris.

Diamond, W. P. and Levine, J. R. (1981) Direct method determination of the gas content of coal: procedures and results. *US Bureau of Mines Rep. 8515*, 1–36.

Dunmore, R. (1980) Towards a method of prediction of firedamp emission for British coal mines. *Proc. 2nd Int. Mine Ventilation Congr.*, *Reno, NV*, 351–64.

Ely, K. W. and Bethard, R. C. (1989) Controlling coal mine methane safety hazards through vertical and horizontal degasification operations. *Proc. 4th US Mine Ventilation Symp.*, *Berkeley, CA*, 500–6.

Ettinger, I. L., *et al.* (1958) *Systematic Handbook for the Determination of the Methane Content of Coal Seams from the Seam Pressure of the Gas and the Methane Capacity of the Coal*, Institute of Mining Academy, Moscow, (NCB translation A1606/SEH).

Ettinger, I. L., *et al.* (1986) Specification of gas dynamics in coal seams with the application of methane diffusion in pieces. In *Urgent Problems of Mine Aerogas Dynamics*, USSR Academy of Sciences, Moscow, 88–93.

Evans, H. and Brown, K. M. (1973) Coal structures in outbursts of coal and firedamp conditions. *Min. Eng.* **173**(148), 171–9.

Gan, H., Nandi, S. P. and Walker, P. L. (1972) Nature of the porosity in American coals. *Fuel* **51**, 272–7.

Gawugal, J. (1979) Flow of gas through stressed carboniferous strate. *PhD Thesis*, University of Nottingham.

GRI (1988) *Technology Profile*, Gas Research Institute, Chicago, IL.

Harpalani, S. (1984) Gas flow through stressed coal. *PhD Thesis*, University of California, Berkeley.

Hucka, V. J. and Lisner, U. W. (1983) *In situ* determination of methane gas in Utah coal mines (a case history). *Proc. SME-AIME Meet., Atlanta, GA*, 1–9.

Janas, H. (1980) Bestimmungen des desorbierbaren methaninhaltes an kohlenproben aus exploratonsbohrunge. Unpublished report.

Jolly, D. C., Morris, L. H. and Hinsley, F. B. (1968) An investigation into the relationship between the methane sorption capacity of coal and gas pressure. *Min. Eng.* **127**(94), 539–48.

King, G. R. and Ertekin, T. (1988) Comparative evaluation of vertical and horizontal drainage wells for the degasification of coal seams. *SPE Reserv. Eng.* (May), 720–34.

Klinkenberg, L. J. (1941) The permeability of porous media to liquids and gases. *API Drill Prod. Pract.*, 200–13.

Lama, R. D. and Nguyen, V. U. (1987) A model for determination of methane flow parameters in coal from desorption tests. *APCOM '87. Proc. 20th Int. Symp. on the Application of Computers and Mathematics in the Minerals Industry, Johannesburg*, 275–82.

Langmuir, I. (1916) The constitution and fundamental properties of solids and liquids. *J. Am. Chem. Soc.* **38**, 2221–95.

Langmuir, I. (1918) The adsorption of gases on plane surfaces of glass, mica and platinum. *J. Am. Chem. Soc.* **40**, 1361–403.

Mavor, M. J. and Schwoebel, J. J. (1991) Simulation based selection of underground coal mine degasification methods. *Proc. 5th US Mine Ventilation Symp., West Virginia*, 144–55.

Mills, R. A. and Stevenson, J. W. (1989) Improved mine safety and productivity through a methane drainage system. *Proc. 4th US Mine Ventilation Symp., Berkeley, CA*, 477–83.

Morris, I. H. (1974) Substantial spontaneous firedamp emissions. *Min. Eng.* **133**(163), 407–21.

Obert, L. and Duvall, W. I. (1967) *Rock Mechanics and Design of Structures in rock*, Wiley, New York, p. 333.

Oldroyd, G. C., *et al.* (1971) Investigations into sudden abnormal emissions of firedamp from the floor strata of the Silkstone Seam at Cortonwood Colliery. *Min. Eng.* **130**(129), 577–91.

O'Shaughnessy, S. M. (1980) The computer simulation of methane flow through strata adjacent to a working longwall coalface. *PhD Thesis*, University of Nottingham.

Owen, W. L. and Esbeck, D. W. (1988) *Industrial Gas Turbine Systems for Gob Gas to Energy Projects*, Solar Tubrines Inc., San Diego, CA, 1–14.

Radchenko, S. A. (1981) The time of diffusion relaxation as the characteristics of coal disturbance. In *The Main Questions of the Complex Development of the Deposits of Solid Mineral Resources*, USSR Academy of Sciences, Moscow, 43–118.

Raine, E. J. (1960) Layering of firedamp in longwall workings. *Min. Eng.* **119**(10), 579–91.

Richmond, J. K., *et al.* (1983) Historical summary of coal mine explosions in the United States, 1959–81. *US Bureau of Mines Information Circular 8909*, 1–51.

Robinson, G., *et al.* (1981) Underground environmental planning at Boulby Mine, Cleveland Potash, Ltd., England. *Trans. Inst. Min. Metall.* (July) (also published in *J. Mine Vent. Soc. S. Afr.* **35**(9), (1982), 73–88.

Schatzel, S. J. and Dunsbier, M. S. (1989) Roof outbursting at a Canadian bedded salt mine. *Proc. 4th US Mine Ventilation Symp., Berkeley, CA*, 491–9.

Schwoebel, J. J. (1987) Coalbed methane: conversion of a liability to an asset. *Min. Eng.* (April), 270–4.

Sheng, J. and Otuonye, F. (1988) A model of gas flow during outbursts. *Proc. 4th Int. Mine Ventilation Congr., Brisbane,* 191–7.

Smith, D. M. and Williams, F. L. (1984a) Diffusional effects in the recovery of methane from coalbeds. *J. Soc. Petrol. Eng.* **24**(5), 529–35.

Smith, D. M. and Williams, F. L. (1984b) Diffusion models for gas production from coal—application to methane content determination. *Fuel* **63**, 251–5.

Somerton, W., *et al.* (1974) *Effect of Stress on Permeability of Coal, Berkeley, CA* (US Bureau of Mines Contract H0122027).

Thakur, P. C. and Dahl, H. D. (1982) Methane drainage. In H. L. Hartman (ed.), *Mine Ventilation and Air conditioning*, Chapter 4, 69–83.

US Bureau of Mines (1980) *Creating a Safer Environment in US Coal Mines*, Bureau of Mines Methane Control Program, US Government Printing Office, Washington, DC.

Williams, R. and Morris, I. H. (1972) Emissions and outbursts in coal mining. *Proc. Symp. on Environmental Engineering in Coal Mining,* Institute of Mining Engineers, Harrogate, 101–16.

Wolstenholme, E. F., *et al.* (1969) Movement of firedamp within the floor strata of a coal seam liable to outbursts. *Min. Eng.* **128**(105), 525–39.

FURTHER READING

Harpalani, S. and McPherson, M. J. (1984) The effect of gas evacuation on coal permeability test specimens. *Int. J. Rock Mech. Min. Sci.* **21**(3), 161–4.

James, T. E. and Purdy, J. L. (1962) Experiments with methane layers in a mine roadway. *Min. Eng.* **121**(21), 561–9.

King, G. R. and Ertekin, T. (1989) State of the art in modeling of unconventional gas recovery. *SPE (18947) Symp.,* Denver, CO.

13

Radiation and radon gas

13.1 INTRODUCTION

The element uranium is widely distributed within the crust and oceans of the earth. It has been estimated that crustal rocks contain an average of some 4 g of uranium per tonne. The structure of the uranium atom is unstable; emission of subatomic particles from the nucleus causes uranium to change or **decay** into a new element, thorium. The process of radioactive decay continues down through a series of elements until it reaches a stable form of lead. This process has existed on earth from before the crust was formed. All forms of life on earth have evolved and exist within a constant bombardment of natural radiation including that from the uranium series of elements.

Although most of the uranium series of elements are solids, one known as **radon**, Rn, is a gas. This facilitates escape, or **emanation**, of radon from mineral crystals into the pore structure of the rock from where it may migrate via a fracture network or interconnected interstices towards the free atmosphere. However, the liberated radon itself decays into microscopic solid particles known as the **radon daughters** which may adhere to dust or other aerosol particulates, or remain suspended as free ions in the air. If the rate of radon emanation into mine workings is high or the space inadequately ventilated, then the radioactivity caused by continued decay of radon and its daughters will reach levels that are hazardous to health. This is likely to occur within closed environments in close proximity to rocks that contain traces of uranium ores. Radon is a potential problem in any mine and tests for its presence should be carried out early in the development of a new mineral deposit. The hazard can also exist in the basements of surface buildings.

As would be expected, the problem of radon is greatest in uranium mines and particular precautions have to be taken in order to protect personnel from the development of lung cancers caused by the inhalation and possible alveolar retention of radon daughters.

In this chapter, we shall outline the uranium decay series and quantify the mechanisms of radon emanation and the growth of radon daughters within the atmospheres of subsurface openings. Methods of measurement will be discussed while the final section is devoted to pragmatic guidelines to be followed in designing a ventilation system for a mine with high radon emanations.

13.2 THE URANIUM SERIES AND RADIOACTIVE DECAY

13.2.1 Atomic structure; alpha, beta and gamma radiation

The current concept of the structure of an atom is that of a nucleus consisting of **protons** of net positive electric charge and **neutrons** of net zero charge, orbited by negatively charged **electrons**. The electrical charge on the nucleus may be expressed in terms of the unit charge of an electron. The **atomic number** of any given element indicates the level of net positive charge on the nucleus. Figure 13.1 gives the atomic numbers of the elements in the uranium series. The **atomic mass** is the nearest integer to the actual mass of the atom expressed relative to the mass of an individual proton or neutron (atomic mass unit).

There are many classes of ionizing radiation. However, three are particularly relevant to decay through the uranium series. An **alpha** (α) particle has a positive charge of 2 units and mass of 4 (helium nucleus). Hence, when an alpha particle is emitted from the nucleus, the atomic number of that element reduces by 2 and the atomic mass reduces by 4. The next element in the decay series is formed. A glance at Fig. 13.1 will reveal the stages at which alpha particles are emitted in the uranium decay chain. Alpha particles have relatively low energy levels. They travel no more than a few centimetres in normal atmospheres and are halted by human skin. However, if alpha particles are emitted within the lung, they can cause cellular alteration within the alveolar walls, leading to possible lung cancer.

A **beta** (β) particle has a negative charge of 1 and a negligible mass (electron). When this is emitted, the atomic mass of the element remains effectively unchanged but the net positive charge on the nucleus (atomic number) increases by one. This can be observed by movement from left to right at atomic masses of 234, 214 and 210 on Fig. 13.1. Beta particles not only cause damage to the lung but can also penetrate human skin and may produce alteration of cell tissue.

Both alpha and beta radiation involve the emission of subatomic particulates. **Gamma** (γ) radiation, however, is a very high frequency electromagnetic wave form (like X-rays) and is deeply penetrating with respect to the human body.

13.2.2 Radioactive decay and half-life

If we commence with a fresh sample of a radioactive substance, then all of the atoms are capable of disintegrating to the next lower element in the chain. However, as this process continues, there are progressively fewer atoms remaining of the original substance. Hence, the rate of decay of that substance will decline exponentially. This may be expressed in terms of the reducing number of atoms that remain unaltered:

$$N = N_0 \exp(-\lambda t) \tag{13.1}$$

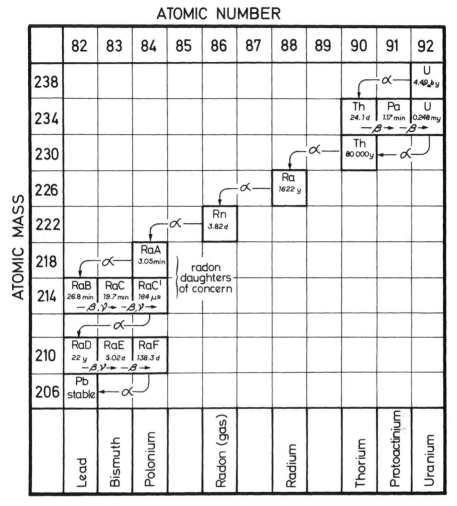

Figure 13.1 The uranium decay series. Radon gas, disintegrating through the radon daughters, RaA, RaB, RaC and RaC′, gives rise to radiation problems in some mines.

where N = number of unaltered atoms remaining, N_0 = original number of atoms, t = time(s) and λ = the **decay constant** (s^{-1}) for that material and is a measure of the probability of disintegration of any one atom. For radon, $\lambda = 2.1 \times 10^{-6}$ disintegrations (dis) per second.

The rate at which atoms disintegrate, I, is known as the **radioactivity** and depends on the number of unaltered atoms remaining and the probability of disintegration:

$$I = \lambda N \quad \text{dis/s} \tag{13.2}$$

However, the activity may also be expressed as

$$I = -\frac{dN}{dt} \quad \text{dis/s} \tag{13.3}$$

or, from equation (13.1)

$$I = -\frac{d}{dt} \, [N_0 \exp(-\lambda t)]$$

$$= \lambda N_0 \exp(-\lambda t)$$

However, the initial activity $I_0 = \lambda N_0$ from equation (13.2). Hence,

$$I = I_0 \exp(-\lambda t) \quad \text{dis/s} \tag{13.4}$$

showing that the radioactivity also decays exponentially with time. Figure 13.2 illustrates a decay curve. By measuring the activity given by a radioactive substance while removing the decay products, the corresponding curve can be plotted and the value of λ determined.

The decay curve approaches zero activity exponentially; hence, the theoretical full

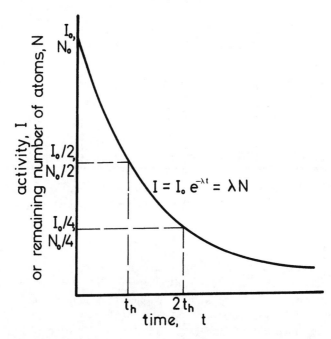

Figure 13.2 The activity, I, and number of atoms, N, of the original substance both decay exponentially with time. The half-life, t_h, occurs when I and N are each half of their original values.

lifespan of a radioactive element is infinite. A more useful indicator of the aging process is the **half-life** of the element. This is defined as the time taken for half of the original atoms to decay. As $I = \lambda N$, the activity also reduces to half its original value at this same time. A simple relationship exists between the half-life, t_h, and the decay constant, λ. Initially,

$$I = I_0$$

At the half-life

$$I = \frac{I_0}{2} = I_0 \exp(-\lambda t_h)$$

Hence,

$$\tfrac{1}{2} = \exp(-\lambda t_h)$$

or

$$-\lambda t_h = \ln(\tfrac{1}{2}) = -0.6931$$

giving

$$t_h = \frac{0.6931}{\lambda} \quad \text{s} \tag{13.5}$$

For radon, $\lambda = 2.1 \times 10^{-6}$ dis/s, giving

$$t_h \,(\text{radon}) = \frac{0.6931}{2.1 \times 10^{-6}} = 0.33 \times 10^6 \quad \text{s}$$

or 3.82 days

The half-lives of the other elements in the uranium decay series are given in Fig. 13.1 and vary from ^{238}U (4.49 billion years) to RaC' (164 μs).

The problems of radon in mines occur not only because it is a gas and can be emitted into the ventilating airstream, but also from a consideration of half-lives. Radon gas is the decay product of radium, ^{226}Ra. This has a half-life of 1622 years, i.e. extremely long relative to the time taken for a ventilating airstream to traverse through a mine. Hence, the source of radon is effectively infinite. With a half-life of only 3.82 days, some of the radon will decay before it leaves the mine. More importantly, the short half-lives of the radon daughters, RaA, RaB, RaC and RaC', indicate that they will disintegrate readily, emitting alpha, beta and gamma radiation. These observations have an important bearing on the design of ventilation systems for uranium mines.

13.2.3 Units of radioactivity

In the previous section we referred to the level of radioactivity in terms of atomic disintegrations per second, dis/s. This unit is sometimes called the **becquerel** (Bq) after the French physicist who found in 1896 that uranium salts are radioactive:

$$1\,Bq = 1\,dis/s \tag{13.6}$$

A less rational but widely used unit of radioactivity is the **curie** named after Marie (1867–1934) and Pierre (1859–1906) Curie, who, in France, first separated and identified radium and a number of other radioactive elements.

One curie, Ci, approximates to the activity of 1 g of radium or, more precisely, a source that is disintegrating at a rate of 3.700×10^{10} atoms per second. Owing to the magnitude of this latter value, submultiples of the unit are normally employed:

$$1\,\text{microcurie} = 1\,\mu Ci = 10^{-6}\ Ci = 37\,000\,dis/s$$

$$1\,\text{picocurie}\ \ = 1\,pCi = 10^{-12} Ci = 0.037\,dis/s \tag{13.7}$$

Hence,

$$1\,pCi = 0.037\,Bq \tag{13.8}$$

or

$$1\,Bq = 27\,pCi \tag{13.9}$$

Analyses on radioactivity may be conducted in terms of either becquerels or curies.

Prior to the establishment of current knowledge on the effects of radioactivity on the human body, it was thought that an average level of 100 pCi/l from radon daughters was safe. This has since been reduced to a third of that value. However, 100 pCi/l became known as an acceptable working level. The term was truncated to working level, WL, and is now used widely with respect to radon daughters.*

Ionizing electromagnetic radiation produced by gamma emissions is measured in röentgens (after Wilhelm Röentgen of Germany who discovered X-rays in 1895). The Röentgen is defined formally as the level of X- or gamma radiation that produces 1 electrostatic unit of charge[†] per 0.001 293 g of air (1 cm^3 at 101.324 kPa and 0 °C). In order to apply this in terms of the effect on human bodies, the Rem (Röentgen equivalent man) is employed. This is the dosage of ionizing radiation that will cause the same biological effect as 1 Röentgen of X- or gamma rays. **Dose rates** are quoted in mRem/hour. Hence, a dose rate of 50 mRem/h would produce a **dosage** or **dose equivalent** of 25 mRem after half an hour.

The röentgen and the Rem are the most commonly used units for ionizing radiation and biological dosage. They are not, however, SI units. The rational SI equivalents are

$$1\,C/kg\,(\text{coulomb per kilogram}) = 3876\,\text{röentgens}$$

$$1\,Sv\,(\text{sievert}) = 1\,J/kg = 100\,\text{Rems}$$

As a sievert is a very large unit, radiation doses are quoted in millisieverts (mSv). One chest X-ray is equivalent to about 0.2 mSv.

---

*The term working level is also defined as that concentration of short-lived radon daughters which represents 1.3×10^5 MeV of potential α particle energy while decaying to the stable ^{210}Pb.

[†]1 electrostatic unit of charge is equivalent to 2.083×10^9 ion pairs.

13.3   RADON AND ITS DAUGHTERS

Radon gas emanates from the crystalline structure of minerals into the pores and fracture networks of rocks. It migrates through the strata by a combination of diffusion and pressure gradient towards mine openings. Radioactive decay produces the solid particulates of the radon daughters as given in Fig. 13.1:

				Half-life
Polonium,	^{218}Po,	or radium A,	RaA	3.05 min
Lead,	^{214}Pb,	or radium B,	RaB	26.8 min
Bismuth,	^{214}Bi,	or radium C,	RaC	19.7 min
Polonium,	^{214}Po,	or radium C',	RaC'	164 μs
Lead,	^{210}Pb,	or radium D,	RaD	22 years

As the half-life of RaD is 22 years, we need concern ourselves only with the radon series down to RaC'. Furthermore, RaC' has the extremely short half-life of 164 μs. Hence, the effect of its decay is normally coupled with that of RaC.

The solid particulates of radon daughters that form during migration of the gas through strata are likely to plate onto the mineral surfaces and be retained within the rock. However, the remaining radon will continue to decay after it has been emitted into a mine opening. The radon daughters will then adhere to aerosol particles or remain as free ions within the airstream.

In this section, we shall examine the migration of radon through the rock and the growth of radon daughters within a ventilating airflow.

13.3.1   Emanation of radon

When an alpha particle is projected from an atom of radium, the resulting atom of radon recoils through a distance of some 3×10^{-8} m in minerals and 6×10^{-5} m in air (Thompkins, 1982). Furthermore, the diffusion coefficient for radon within mineral crystals is very small. Hence, although movements of radon atoms will occur within the crystals, the distances of individual motion are small in comparison with most mineral grain sizes; the probability of a single radon atom escaping into a pore is, therefore, also small. Nevertheless, a sufficient number do escape to give rise to radon problems in uranium mines.

The migration of radon through the pore and fracture network of the strata may be analysed on the basis of Fick's laws of diffusion modified for radon production and decay (Bates and Edwards, 1980). This leads to an approximate equation that describes the radon concentration, C, within the pores with respect to distance into the rock:

$$C = C_\infty \left[1 - \exp\left(-x \sqrt{\frac{\lambda \phi}{D}} \right) \right] \quad \text{pCi/(m}^3 \text{ of space)} \qquad (13.10)$$

where C_∞ = concentration at infinite distance into rock (pCi/m³), x = distance into rock from free surface (m), λ = radon decay constant (2.1×10^{-6} Bq), ϕ = rock porosity (fraction) and D = diffusion coefficient for rock (m²/s).

The units of radon concentration require a little explanation. The normal volumetric concentration commonly used for other gases would give excessively low values for radon. It is more convenient to express radon concentration in terms of the level of radioactivity (pCi or Bq) emitted by each m³ of the radon:air (or radon:water) mixture.

Equation (13.10) is based on an assumed radon concentration of zero at the open rock surface. The actual emanation at the surface can be measured directly or calculated as (Bates and Edwards, 1980)

$$J = C_\infty \sqrt{\lambda D \phi} \quad \frac{\text{pCi}}{\text{m}^2 \text{s}} \tag{13.11}$$

The maximum value of radon concentration in the rock, C_∞, may be determined from

$$C_\infty = \frac{B}{\lambda \phi} \quad \frac{\text{pCi}}{\text{m}^3} \tag{13.12}$$

Combining equations (13.11) and (13.12) gives

$$J = B \sqrt{\frac{D}{\lambda \phi}} \quad \text{pCi/(m}^2 \text{s)} \tag{13.13}$$

where B is the rate of emanation from unit volume of rock (pCi/(m³ s)) (sometimes known as emanating power) and can be measured from samples of the rock. Values of both B and J for differing rocks vary by several orders of magnitude.

Table 13.1 Coefficients of diffusion, D, for radon in various porous media (after Thompkins, 1985)

Medium	Coefficient of diffusion (m²/s)
Rocks	
Dense rock	0.05×10^{-6}
Porosity 6.2%	0.2×10^{-6}
Porosity 7.4%	0.27×10^{-6}
Porosity 12.5%	0.5×10^{-6}
Porosity 25%	3×10^{-6}
Air	$(10–12) \times 10^{-6}$
Water	0.0113×10^{-6}
Alluvial soil	$(3.6–4.5) \times 10^{-6}$
Concrete	$(0.0017–0.003) \times 10^{-6}$

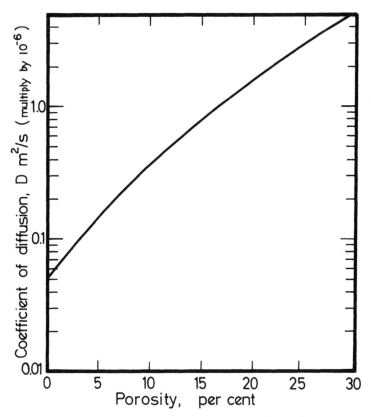

Figure 13.3 Guideline to the approximate coefficient of diffusion for radon in rocks of known porosity. Actual coefficients of diffusion depend on the type and state of the pore fluid.

A guide to coefficients of diffusion for radon in a range of materials is given in Table 13.1. Figure 13.3 may be used as a guideline to estimate a coefficient of diffusion for radon in rocks of known porosity. However, it should be noted that the coefficient of diffusion varies widely with the type and condition of pore fluid.

Figure 13.4 illustrates the variation of radon concentration within the pores for the four rocks of specified porosities and corresponding coefficients of diffusion given in Table 13.1. These curves indicate that peak emanations of radon are likely to occur when rock is broken from the orebody and reduced to fragments of about 10 cm in size. However, below that dimension there is relatively little change in radon concentration within the pores, particularly for the higher porosity rocks.

13.3.2 Growth of radon daughters

Radon is emitted into mine openings not only directly from the surrounding rock surfaces but also from old workings and other zones of voidage. Hence, the total

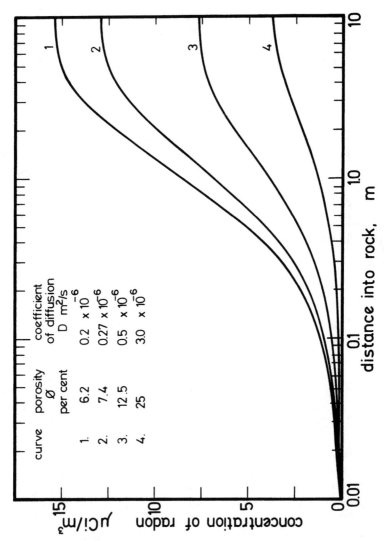

Figure 13.4 Examples of the variation in radon concentration with respect to distance into the rock, based on diffusion flow, equation (13.10). $B = 2 \, \text{pCi}/(\text{m}^3 \, \text{s})$ and $\lambda = 2.1 \times 10^{-6} \, \text{Bq}$.

emanations will consist of a mixture of the gas and its particulate daughters. In order to analyse the growth of the daughters, consider an imaginary experiment. We commence with 1 l of filtered air that contains a radon concentration of 100 pCi/l. The radon will immediately begin to suffer disintegration into RaA. This has a half-life of only 3.05 min. Hence, the second radon daughter, RaB, soon appears. However, this has a half-life of 26.8 min and concentrations of the RaC (and RaC′) do not become significant until some 20 min from the start of the experiment. We now have a situation in which the initial amount of radon gas is diminishing slowly (half-life of 3.82 days) but each of the daughters is simultaneously being generated and decaying on a shorter time scale. With its relatively small half-life, the concentration of RaA reaches a state of dynamic equilibrium within a few minutes.

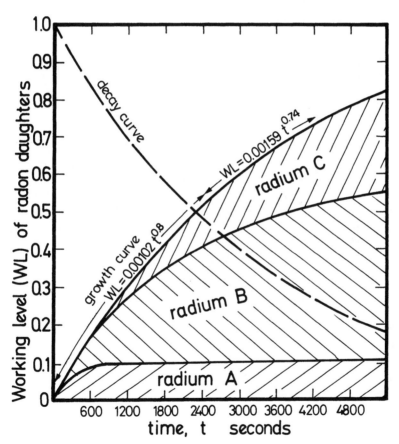

Figure 13.5 The growth curve shows the development of radon daughters from a radon concentration of 100 pCi/l. The decay curve shows the fall in radon daughters from an initial working level of 1.0 in the absence of replenishing radon and assuming that the daughters are in near equilibrium with their radon parent. The decay curve is a mirror image of the growth curve.

Figure 13.5 illustrates the growth of the radon daughters from a radon concentration of 100 pCi/l. Full (secular) equilibrium is reached in some 30 h when the air is said to have fully 'aged'. However, 80% of the ageing occurs within 90 min. The concept of the 'age' of the air is most important in understanding the radioactive decay of radon. At secular equilibrium, the number of atoms in each radon daughter is proportional to its half-life. The age of the air can be estimated from the growth curve of Fig. 13.5 for a given working level of radon daughters and having corrected for the actual concentration of radon gas present. (The latter may be eliminated by using Fig. 13.9 which gives the growth in total radon daughters for a range of radon gas concentrations.)

With a half-life of 3.82 days, the radioactivity caused by the decay of radon during the short time it spends within the human lung is very limited. However, the shorter half-lives of the radon daughters cause them to be much more prolific emitters within the respiratory system. Furthermore, as they are particulates, they may adhere to mucous membranes and be retained within the lung tissue. It is, therefore, the daughters of radon rather than the gas itself that are dangerous to health.

As it is the decay of the radon daughters that causes the health hazard, it is clear from Fig. 13.5 that airflows should be sufficiently vigorous to remove radon from mine workings as rapidly as possible and before significant ageing has occurred. It is also clear that near stagnant air in old workings will be fully aged, i.e. contain the maximum concentrations of radon daughters, and may cause serious escalations in working levels if leakage occurs from those old workings into active airways.

Although Fig. 13.5 illustrates the growth of each radon daughter, it will be recalled that each daughter is simultaneously suffering disintegration. Hence, a mirror image of Fig. 13.5 would indicate the corresponding decay curves. At full secular equilibrium, the growth and decay curves for each radon daughter would be horizontal lines, equal in magnitude but opposite in sign. For clarity, the decay curve for total radon daughters only is shown on Fig. 13.5.

13.3.3   Threshold limit values

Lung cancers occur throughout the general population but are exacerbated by the inhalation of some types of dust particles and products of combustion. However, the higher incidence of lung cancer in uranium mine workers is attributed to radon daughters. The current method of dosage assessment is based on cumulative exposure. Any unit of time may be employed. However, the most widespread measure of cumulative exposure is the working level month, WLM, defined as an exposure of 1 WL for a period of 1 month.

Assuming that an average of 170 h are spent in the workplace each month, the cumulative exposure of any individual may be calculated as

$$\frac{\sum(\text{WL} \times \text{hours of exposure})}{170} \quad \text{WLM}$$

Example   During the 40 h a miner spends underground during a week, radio-
logical monitoring indicates the following levels and periods of exposure to
radon daughters:

$$20 \text{ h at } 0.1 \text{ WL}$$
$$15 \text{ h at } 0.2 \text{ WL}$$
$$5 \text{ h at } 0.4 \text{ WL}$$

Determine the cumulative exposure in WLM for that week.

Solution

$$\text{cumulative exposure} = \frac{20 \times 0.1 + 15 \times 0.2 + 5 \times 0.4}{170}$$

$$= 0.041 \text{ WLM}$$

The most commonly accepted TLV (threshold limit value) is that no person should
be subjected to a cumulative exposure exceeding 4 WLM in one year. Legislation
may also require that there should be an upper limit of 2 WLM in any consecutive
three months and an instantaneous ceiling limit level of 1.0 WL.

The TLV of 4 WLM per year implies an average of $4/12 = 0.33$ WLM per month,
i.e. an average radiation level of 0.33 WL. Legislation may require mandatory
reporting and increased rates of sampling at higher levels. However, the aim should
be for lower actual levels in the active work areas of a mine as leakage from old
workings can cause rapid escalation of working levels in return airways. (National
or state legislation may also mandate notification to inspection agencies at very low
levels of radiation (Howes, 1990).)

There remains considerable doubt on the long-term effects of exposure to low
levels of radioactivity. It should be expected that threshold limit values will be
decreased yet further. In the meantime, the International Commission on Radio-
logical Protection has recommended that, in addition to remaining within specified
threshold limit values, radiation levels should be maintained 'as low as reasonably
achievable'. This is often termed the ALARA principle. The practical application of
this rather vague phrase may involve mine management being required to notify the
inspectorate when a specified fraction of the exposure TLV is exceeded for any
worker, and, furthermore, to demonstrate that all reasonable steps have been taken
to alleviate the problem.

The threshold limit values for gamma (ionizing) radiation are, typically, 5 Rems
per year. Where average gamma radiation measurements are in excess of 2 milli-
röentgens per hour, personal gamma radiation dosemeters may be mandated for all
personnel and records of cumulative exposure maintained for each individual
(e.g. CFR (1990), section 57.5047).

13.4 PREDICTION OF LEVELS OF RADIATION

13.4.1 Emanation rate

In order to predict the levels of radiation to be expected in any active area of a mine, the emanation rates, J (pCi/(m^2 s)), from exposed surfaces and/or emanating powers, B (pCi/(m^3 s)), from drill cores or other solid samples must first be measured. The relationship between the two is given by equation (13.13). Surface emanation rates can be measured *in situ* by attaching steel collection receptacles tightly to a rock face (Archibald *et al*, 1980) or by excavating chambers into the rock. Multiple measurements made on a variety of ores and waste rock allow tables of surface emanation rates, J, and emanating powers, B, to be assembled for a given mine or rock types.

In order to determine the total rate of emanation in a specified zone, the surface area of all exposed rock faces, A (m^2), and the volumes of broken ore, V (m^3), must be assessed and multiplied by the relevant values of J and B respectively.

Example In a given stope, 350 m^2 of ore and 300 m^2 of waste rock surface are exposed. The stope contains 320 t of fragmented ore. If the density of the ore is 2000 kg/m^3, determine the rate of emanation into the stope given that

$$J(\text{ore}) = 500 \, \text{pCi/(m}^2 \, \text{s})$$

$$B(\text{ore}) = 600 \, \text{pCi/(m}^3 \, \text{s})$$

$$J(\text{waste}) = 100 \, \text{pCi/(m}^2 \, \text{s})$$

Solution
Radon from:

$$\text{ore wall surface} = 350 \times 500 = 175\,000 \, \text{pCi/s}$$

$$\text{waste wall surface} = 300 \times 100 = \quad 30\,000$$

$$\text{Total wall surface} = \qquad\qquad\quad 205\,000 \, \text{pCi/s}$$

$$\text{volume of ore fragments} = \frac{320\,000 \, \text{kg m}^3}{2000 \quad \text{kg}}$$

$$= 160 \, \text{m}^3$$

$$\text{radon from fragmented ore} = 160 \times 600 = \quad 96\,000 \, \text{pCi/s}$$

$$\text{total emanation of radon} = 250\,000 + 96\,000 = 301\,000 \, \text{pCi/s}$$

13.4.2 Changes in working levels of radon daughters

A commonly accepted method of calculating the variation in natural radioactivity due to radon daughters along an airway was developed by Schroeder and Evans at

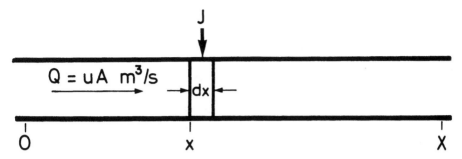

Figure 13.6 Radon emanates from rock surfaces at a rate of J pCi/(m^{2}s).

the Massachusetts Institute of Technology in 1969. Consider the length of airway shown in Fig. 13.6. The working level of radon daughters at the exit from the airway, distance X (m) from the entrance, arises from three factors.

1. decay of the radon gas already in the air at the entry point,
2. decay of the radon that emanates from the walls or fragmented rock within the airway itself, and
3. continued ageing of radon daughters that already exist in the airway at the entry point.

Each of these three behaves independently of the others. They may, therefore, be determined separately and summed to give the working level of radon daughters at exit. Let us consider each component in turn.

Decay of radon existing at entry

This can be determined from Fig. 13.5. Curve fitting to the total working level (growth) curve on this figure gives

$$\text{WL} = 102 \times 10^{-5} t^{0.8} \quad \text{working levels} \tag{13.14}$$

for $t < 2400$ s (40 min), where $\text{WL} =$ working level due to decay of 100 pCi/l of initial radon and $t =$ travel time for air to traverse the airway(s) (the coarser equation $\text{WL} = 159 \times 10^{-5} t^{0.74}$ may be used for $t < 4200$ s or 70 min). For other concentrations of radon, C (pCi/l), equation (13.14) becomes

$$\text{WL}_1 = \frac{C}{100} \times 102 \times 10^{-5} t^{0.8} \tag{13.15}$$

This equation is the basis of Fig. 13.9.

Example 1 Air with a radon concentration of 20 pCi/l flows into a mine opening. If the travel (residence) time of the airflow along the airway is 400 s, the working level of radon daughters at exit due to the initial radon is

$$WL_1 = \frac{20}{100} \times 102 \times 10^{-5}(400)^{0.8}$$

$$= 0.025\ WL$$

This result may be read directly from Fig. 13.9.
   The example will be continued for the remaining sources of radon daughters.

Decay of radon emanated from rock surfaces into the airway

To facilitate the analysis, it is assumed that radon emanates at a uniform rate of $J(pCi/(m^2 s))$ from all solid surfaces in the opening. Consider the airway shown in Fig. 13.6. The air enters at position O and leaves after traversing a distance of X (m). Over the short length dx, the wall area is per dx (m^2), where per = perimeter (m).
   The radon emitted over this increment is, therefore,

$$J \text{ per } dx \quad pCi/s$$

If the airflow is

$$Q = uA \quad m^3/s$$

where u = mean velocity (m/s) and A = cross-sectional area (m^2), then the radon emanation may be expressed in terms of cubic metres of air, i.e. an increase in *concentration*:

$$\begin{array}{c} \text{concentration from radon} \\ \text{emitted in increment } dx \end{array} = \frac{J \text{ per } dx}{Q} \quad pCi/m^3$$

$$= \frac{J \text{ per } dx}{uA} \quad pCi/m^3 \tag{13.16}$$

The time taken for the air to travel from x to X is

$$t = \frac{X - x}{u} \quad s$$

Equation (13.14) states that from a radon concentration of 100 pCi/l (or 100 000 pCi/m^3) the working level of radon daughters that will develop in this time is

$$102 \times 10^{-5} \frac{(X - x)^{0.8}}{u^{0.8}} \quad WL$$

Hence, from a radon concentration of

$$\frac{J \text{ per } dx}{uA} \quad (pCi/m^3)$$

the working level of radon daughters will grow to

$$\frac{J \, \text{per} \, dx}{100\,000 \, uA} 102 \times 10^{-5} \frac{(X-x)^{0.8}}{u^{0.8}} \, \text{WL}$$

To find the working level of radon daughters at exit due to radon emanations throughout the complete length of airway, WL_2, we must integrate from O to X:

$$WL_2 = 102 \times 10^{-10} \frac{J \, \text{per}}{Au^{1.8}} \int_0^X (X-x)^{0.8} \, dx$$

The integration is accomplished by substitution to give

$$WL_2 = 102 \times 10^{-10} \frac{J \, \text{per}}{Au^{1.8}} \frac{X^{1.8}}{1.8}$$

If the total residence time is $t_r = X/u$, then

$$WL_2 = 56.7 \times 10^{-10} \frac{J \, \text{per}}{A} t_r^{1.8} \tag{13.17}$$

Example 2   The airflow is $24 \, m^3/s$ in an 800 m long airway of perimeter 14 m and cross-sectional area $12 \, m^2$. Calculate the activity of radon daughters at exit if radon is emitted from the surfaces at a rate of $J = 250 \, pCi/(m^2 \, s)$.

Solution   Air velocity,

$$u = \frac{Q}{A} = \frac{24}{12} = 2 \, m/s$$

Residence (travel) time,

$$t_r = \frac{\text{length}}{u} = \frac{800}{2} = 400 \, s$$

Equation (13.17) gives

$$WL_2 = 56.7 \times 10^{-10} \times \frac{250 \times 14}{12} (400)^{1.8}$$

$$= 0.080 \, WL$$

Continued ageing of radon daughters that existed in the air at the entry point

In general, the air that enters at location O on Fig. 13.6 will already contain radon daughters. These will continue to age and, hence, decay to a reduced working level by the time they reach the exit. The fact that the daughters are continuously being replenished from the decay of radon can be ignored as those effects have been considered separately in the previous sections.

The reduction in working levels may be read from the decay curve on Fig. 13.5 and adjusted for the initial working level, WL_{in}, or, as the decay curve is a mirror image of the growth curve, calculated as

$$WL_3 = WL_{in}(1 - 102 \times 10^{-5}t_r^{0.8})\,WL$$

Example 3 If the activity of radon daughters at entry into the airway of Example 2 is 0.05 WL, then, after the residence time of 400 s, this will have decayed to

$$WL_3 = 0.05[1 - 102 \times 10^{-5}(400)^{0.8}]$$

$$= 0.0438\,WL$$

There is, however, a weakness in this technique. It assumes that the radon daughters are in equilibrium with each other and with their parent radon gas. This is not usually the situation in ventilated areas underground. Further analysis of a suggestion made by Schroeder and Evans (1969) leads to an improved estimate:

$$WL_3 = WL_{in}\frac{(t_{in} + t_r)^{0.8} - t_r^{0.8}}{t_{in}^{0.8}} \tag{13.18}$$

where t_{in} = the 'age' of the air at entry.

Example 4 Let us repeat the previous example using the Schroeder and Evans technique, given that the radon concentration is $C = 20\,pCi/l$.

An estimate of the age of the air at entry, t_{in}, is given by an inversion of equaiton (13.15):

$$t_{in} = \left(WL_{in}\frac{100}{C}\frac{1}{102 \times 10^{-5}}\right)^{1/0.8} \tag{13.19}$$

$$= \left(0.05 \times \frac{100}{20}\frac{1}{102 \times 10^{-5}}\right)^{1/0.8}$$

$$= 970\,s$$

(This result can be estimated from Fig. 13.9 at WL = 0.05 and radon concentration = 20 pCi/l.) Equation (13.18) then gives

$$WL_3 = 0.05\frac{(970 + 400)^{0.8} - (400)^{0.8}}{(970)^{0.8}}$$

$$= 0.0413\,WL$$

To complete the series of examples given in this section, the total working level of radon daughters exiting the airway is

$$WL_{out} = WL_1 + WL_2 + WL_3$$

$$= 0.025 + 0.080 + 0.041$$

$$= 0.146\,WL$$

13.5 METHODS OF MONITORING FOR RADIATION

When radioactive emission strike the atoms of other substances, they produce effects that may include increases in temperature or secondary radiation. These effects can be measured and are a function of the level of the primary emission. In general, there are two types of radiation instrumentation, the thermal and photosensitive detectors.

In thermal detectors, the radiation is directed on to the hot junctions of a series of thermocouples (thermopile) or a resistance thermometer within an evacuated chamber. The increase in temperature of these sensors produces an electrical output that is representative of the level of radiation. Another type employs a sensitive gas thermometer. Thermal detectors tend to be fragile and are less suitable for portable instruments.

The most widely used principle of radiation detection in subsurface openings is photosensitivity. The radiation is aimed at a material that emits photons (quanta of light) when irradiated. Zinc sulphide is commonly employed.

The photons are directed towards a photomultiplier tube (PMT) where the light is amplified and converted to electrical pulses for counting and display.

13.5.1 Measurement of radon daughters

A number of instruments have been devised to measure the working levels of radon daughters (Williamson, 1988). In the Kusnetz (1956) method, the air is pumped for 5 min at a steady rate of 2 to 10 l/min through a filter of pore size less than 0.8 μm. The radon daughters collected on the filter are allowed to age for a further 40 to 90 min and are then exposed to a photomultiplier tube. The pulses of output energy are counted over a period that depends upon the level of activity but should be small compared with the delay period, typically 1 to 2 minutes (Calizaya, 1991). The concentration of radon daughters in working levels is then given as

$$WL = \frac{C \times CE}{TF \times V} \qquad (13.20)$$

where C = measured count rate (counts/min), CE = counter efficiency (instrument factor), TF = time factor corresponding to the 40 to 90 min delay (Fig. 3.7) between the end of sampling and the midpoint of the counting interval and V = sample volume (l).

A disadvantage of the Kusnetz and similar methods is the delay between sampling and measurement. This limits the number of samples that can be taken in any one shift. 'Instant' radiation meters reduce the sampling and measurement cycle to a few minutes but may suffer from reduced accuracy, particularly at low levels of activity (Williamson, 1988). These instruments also involve the collection of radon daughters on filters and employ photomultiplier tubes and display units to indicate count rates. The Instant Working Level Meter (IWLM) gives gamma radiation in mRem/h as well as separate counts for alpha and beta radiation.

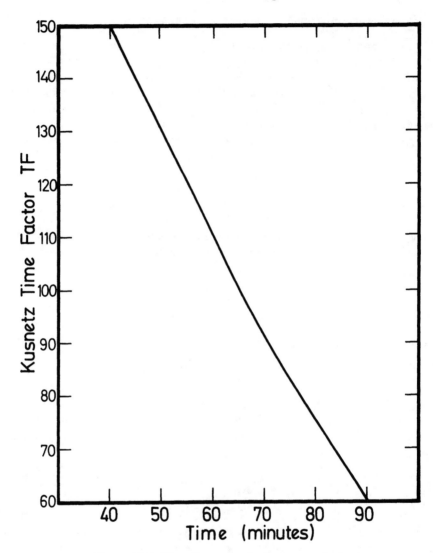

Figure 13.7   Time factors for the Kusnetz method.

13.5.2   Measurement of radon concentration

A Lucas flask is a lucite container whose sides and top are coated with zinc sulphide. To measure radon concentration the flask is first evacuated by a vacuum pump. Sample air is then admitted via a valve and filtered to remove radon daughters. The sample is allowed to age for about 3 h in order to achieve secular equilibrium. A window at the bottom of the flask is then attached to a photomultiplier unit and a reading taken of the output count rate. The radon concentration is given as

$$\frac{(C - b)\, \text{TF}}{\text{FE} \times 2.22 \times 3 \times \text{CE} \times V}\ \text{pCi/l} \qquad (13.21)$$

where C = counts per minute, b = counts per minute due to background radiation from the ambient surroundings, TF = time factor (varies from 0.9762 at 2.5 h to 1.0051 at 3.5 h of elapsed time), FE = flask efficiency (given by flask manufacturer), 2.22 = conversion of counts per minute to pCi (60×0.037 from equation (13.8)), CE = counter efficiency (instrument factor) and V = volume of flask (l). The remaining constant of 3 arises from the fact that alpha emission occurs at three levels during decay from radon to RaD. This is shown on Fig. 13.1. At secular equilibrium, the alpha emissions at each level are equal. As the instrument detects total alpha activity the result must be divided by 3 to give the concentration of radon alone (Calizaya, 1985).

13.5.3 Personal dosemeters

Radiation badges or personal dosemeters are designed to be attached to the clothing and provide a measure of the cumulative radiation dosage to which the wearer has been subjected. Thermoluminescent dosemeters (TLDs) consist of four luminescent phosphors. At periods of one to three months each badge is processed by heating it to a given temperature. The amount of light emitted from each phosphor indicates the average levels and types of radiation to which the badge has been exposed. Another type of radon detector employs an element consisting of poly-allyl diglycol carbonate (PADC detector). A difficulty with personal detectors is that they may be incapable of distinguishing between the radon daughters which are the main radiation hazard in mines and other less harmful forms of radiation including the radon gas itself (Howes, 1990). They are also susceptible to changes in atmospheric pressure, temperature and humidity that are characteristic of subsurface environments.

Miniature versions of the pump and filtration units may prove to be more reliable than current radiation badges. However, these are likely to be cumbersome as well as expensive for routine use. To this time, personal dosemeters have not found widespread use in subsurface workings.

13.6 CONTROL OF RADIATION IN SUBSURFACE OPENINGS

The control of radon and its daughters in underground mines should be addressed during the design of the mine layout, choice of mining method and in selecting mineral transportation routes as well as in planning the ventilation system. In this section, we shall discuss the measures that may be taken to reduce the hazard of radiation in mines. The guidelines that are suggested have been established through a combination of practical observations and theoretical analysis. Although these guidelines can be followed without regard to theoretical background, their success is better assured if the ventilation engineer is familiar with the earlier sections in this chapter. The concepts of ageing and residence time are particularly important. It follows that the need for rapid removal of radon and its daughters results in higher airflow requirements in uranium mines than for most other subsurface openings. The high operating costs that can ensue make it particularly important to design the ventilation system with a view to high efficiency and to employ the techniques of computer-assisted network analysis (Chapter 7). A further prerequisite is to obtain data on the geology of the area, rock surface emanation (J) rates and emanating powers of fragmented ore (B) (section 13.3.1).

13.6.1 Ventilation systems for uranium mines

Particular regard should be paid to the locations and dimensions of intake airways in uranium mines. The purpose is to deliver the intake air to stoping areas as free as practicable from radon or its daughters. There are three methods of achieving this objective. First, the intakes should be driven in the native rock and, as far as possible, not within the orebody. Such airways will be less subject to emanations of radon from the rock surfaces.

Secondly, the residence times of air in intake airways should be kept to a minimum. Air velocity limits normally accepted for the purposes of economics and dust control (section 9.3.6) are frequently exceeded in the intakes of uranium mines. Ventilation requirements for uranium mines are usually much higher than for other mines. Thirdly, where long intake airways are unavoidable, then even higher velocities may be necessary and consideration may be given to the use of airway liners. These are discussed further in section 13.6.7.

Within the stoping areas, emphasis should, again, be placed on rapid air changes. Series ventilation should be avoided and when booster fans are used, particular care should be taken in the choice of their locations and duties in order to minimize recirculation. Uranium mines are a case in which systems of controlled partial recirculation should not be employed.

Pressure differentials across sealed old workings should be in a direction such that any leakage will pass into return airways and not into intakes. The 'dirty pipe' principle may be used to advantage in the design of ventilation systems for uranium mines (section 18.3.1). The number of personnel required to work or travel in the return airways of uranium mines should be kept to a minimum.

In order to provide 'young' air to the faces of headings in uranium mines, it is preferable to employ forcing systems of auxiliary ventilation. An exhaust overlap duct and filter may be added to deal with dust problems (sections 4.4.2).

13.6.2 Dilution and mixing processes

At a constant rate of emission, the rise in concentration of non-radioactive gases is inversely proportional to the rate of throughflow of fresh air (section 9.3.1). This is not the case for radon daughters because of the ongoing effects of radioactive decay. Equation (13.17) indicates that, if an airway or stope is supplied with uncontaminated air and the rate of radon emanation remains constant, then the exit working level of radon daughters is proportional to the residence time raised to the power 1.8, i.e.

$$WL \propto t_r^{1.8} \tag{13.22}$$

where $\propto$ means 'proportional to'. This indicates that if the airflow is halved and, hence, the residence time doubled, then the exit working level of radon daughters will increase by a factor of $2^{1.8} = 3.48$.

A more general relationship is gained by substituting

$$Q \propto \frac{1}{t_r}$$

for a given airway geometry where $Q =$ airflow (m³/s), giving

$$WL \propto \frac{1}{Q^{1.8}} \tag{13.23}$$

Then

$$\frac{WL_1}{WL_2} = \left(\frac{Q_2}{Q_1}\right)^{1.8} \tag{13.24}$$

Example 1 A mine opening is ventilated by an airflow of 10 m³/s. The exit concentration of radon daughters is 0.9 WL. If this is to be reduced to 0.33 WL, determine the required airflow.

Solution From equation (13.24)

$$Q_2 = Q_1 \left(\frac{WL_1}{WL_2}\right)^{1/1.8}$$

$$= 10 \left(\frac{0.9}{0.33}\right)^{1/1.8}$$

$$= 17.46 \, m^3/s$$

Example 2 The radon daughter concentration leaving a mine section is 0.3 WL when the airflow is 15 m³/s. A temporary obstruction caused by stocked materials reduces the airflow to 5 m³/s. Determine the effect on the radon daughter concentration.

Solution Equation (13.24) gives

$$WL_2 = WL_1 \left(\frac{Q_1}{Q_2}\right)^{1.8}$$

$$= 0.3 \left(\frac{15}{5}\right)^{1.8} = 2.17 \, WL$$

This is a dangerous concentration of radon daughters and illustrates the importance of maintaining adequate airflows at all times in a uranium mine.

It should be recalled that equation (13.24), upon which this method of estimating the effects of airflow is based, assumes that the air at entry to the opening is uncontaminated. This may not be the situation in practice and, indeed, experience has shown that the formula often underestimates the amount of air required (Rock and Walker, 1970).

When two airstreams of differing concentrations of radon daughters are mixed, then the resulting concentration is given simply as the weighted mean:

$$WL_{mixture} = \frac{\sum(Q \times WL)}{\sum Q} \qquad (13.25)$$

Example 3 An airflow of $10\,m^3/s$ and radon daughter concentration of 0.25 WL passes a seal from which issues a leakage flow of $0.3\,m^3/s$ at 150 WL. (Near-stagnant air in sealed areas can reach very high concentrations of radon and its daughters.) Determine the radon daughter concentration in the downstream airflow.

Solution From equation (13.25)

$$WL_{mixture} = \frac{10 \times 0.25 + 0.3 \times 150}{10.3}$$

$$= 4.61\,WL$$

This example illustrates the dangerous levels of radiation that can arise from small leakages through abandoned areas.

13.6.3   Radiation surveys

The preceding section makes it clear that modifications to the airflow distribution and small leakages from old workings can have very significant effects on levels of radioactivity in mines subject to radon emanations. In order to ensure the continuity of acceptable conditions and to locate sources of contamination, it is useful to conduct radiation surveys. These involve taking measurements of radon and radon daughters, commencing at points of fresh air entry and tracing the primary ventilation routes through to mine exits. Figure 13.8 illustrates the types of results produced by a radiation survey.

Sampling control stations may be selected at strategic locations in order to establish time-transient trends or to correlate radiation levels with mining activities. Permanent monitoring stations with recording and alarm facilities provide an even greater degree of control (Bates and Franklin, 1977).

The degree of equilibrium between the measured concentrations of radon and radon daughters enables the 'age' of the air to be established (Fig. 13.5 or 13.9 or

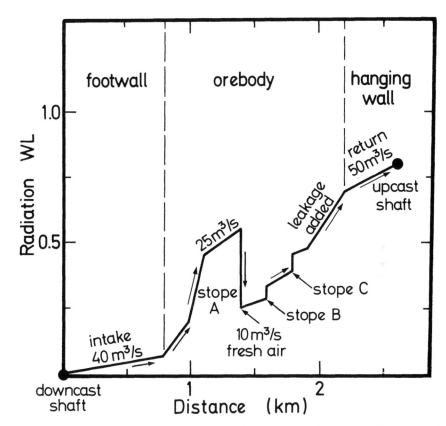

Figure 13.8 Example of a radiation profile produced from a radiation survey along a main ventilation route in a uranium mine.

equation (13.19)). High equilibrium levels ('old' air) measured along the airflow paths are indicative of inadequate ventilation, recirculation or leakage from old workings. Low equilibrium levels ('young' air) but elevated concentrations of radon and radon daughters imply that high rates of radon emanation are occurring.

13.6.4   Mining methods, mineral clearance and backfill

The choices of mine planning, stoping methods and operational procedures have a large influence on the severity of a mine radiation problem that must, subsequently, be handled by the ventilation system. In planning the extraction sequence, stoping areas should progress from the main exhaust airways towards the trunk intake zones. This strategy of retreat mining ensures that worked-out areas and zones of fragmentation and voidage do not contribute towards the radioactive contamination of current workings. Leakages from abandoned areas pass directly into return airways.

Stoping methods should avoid systems that involve large areas of exposed ore,

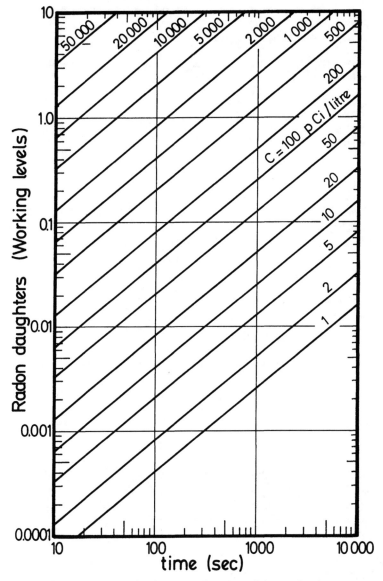

Figure 13.9 Growth of radon daughters as a function of time and radon concentration.

sluggish ventilation or high tonnages of fragmented rock. Hence, open or shrinkage stoping and caving techniques are not advisable in uranium mines. Peak emanations of radon occur during and after blasting. It is particularly important that adequate re-entry periods are employed for the clearance of blasting fumes and the associated radon daughters before personnel are allowed to return to the workings.

Piles of fragmented ore may produce high radiation levels, especially when they are disturbed by mucking operations. This can result in very large variations in radon daughter concentrations during a mining cycle. The broken ore should be transported from the mine as rapidly as practicable. During mineral transportation, air velocities over the broken mineral should be minimized. Where some aeration is unavoidable such as in orepasses or at transfer points, then consideration should be given to the use of exhaust hoods or air byepasses to route the contaminated air directly into a return airway. The number of ore handling operations within the mine should be kept as low as possible and out of main intake airways.

Haulage airways should be well maintained in order to avoid unnecessary comminution of the ore or spillage during transportation. Furthermore, the production of dust particles from uranium-bearing rocks will cause increased levels of radiation (Bigu and Grenier, 1985). The settlement of such dust particles within subsurface airways produces an escalating source of radon. Dust suppression by water sprays or filters is particularly important in uranium mines (however, refer to section 13.6.6 for the effects of water vapour on emanating surfaces).

Localized peaks of radon emanation may occur during drilling operations either for blasting or for orebody investigation. Consideration should be given to the location of machine operators. Long exploration holes should be sealed. Blasting patterns for developments should be selected to minimize the overbreak envelope of fractured rock around the opening. The induced fracture network within this envelope produces additional surface area for radon emanation as well as enhancing the inflow of radon-contaminated groundwater. Good strata control techniques including rock bolting and other support methods will help to minimize radon emanations.

In addition to controlling ground movement, the employment of backfill material will reduce leakage flows through old workings. Both of these features are particularly important in uranium mines. However, tests should be carried out on the radon emanation characteristics of the fill material itself, particularly when it contains mill tailings. Freshly placed wet fill may emanate radon gas at about twice the rate of the consolidated fill (Bates and Franklin, 1977; Thompkins, 1982). The addition of cement to the fill material may result in a reduction in the radon emanation rate. If backfilling operations produce a significant amount of radon, then care should be taken to ensure that the contaminated air is exhausted into return airways.

13.6.5 Contamination from abandoned workings

Example 3 in section 13.6.2 indicated the high level of radioactive contamination that can occur in uranium mine ventilation systems when slight leakage occurs from abandoned areas or unventilated blind headings. The near-stagnant air in such zones achieves a high concentration of radon at secular equilibrium with the radon daughters (fully aged). Radon concentrations of many thousands of pCi/l may occur behind seals in uranium mines. It becomes particularly important that barrier seals in uranium mines should be constructed and maintained to a high standard. The faces of stoppings and adjoining rock walls may be coated with additional sealant material (section 13.6.7). However, minor transients in atmospheric pressure can cause 'breathing' through seals and stoppings, resulting in peak emanations of radon and radon daughters during periods of falling barometric pressure (section 4.2.2).

Such effects occur not only from sealed areas of the mine but also from fracture networks and other voidages within the strata. Attempts have been made to modify mine atmospheric pressures by fan control in uranium mines such that air pressures are elevated during working shifts and depressed when few or no persons are underground (Schroeder *et al.*, 1966; Bates and Franklin, 1977). Leakages from sealed areas can be controlled by pressure balance chambers (section 21.5.5) which maintain pressure differentials across seals at near zero. Another technique is to employ a bleed pipe that connects the sealed area to a main return airway or through a vertical borehole to surface. The sealed zone can be maintained at subatmospheric pressure by employing a low capacity extractor pump or fan within the bleed pipe. Leakage into the sealed area then remains safely in an inward direction. In the absence of these pressure control techniques, a small pressure differential should be maintained across the sealed area such that any leakage that occurs will be into return airways.

13.6.6 The influence of water

The emanation of radon from mineral crystals into the pores of a rock will be essentially the same whether the interstices are filled with air or water. However, any migration of the groundwater can provide a transport mechanism for the dissolved radon that is more efficient than diffusion of the gas through a dry rock. When the water reaches a mine opening it will yield up its dissolved radon very readily, particularly if the water is aerated by spraying or dripping into the airway. It is prudent to capture such water into pipes as soon as possible in uranium mines and minimize its exposure to the air.

In permeable wet strata, the emanation of radon can be reduced significantly by pre-draining the area. This may be achieved by pumping from a ring of drainage boreholes. Better results can be obtained by driving drainage levels below the stoping areas prior to mining. The effectiveness of this technique can be further enhanced by boreholes drilled into the strata from the drainage levels. In severe cases, the flow of water into development headings can be reduced by grouting.

Radioactive decay of the radon occurs whether it is contained within air or water. Hence, the concentration of radon within groundwater depends on the elapsed time since the gas was emitted from mineral crystals into the water-filled interstices. Mine water that contains dissolved radon should not be used for dust suppression sprays. However, if it is first brought to the mine surface and aerated, then the radon content may be diminished to a level that renders the water suitable for dust suppression.

Experimental observations indicate that, commencing with dry rock, radon emanations can increase dramatically as the moisture content of the rock increases (Bates and Franklin, 1977). A similar effect occurs from raising the moisture content of the air that is in proximity to rock surfaces. However, as the rock or air approaches saturation, the effect diminishes and radon emanations fall when liquid water appears on the rock surface. The mechanisms that produce these phenomena appear not to be clearly understood. It is thought that displacement of adsorbed radon by water molecules on mineral surfaces may explain the initial increase in emanation rates, while the later inhibition of radon release may be due to the interstices near the surface becoming filled with liquid water.

13.6.7 Air filters and rock surface liners

As radon daughters are particulates, it is possible to remove a large proportion of them by passing the air through high efficiency dust filters. These should be capable of removing at least 95% of particles 0.3 μm in size. As such filters are relatively expensive and can rapidly become blocked in mining conditions, fibreglass prefilters may be used to remove the coarser particles and, hence, improve the life of the high efficiency filters (Rock and Walker, 1970). The latter are also affected adversely by humid conditions. The pressure drop across filters can be monitored to indicate when renewal or cleaning has become necessary.

The major drawback to filters is that they do not remove the radon gas that continues to replenish radon daughters. The filtered air must be supplied quickly to the personnel who are to be protected. Figure 13.9 or equation (13.15) shows that even if perfect filtration of particulates is achieved, a radon concentration of 100 pCi/l will generate 0.3 WL of radon daughters in 1218 s (20.3 min). However, if the radon concentration is 500 pCi/l, then 0.3 WL of radon daughters will appear in only 163 s (2.7 min) after filtration. Filters for radon daughters are perhaps most effective for forcing duct systems supplying rejuvenated air to headings.

Activated charcoal can remove radon gas from air. At the present time, large-scale applications appear to be impractical. However, gas masks with activated charcoal filters can be used to protect personnel who are required to venture into high radon and radon daughter concentrations for a short time.

A number of trials have been carried out into the use of sprayed coatings and film membranes to reduce radon emanations into mine openings. These are unlikely to replace good ventilation as the primary means of combatting the radon problem. However, they may have an application in long intakes driven in high grade ore or permanent work places such as workshops. Liquid sprays are suitable for application on rock surfaces while membranes may be attached to the faces of stoppings. Grouting of the rock envelope with or without rock bolting can also be effective in reducing inflows of both radon and water.

Tests conducted by the US Bureau of Mines investigated the ability of a range of polymer sealants to resist the passage of radon (Bates and Franklin, 1977). The permeability of a material with respect to air has little bearing on its resistance to radon gas. The latter is a monatomic gas that will diffuse through most substances. Furthermore, any lining material intended for use in mines should have low toxicity and flammability both during application and after curing has been completed. Additionally, it should be convenient to apply, inexpensive and not productive of smoke or toxic fumes when heated. These stringent requirements limit the use of radon barriers in practice.

The US Bureau of Mines tests indicated that water-based epoxies were suitable for application to rock surfaces. However, two spray applications are recommended, the first with a low viscosity liquid to penetrate surface interstices, and the second with a thicker fluid to seal visible fractures. Using differing colours for the two applications assists in achieving full coverage.

Further tests in Canada indicated that a polysulphide copolymer spray and aluminized myler sheeting were both capable of reducing radon diffusion through a bulkhead by more than 95% (Archibald and Hackwood, 1985). Additional data on the permeability of membranes to radon are given by Jha *et al.* (1982).

13.6.8 Education and training

Radon gas and radon daughters are particularly insidious hazards. They are invisible, odourless and can be detected only by specialized instruments. Furthermore, they have no short-term observable effects on the human body. It becomes particularly important that all workers in affected environments should be made aware of the health hazards associated with radon and the steps that can be taken to alleviate the problem. Booklets, videotapes and classroom teach-ins are particularly effective. These should emphasize the importance of a brisk throughflow of air and the hidden dangers that may arise from recirculation, damaged stoppings or ventilation doors left open.

REFERENCES

Archibald, J. F. *et al.* (1980) Determination of radiation levels to be encountered in underground and open-pit uranium mines. *2nd Int. Congr. on Mine Ventilation, Reno, NV*, 399–404.

Archibald, J. F. and Hackwood, H. J. (1985) Membrane barriers for radon gas flow restriction. *2nd US Mine Ventilation Symp., Reno, NV.*

Bates, R. C. and Edwards, J. C. (1980) Mathematical modelling of time dependent radon flux problems. *Proc. 2nd Int. Congr. on Mine Ventilation, Reno, NV*, 412–19.

Bates, R. C. and Franklin, J. C. (1977) US Bureau of Mines radiation control research. *Conf. on Uranium Mining Technology, Reno, NV.*

Bigu, J. and Grenier, M. G. (1985) Characterization of radioactive dust in Canadian underground uranium mines. *Proc. 2nd US Mine Ventilation Symp., Reno, NV*, 269–77.

Calizaya, F. (1985) Control study of the evolution of radon and its decay products in radioactive mine environments. *PhD Thesis*, Colorado School of Mines.

Calizaya, F. (1991) Personal communiction.

CFR (1990) *US Code of Federal Regulations*, Vol. 30, *Mineral Resources*, Part 57, *Metallic and Non-metallic Underground Mines*, US Government Printing Office, Washington, DC.

Howes, M. J. (1990) Exposure to radon daughters in Cornish tin mines. *Trans. Inst. Min. Metall.* **99**, A85–A90.

Jha, G., *et al.* (1982) Radon permeability of some membranes. *Health Phys.* **42**(5), 723–5.

Kusnetz, H. L. (1956) Radon daughters in mine atmospheres—a field method for determining concentrations. *Ind. Hyg. Q.* (March), 85–8.

Rock, R. L. and Walker, D. K. (1970) *Controlling Employee Exposure to Alpha Radiation in Underground Uranium Mines*, US Bureau of Mines, US Government Printing Office, Washington, DC.

Schroeder, G. L. and Evans, R. D. (1969) Some basic concepts in uranium mine ventilation. *Trans. AIME* **244**, 301–7.

Schroeder, D. L., *et al.* (1966) Effect of applied pressure on the radon characteristic of an underground mine environment. *Trans. AMIE* **235**, 91–8.

Thompkins, R. W. (1982) Radiation in uranium mines, *CIM Bull.* **75** (845, 846, 847).

Thompkins, R. W. (1985) The safe design of a uranium mine. *Proc. 2nd US Mine Ventilation Symp., Reno, NV*, 289–94.

Williamson, M. J. (1988) The exposure of mining personnel to ionizing radiations in Cornish tin mines *Proc. 4th Inst. Mine Ventilation Congr., Brisbane*, 585–92.